The Philosophy of Biology

An Episodic History

Is life different from the non-living? If so, how? And how, in that case, does biology, as the study of living things, differ from other sciences? These questions lie at the heart of *The Philosophy of Biology*, and are traced through an exploration of episodes in the history of biology and philosophy. The book begins with Aristotle, then moves on to Descartes, comparing his position with that of Harvey. From the eighteenth century, the authors consider Buffon and Kant. From the nineteenth century, the authors examine the Cuvier–Geoffroy debate, pre-Darwinian geology and natural theology, Darwin, and the transition from Darwin to the revival of Mendelism. Two chapters on the twentieth century deal with the Modern Evolutionary Synthesis and such questions as the species problem, the reducibility or otherwise of biology to physics and chemistry, and the problem of biological explanation in terms of function and teleology. The final chapters reflect on the question of human nature and the implications of the philosophy of biology for the philosophy of science in general.

Marjorie Grene is Professor Emeritus of Philosophy at the University of California, Davis, and Adjunct Professor and Honorary Distinguished Professor at Virginia Polytechnic Institute and State University.

David Depew is a Professor in the Department of Communication Studies and in the Project on the Rhetoric of Inquiry at the University of Iowa.

THE EVOLUTION OF MODERN PHILOSOPHY

General Editors

Paul Guyer and Gary Hatfield (*University of Pennsylvania*)

Published

Roberto Torretti: *The Philosophy of Physics*

Forthcoming

Paul Guyer: *Aesthetics*

Gary Hatfield: *The Philosophy of Psychology*

Stephen Darwall: *Ethics*

T. R. Harrison: *Political Philosophy*

William Ewald & Michael J. Hallett: *The Philosophy of Mathematics*

Michael Losonsky: *The Philosophy of Language*

Charles Taliaferro: *The Philosophy of Religion*

Why has philosophy evolved in the way it has? How have its subdisciplines developed and what impact has this development had on the way the subject is now practiced? Each volume of **The Evolution of Modern Philosophy** series focuses on a particular subdiscipline of philosophy and examines how it has evolved into the subject as we now understand it. The volumes are written from the perspective of a practitioner in contemporary philosophy whose point of departure is the question: How did we get from there to here? Cumulatively, the series will constitute a library of modern conceptions of philosophy and will reveal how philosophy does not in fact comprise a set of timeless questions but has rather been shaped by broader intellectual and scientific developments to produce particular fields of inquiry addressing particular issues.

The Philosophy of Biology

An Episodic History

MARJORIE GRENE

Virginia Polytechnic Institute and State University

DAVID DEPEW

University of Iowa

CAMBRIDGE
UNIVERSITY PRESS

PUBLISHED BY THE PRESS SYNDICATE OF THE UNIVERSITY OF CAMBRIDGE
The Pitt Building, Trumpington Street, Cambridge, United Kingdom

CAMBRIDGE UNIVERSITY PRESS
The Edinburgh Building, Cambridge CB2 2RU, UK
40 West 20th Street, New York, NY 10011-4211, USA
477 Williamstown Road, Port Melbourne, VIC 3207, Australia
Ruiz de Alarcón 13, 28014 Madrid, Spain
Dock House, The Waterfront, Cape Town 8001, South Africa

http://www.cambridge.org

First published 2004

Printed in the United States of America

Typeface Sabon 10.25/13 pt.   *System* LaTeX 2_ε [TB]

A catalog record for this book is available from the British Library.

Library of Congress Cataloging in Publication Data
Grene, Marjorie Glicksman, 1910–
The philosophy of biology : an episodic history / Marjorie Grene, David Depew.
p. cm. – (The Evolution of modern philosophy)
Includes bibliographical references (p.).
ISBN 0-521-64371-6 (hb) – ISBN 0-521-64380-5 (pbk.)
1. Biology – Philosophy – History. I. Depew, David J., 1942– II. Title. III. Series.
QH331.G736 2004
570'.1–dc22   2003055891

ISBN 0 521 64371 6 hardback
ISBN 0 521 64380 5 paperback

Contents

Contents

List of Figures

ix

Notes on Citations and References

- For most authors of a single work or a small body of work, we have used the standard citation form in the text and in the references: for example – Adams 1980.

- For certain authors of a substantial, well-edited body of work, we have used a standard reference system or abbreviations, as noted here, to cite the editions in which these authors' writings can be found.

Aristotle
References to the Greek text of the various treatises of Aristotle – *History of Animals, On the Soul, Politics*, and so on – are to the page numbers of I. Bekker's text. Most of these now standard texts can be found in editions published by Oxford University Press. Unless otherwise specified, English translations of Aristotle's works are from Barnes 1984.

Buffon
Buffon's works are referenced first to the original French publication date and place, and then, in most cases, to page numbers in Buffon's *Oeuvres philosophiques* (edited by Jean Piveteau, 1954), which we have abbreviated as OP. English translations are our own, except those taken, as noted, from Lyon and Sloan 1981.

Darwin
Darwin's works are generally cited in their first editions. The first edition of the *Origin of Species* is abbreviated *Origin*. Publications of Darwin's notebooks, letters, marginalia, and so on will be found under

Darwin, not under the names of their editors. Unpublished manuscripts found in the Darwin Research Archive at Cambridge are abbreviated as DAR.

Descartes

The abbreviation AT refers to the French text of Descartes's *Oeuvres* edited by C. Adams and P. Tannery. CSMK refers to English translations found in *The Philosophical Writings of Descartes*, edited by J. Cottingham et al.

Kant

The abbreviation Ak refers to the so-called Akademie edition of Kant's works published by the Deutsche Akademie der Wissenschaften, 1908–1913. Kant's various works are cited in the text first by title and date, then by paragraph (where the original text is so divided), then by volume and page number in Ak).

Titles are cited in English in the text and in both English and German in the references – for example, Kant (1781), *Critique of Judgment (Kritik des Urteilskraft)*. Translations of the *Critique of Pure Reason* are by Norman Kemp Smith (Kant 1929) unless otherwise noted. Translations of the *Critique of Judgment* are by W. Pluhar (Kant 1987).

Acknowledgments

In this book, we deal with the relationships between philosophy and biology at various times and places in our tradition. As befits a contribution to a series on the evolution of modern philosophy, it is a history of the philosophy of biology rather than a philosophy of biology in the usual sense. Although we deal with recent and even contemporary issues in the last four chapters, our approach is primarily historical rather than "systematic." Philosophical issues are raised in close connection with actual biological discoveries and theorizing.

It will be clear from our text that we are much indebted to others who have worked in this area in a similar spirit, at once philosophical, historical, and biologically alert – particularly Jean Gayon, Jon Hodge, and Phillip Sloan – as well as to Richard Burian, who has read and criticized some of our chapters. Regrettably, the manuscript had been nearly completed before the untimely death of Stephen Jay Gould and the appearance of his last and massive work; we were able to take account only partially of his arguments on aspects of our subject.

We are grateful to our manuscript editor, Ronald Cohen, for his encouragement and many helpful suggestions.

We wish to acknowledge the help of our research assistants at the University of Iowa – John Franzen and Rosemary Steck – and of Kai Weatherman of the University of Iowa Libraries. We also gratefully acknowledge support from the Arts and Humanities Initiative at the University of Iowa. We appreciate the help and hospitality of the Philosophy Department of Virginia Polytechnic Institute and State University.

We acknowledge the kind permission of the University of Illinois Press to reprint a figure from Willi Hennig's *Phylogenetic Systematics*

(Urbana: University of Illinois Press, 1966, 2d. ed., 1979, p. 89); and of Harvard University Press to reprint a figure from Ernst Mayr's *Toward a New Philosophy of Biology* (Cambridge, MA: Harvard University Press, 1988, p. 279).

Preface

There would never be, Immanuel Kant assured his readers, a "Newton of a blade of grass." Living things, he believed, are examples of "natural purposes," entities organized so purposefully that we cannot explain them altogether through the blind causality we apply to inanimate nature. At the same time, Kant argues that if living beings are organized purposely, or on purpose, rather than just purposefully, we cannot know it. There seems to be something special about things that are alive that exempts them from Newtonian mastery.

Something like this, although not quite in Kantian terms, has been the view of many natural historians, physicians, and comparative anatomists, as well as philosophers, in our tradition. Others, notably Descartes and his followers, as well as more recent "reductionist" thinkers, have denied that any such difference exists.

Yet even among those who stress the uniqueness of life, a number have appeared, at least implicitly, to welcome the accolade of "Newton of a blade of grass," whether for themselves or others. Georges Cuvier seems to have been happy to assume that title, though it was animals, not plants, that he studied. But he would also have been willing to claim the crown for Master Aristotle. Some thought Cuvier's rival, Etienne Geoffroy Saint-Hilaire, more worthy of that honor. And of course many have since found that it was Charles Darwin who gave the study of life such a new and scientifically satisfying solution that he truly deserved the title "Newton of a blade of grass." Still, all of these claimants would at least have agreed with Kant that the subject matter of biology has something about it that is not quite the same as physics.

Is life different from the non-living? If so, how? And how, in that case, does biology, as the study of living things, differ from other sciences? Or does it after all? That is the basic cluster of questions we have focused on in what follows. The result is a study of figures and of episodes in the history of biology, as well as of philosophy, that seem to us to illuminate these basic problems in one way or another. Thus, while we are dealing with *some* interactions between biology and philosophy, we are far from attempting a survey of such events. That would be both tedious and beyond our competence. For example, we touch only incidentally on the preformation/epigenesis debate; we come close to ignoring the coming of the cell theory; we overlook Pasteur and the question of spontaneous generation, as well as Claude Bernard and the analysis of experimental method. Indeed, our treatment of the nineteenth century focuses, as we admit, on Darwin and some of the events that precede and follow the development of his theory. Linnaeus appears, parenthetically, in our chapter on Buffon, and Lamarck receives equally short shrift in the context of the Paris Museum of Natural History in the early nineteenth century and the Cuvier–Geoffroy debate. Nor, it should be added, despite the appearance of Harvey, whom we treat as a physiologist rather than a physician, have we dealt even episodically with the very complex, and different, subject of the history of medicine. Yet our hope is that by this highly selective, if not idiosyncratic, procedure we can illuminate some facets of the prehistory of the philosophy of biology as it has recently developed.

We begin in antiquity, in particular with Aristotle, whose views we explore in Chapter I for at least three reasons. First, Descartes's would-be reform of the foundations of biology is intelligible only in the context of the Scholastic, at least remotely Aristotelian, tradition in which he was educated. Indeed, the Scholastic tradition itself reaches through to Cesalpino and into the beginnings of modern taxonomy. Second, Descartes's position is especially clear, as he himself stresses, in his arguments against Harvey's view of the motion of the heart, which we review. And Harvey, for his part, was a profoundly Aristotelian thinker, not in following the Scholastic model, but as an investigator with a deeply rooted interest in the functional particularities of living beings. Finally, in the nineteenth century, we find Cuvier attributing to Aristotle the foundation of comparative anatomy, insisting that in his own far-reaching study of the animal kingdom he was only carrying on the work for which Aristotle had laid down the foundation and the method. Geoffroy, of course, disagreed; we needed a new beginning, he believed. But even if he was right, we would want to know what it was he wanted to displace or transcend. Quite apart from those historical reasons, moreover, Aristotle, as the one great philosopher in our tradition who was also a great biologist, deserves to stand at the head of our study. So we begin by looking at his biology and, briefly, at some of the changes in biological philosophy that followed in the Hellenistic period.

From Aristotle and his successors we move abruptly to the origin of modern mechanism in the person of Descartes, comparing his position on the motion of the heart with that of Harvey. Setting the Cartesian enterprise within its Scholastic context, we see it reducing to local motion the four Aristotelian kinds of change: substantial, quantitative, qualitative, and local. Apart from God and mind, there is just spread-outness, and those things that look in some ways a little like us are only bits of matter ingeniously engineered by their creator. In effect, the notorious doctrine of the beast-machine does away with life. At the same time, Descartes, accepting the new doctrine of the circulation of the blood credited to Harvey, boasts that his own very different account of cardiac motion is superior to that of the English physician, precisely because it is more mechanical. The heart is a furnace in which the entering blood is rarefied, so that it pushes its way out into the aorta. Harvey's account is less perspicuous, and, what is worse, it reverses the traditional order of diastole and systole in a manner difficult for well-trained physicians to accept. In fact, Descartes's doctrine was commonly accepted for some time to come.

In the eighteenth century, we focus, first, on the work of the Comte de Buffon, the great natural historian whose work spans most of the century. Here we have clearly left the Scholastic tradition for a period in which it is the Newtonian heritage that has come to dominate – in Buffon's case in the guise of the seeming inductivism of Newton's method. We find the inverse-square law prevailing everywhere, but do not ask why. We simply face reality and accept it. What we call "causes" are in fact only carefully generalized effects. Buffon sharply contrasts the abstractions of mathematics with what he calls "physical truth," which provides us with the best certainty we can attain. True, we must sometimes hypothesize, but then only carefully, in close touch, Buffon hopes, with concrete reality. Given this methodology, we look at Buffon's work in several areas: his radical skepticism about taxonomy, in contrast to the widely accepted Linnaean system; his concept of the species as a historical – though permanent – entity; his view of generation, which is opposed to the popular notion of "emboitement," or preformed encasement, but involves so-called organic molecules and internal molds, as well as special forces that maintain each species; and, finally, Buffon's sense of the uniqueness of the living, which puts him into some relation to the emergence of a vitalist philosophy.

Toward the close of the (eighteenth) century we have, further, Kant's "Critique of Teleological Judgment," an account, as we have seen, of organisms as "natural purposes," in which all parts are harmoniously both ends and means (Kant 1793). In this work, Kant was both influenced by, and in his turn, an influence on work in the unified, or unifying, set of inquiries that was beginning to be called "biology," especially at the University of Göttingen. In Kant, the conflict between mechanistic physics and the purposiveness of life forms is at its most acute – so acute, he held, that we will never resolve it.

The early nineteenth century is marked by the preeminence of the Museum of Natural History in Paris. It was a time of explosive development in comparative anatomy. In particular, we consider the famous controversy between two leading figures there, Georges Cuvier and Etienne Geoffroy Saint-Hilaire. For Cuvier, it was the details of comparative anatomy that mattered. Geoffroy, in contrast, was always in search of constancies, of one overarching plan that would explain the morphological relations (though not primarily genealogical relations!) between animals of seemingly different types. By the time Darwin was formulating his theory, both of these perspectives had come to be accepted alongside one another, although in the *Origin of Species*, Cuvier's "conditions of existence" were ranked as more fundamental than Geoffroy's "unity of type." We explore the Cuvier–Geoffroy controversy in Chapter 5.

Two interconnected lines of thought in Britain remain to be considered as background to Darwin's theorizing. On the one hand, there is the development of geology, which revealed extinctions – and plenty of them. The development of geology gave rise to the important debate between uniformitarians and catastrophists about the proper reading of the earth's history. Their divergence reflected a wider disagreement in the philosophy of science between John Herschel's conception of the identification of "true causes" as the heart of the inductive process, and William Whewell's perhaps more conservative notion of the "consilience of inductions." At the same time, a traditional belief in the fixity of species was shaken, in some quarters by the work of Lamarck, but (in Britain at any rate) more shockingly and at a more popular level by the publication of the notorious *Vestiges of Creation* in 1844. All these developments, finally, were closely related to challenges posed to, and defenses attempted for, a deeply rooted British trust in the tradition of natural theology, as exemplified most conspicuously in the work of William Paley.

When it comes to Darwin himself, we first consider the road to his theory, which is somewhat more circuitous than used to be thought. We then single out some of the epistemic presuppositions of the method he followed when he finally gave an "abstract" of his theory to the public in 1859, and we notice the implications for taxonomy of the understanding of species that follows from that theory. It may look as if the author of the *Origin of Species* has done away with species; in fact, he has given them a new genealogical reality. Finally, there is the question of man's place in nature, with which Darwin dealt privately in his *Notebooks*, but made public in 1871 in *The Descent of Man*, as well as in his book of 1872 on the emotions.

If Darwin changed the tenor of biology, he did not in fact convert his contemporaries and successors wholesale to the acceptance of his theory of natural selection. In looking at the period from Darwin to the rise of genetics, we note, first, some workers who accepted selection as central to evolution. But then we also consider the work of Haeckel and his "biogenetic law," which made the search for phylogenies the primary focus of evolutionary speculation (Haeckel 1866). The work of Francis Galton developed partly out of his disappointment in se- lectionist explanation, but contributed to the advance of evolutionary theory through his introduction of statistical methods. These were car- ried on, under a phenomenalist banner, by the biometricians Weldon and Pearson. Contrasted with their stress on continuity in the natu- ral world was the insistence on discontinuity put forward by Bateson and the first "geneticists," who, like De Vries with his mutation theory, found Mendel's work, "rediscovered" in 1900, opposing the notion of small gradual variations leading slowly and smoothly to new varieties or species.

It was the founding of the "evolutionary synthesis" in the 1920s and 1930s, whose history we consider in Chapter 9, that dissolved this seem- ing opposition, although, more broadly, the synthesis has been taken to mean the coalescence, not just of two fields – genetics and Darwinism – but of a number of fields: systematics, paleontology, and botany, as well as genetics and the theory of selection. We examine the synthesis in a number of its architects, such as Dobzhansky, Mayr, and Wright, and then note some of the recent challenges to it: from some paleontologists, for example, who have objected to the reduction of macro-evolution (evolution at and above the species level) to subspecific micro-evolution;

and from developmental biologists, especially in recent attempts to read some features of development into evolutionary theory.

Although the analysis of evolutionary theory is certainly central to the philosophy of biology as it developed in the second half of the twentieth century, there have also been other conspicuous areas of debate. The species problem is still hotly debated. After Darwin, can we call species real? And if so, what are they? An authoritative view has been that of Ernst Mayr's "biological species concept," according to which species are potentially interbreeding populations. But there has also been opposition to this notion. One question vigorously debated concerns the ontological status of species, again in the wake of evolutionary theory: Are they classes (which, some argue, cannot be altered) or individuals? Further, whatever species are, there is the problem of classifying them. Numerical taxonomists wanted a purely conventional basis for their science. Various sorts of "cladists" want a method based somehow on Willi Hennig's "phylogenetic systematics." Still other, evolutionary, taxonomists claim that classification by splitting alone will not do; evolutionary distance, change in life style, and so on, should be taken into account.

A second group of problems concerns the reduction, or otherwise, of biology to physics and chemistry. Early in the century, there were still vitalists opposing programmatic mechanists. Although confrontation in those terms seems to be over, biologists and biologically concerned philosophers still take diverse stands on the question, permitting, or opposing, various degrees of reduction – theoretical, ontological, or methodological.

Finally, there is the problem of biological explanation. If it is somehow *sui generis*, is it, as many used to think, teleological in character? There is still lively discussion about this question, especially in relation to the concept of function and its connection, if any, with teleology of some sort. Here we consider the two major contenders, "etiological" or "selected effect" functions, and "causal role" functions, as well as some variants on them.

As the title of our book indicates, the topics we are concerned with have chiefly to do with matters in biology in general – reflections on understanding life – rather than with human concerns, with questions having to do with our life in particular. However, in Chapter 11, we do look, if sketchily, at questions connected with human nature. First, there is the question of human origins, especially with respect to the unity of the species. Second, there is the old worry of nature versus nurture. Third, we ask about some of the characteristics we alone seem to possess, such as language and mind. Finally, we discuss briefly some of the implications of the Human Genome Project and of the recent development of technology that permits direct manipulation of our genetic material.

In conclusion, in Chapter 12, we reflect briefly on some implications of the philosophy of biology for the philosophy of science in general.

✦

CHAPTER 1

Aristotle and After

Beginning with Aristotle

It is no longer possible to begin an account of modern philosophy of physics in modernity itself; one must go back at least to the Middle Ages. In the case of philosophical thought about living things, however, or what has recently come to be called philosophy of biology, one must go back even further – to the figure of Aristotle, who lived in the fourth century B.C.E. (384–323). For one thing, Aristotle is the only major philosopher in our tradition who is also a major biologist. One cannot read him for any length of time without seeing that his central philosophical concerns were closely related to his biological interests. Moreover, Aristotle first raised the questions that have preoccupied philosophers of biology ever since: arbitrary imposition versus "cutting nature at the joints" when it comes to naming traits and classifying kinds of organisms; purposive function versus haphazardness and accident in the distribution of traits to various kinds; mechanistic reduction versus teleology or goal-orientation in the process of embryogenesis. These topics are all explicitly formulated in Aristotle's biological treatises, which comprise no less than a quarter of the corpus of his writings that have come down to us.

We must begin with Aristotle, however, not only because we find him raising issues that recur, but because Aristotle's biological way of thought forms the background of subsequent philosophy of biology. For its part, modern physics, as is well known, began by rejecting not just scholastic Aristotelianism, but the fundamental principles of Aristotle's physics itself. We can say, generalizing rather crudely, that the late Scholastics, or "Aristotelians," had forgotten Aristotle's biology, and the way it concretely informed and was informed by his metaphysics, in favor of Christianized versions of his physics and metaphysics. We shall look at this problem in the next chapter, since modern mechanists, beginning with Descartes, took off from there. It is important to recognize, however, that the development of modern biology did *not* follow this pattern. Indeed, biologists who worked after Descartes made increasingly systematic use of the concepts of end and form in their explanations of living things, and the name of Aristotle was often spoken reverently among them. Modern biologists have, in fact, returned again and again to Aristotle as their master.[1]

In this chapter, we explicate the conceptual structure of Aristotle's program for biological research, and the ways in which that program informed and was informed by his logical, methodological, and metaphysical doctrines. Aristotle wanted this program to be completed by the Lyceum, which he founded. Very early on, however, perhaps as soon as his immediate disciple Theophrastus of Eresus ceased working, the sharp edges of Aristotle's philosophy of biology became blurred. This did not keep Aristotle's biological works from inspiring much creative thought about living things, especially after his texts were republished in the Italian Renaissance. Clearly, however, even the best of this work, such as William Harvey's, was almost never carried out under Aristotle's precise conception of what a philosophically informed biology should look like. Much had been transformed throughout the long tradition.

Hippocratic Medicine and Aristotelian Biology

Aristotle was not the first Greek to have left written reflections on living things. As the son of a doctor – his father was physician to the Macedonian court – he was clearly familiar with those literate practitioners of the medical art, the Hippocratics. The Hippocratics held that the two basic opposites – hot-cold and wet-dry – can be combined in four ways, producing the elements of earth (dry-cold), air (hot-wet), fire (hot-dry), and water (wet-cold). According to a view developed by some of them, these elements give rise to the four bodily humors – black bile (earth, located in the spleen), blood (fire, located in the heart, thought to be the source of life and hence hotter than the rest of the body), yellow bile (air, located in the gall bladder), and phlegm (water, located in the

---

[1] We refer here to figures as diverse as William Harvey, George-Louis Leclerq Comte de Buffon, Georges Cuvier, John Hunter, and Richard Owen, many of whom will be discussed in later chapters.

lungs) (Hippocrates, "The Nature of Man," section 4, in Hippocrates 1978, p. 262). The point of medical practice was to maintain, and whenever necessary restore, the right blending among these sometimes competing humors. In the most highly articulated versions of Hippocratic thought, the humors were in turn believed to correspond to four temperaments – melancholic, sanguine, choleric, and phlegmatic – which also corresponded to the four seasons. Unsurprisingly, in view of this picture, the Hippocratics were sensitive to the effect of diet and environment on health and of climate on the character of populations. In the Hippocratic treatise on *Airs, Waters, and Places*, for example, we learn that on the mainland of Asia Minor, "the people are milder and less passionate" than in Europe because Asia "lies equally distant from the rising of the sun in the summer and winter," and "luxuriance and ease of cultivation are to be found most often where there are no violent extremes, but when a temperate climate prevails" (Hippocrates, "Airs, Waters, and Places," section 12, in Hippocrates 1978, p. 159).

One can find plenty of claims in Aristotle that sound Hippocratic enough. Aristotle's theory of elements, for example, is very much like theirs; the elements are not hard, entity-like substances, but phases of a self-perpetuating cyclical process in which the opposites – hot-cold and wet-dry – necessarily and predictably give way to one another (*On Generation and Corruption* 337 a 1–15). Aristotle was also conscious of how diet and climate affect character, as when he characterizes "Asians" as lacking in aggression due to the heat of their climate, and says that northern barbarians, coping as they must with extreme cold, are excessively aggressive (*Politics* 1327 b 19–32). In his *Ethics*, too, Aristotle's stress on finding a virtuous mean between opposing passions, and on finding it in a way that is uniquely appropriate to the individual, fits in with the Hippocratic approach to medicine.

Yet in spite of many stray remarks suggesting off-hand familiarity, if not complete agreement, with Hippocratic views, the spirit of Aristotle's approach to living things differs entirely from those of the Hippocratics. Although it may be said that, in a general way, the Hippocratics projected a certain theoretical framework, they did so in a pragmatic rather than a dogmatic spirit. The dominant tone of the best of their writings is one of suspicion about applying reasoning from theoretical postulates to particular cases, after the fashion of the pre-Socratic natural philosophers. "I am utterly at a loss," writes the author of the fifth-century treatise, "to know how those who prefer hypothetical arguments and reduce the science of medicine to a simple matter of 'postulates' (*hypotheses*) could ever cure anyone" (Hippocrates, "Tradition in Medicine" [also known as "On Ancient Medicine"], in Hippocrates 1978, p. 7. The Hippocratics, in sum, were proud practitioners of the *art* of medicine, not devotees of a theoretical *science*. The focus of their writings, which were collected over a period of several centuries, was on urging their would-be adepts to cultivate skills that would enable doctors to remain true to the internal norms governing their art. That is the thrust of the famous Hippocratic Oath.

Now it is certainly true that Aristotle recognized medicine as an art, which if practiced skillfully would both require and exhibit judgment (of a sort different from the deliberative wisdom of the citizen-politician [*phronēsis*], but no less focused on how to deal with the contingencies of particular cases) (*Nicomachean Ethics* 1140 a 1–24). Indeed, Aristotle regarded medicine as the very paradigm of a craft or *technē*; and, like both Plato and the Hippocratics, he was at pains to distinguish genuine crafts such as medicine from mere empirical knacks. In the work of Aristotle and his school, however, we find for the first time a sustained effort to pursue biological inquiry (collection, description, explanation) for its own sake rather than for practical benefit. *Aristotle was the first theoretical biologist.* This drive toward theory means that in Aristotle, problems we now recognize as scientific were penetrated at every point by questions that he recognized as philosophical – and that we should, too. Theory (literally "vision" or observation) *means* philosophical insight.

Aristotle's orientation to theory leads him to judge that reasoning from hypotheses, the very process eschewed by the Hippocratics, can be helpful in searching for the indemonstrable, but certain, first principles from which the propositions constituting a science follow (*Posterior Analytics* 92 a 7–32). Presumably, the Hippocratics should not object to that. For unlike both the craft-knowledge they prized and the political-ethical activity of citizens, science is not concerned with particular cases, as doctors and politicians are, with all the uncertainty that attends these cases. It is concerned instead with what happens "always or for the most part." Aristotle divides the work of theoretical inquiry into an inductive procedure (*epagōgē*), which leads to the establishment of explanatory first principles, and a demonstrative procedure, which solves the problems encountered on the way toward principles by deducing their correct answers from these principles once they are found (*apodeixis*). Having said this, it is important for us to note that, as Aristotle understands scientific knowledge, the principles governing what happens always, and even for the most part, are not arrived at by simple enumerative induction, as in proverbial nose-counting exercises like that about white swans. Instead, the upward path involves properly dividing the subject matter until its proper elements and its essential definition are identified, often by sorting through what is plausible and implausible in the views of predecessors. Properly conducted, inquiry of this sort will arrive at principles that are true, primary, immediate, better known than, prior to, and causative of the conclusions drawn from them (*Posterior Analytics* 71 b 16–20). Although these principles are as certain as certain can be, only the second, downward leg of the process of inquiry constitutes demonstrative scientific knowledge (*epistēmē*) as such. The sciences of nature, including what is now called biology, are for Aristotle demonstrative sciences in just this sense. They are presumed to have their own first principles, from which universally valid and sound conclusions about living things necessarily follow.

Biology Within the Bounds of Physics

In describing Aristotle's program of theoretical biology, we must first recognize that he had no word for "biology." That term was coined toward the end of the eighteenth century (see Chapter 4). For Aristotle, on the contrary, what *we* recognize as biology was part, indeed a central part, of the science of natural philosophy or physics. Clearly, Aristotle had a wider notion of the study of "physics" (nature) than has become conventional in modern times.

For Aristotle, physics, which in Greek means "things that grow or develop" (*phuomena*), is the study of any and all beings that have within themselves a non-incidental source of motion and of rest (*Physics* 192 b 12–15; 20–23; 199 b 15). All such beings are substances, the individuated entities that collectively make up the world (*Physics* 192 b 32–34). Some of these substances – the ones physics studies – come into being and pass away. This process constitutes substantial change. Moreover, almost all substances – at least all of the perishable ones – are able, while they exist, to remain themselves by means of various processes of change – qualitative, quantitative, locomotive – in which they acquire or lose properties. These are non-substantial, or in Aristotle's terminology, "incidental," changes; they are less fundamental than the generation or extinction ("corruption") of a substantial unity itself. That there are many (relatively) independently existing entities or substances, whose natures our minds are suited to understand, is absolutely basic to Aristotle's view of things. These are the everyday things we see around us: plants, animals, and, among animals, ourselves. Substances that remain the same things through incidental changes are said to have a nature (*phusis*) and to change by nature (*phusei*) (*Physics* 192 b 33–193 a 1).

To understand more precisely what Aristotle means by "nature," it helps to see that what happens naturally is contrasted in various places in his works with three other sorts of things:

1. Natural philosophy (physics) is contrasted in the first instance with the study of substances that do not move at all, even if they move other things (*Metaphysics* 1026 a 10–20). What Aristotle has in mind here is the outermost sphere of the *kosmos*, which for him is also the divine self-understanding of the eternal world-order itself, to which finite things are both oriented and subordinated. This substance is the ultimate subject of "first philosophy," or what Aristotle calls "theology" (of a highly rationalized sort by typical Greek standards) (*Metaphysics* 1026 a 19–20). The sphere of physics, by contrast, is "second philosophy."

2. What happens by nature is also contrasted with what happens by art or craft (*technē*). What happens by art comes into being not naturally, but by way of a source external to itself – namely, the thought in the mind and the artfulness in the hands of an artificer or practitioner. In contrast to the materials from which they are made, for example, "a bedstead or a coat or anything else of that sort...has within itself no internal impulse to change" (*Physics* 192 b 16–18, revised Oxford translation, amended). We can see from this example what Aristotle means by a non-incidental source of change. A bedstead can change incidentally when the wood from which it is made grows brittle and needs to be glued, or when, like Antiphon's bed, it rots and sprouts branches (*Physics* 193 a 12–16). But this does not happen insofar as it is a *bed*. It happens insofar as it is *wooden* (*Physics* 193 b 8–11). In part because of their external source of motion, products of art are not sufficiently integrated to count as substances. In artefacts, as the example of the bed shows, matter (the stuff of which something is composed) and the form (the kind of thing it is) do not fully fuse. In natural entities, matter and form are not so separate.

Granted, there is an important analogy for Aristotle between what comes to be by art and what comes to be by nature (*Physics* 199 a 9–19). In both processes, as in both kinds of entities, Aristotle distinguishes four "causes" – one might say four reasons why a thing is as it is. Its matter and form are two of these, which are always distinguishable when we look at a substance (or an artefact) in cross-section, so to speak, at a given period of its existence. When we consider its life-history over time, however, we find two more correlative explanatory factors: the efficient or moving cause, which names the agency by which a thing comes into existence, and the final cause, which refers to the end for which it comes into existence or its terminal point of development. In the case of natural substances, as we will see, form, end, and efficient cause are often identified. Matter, which for Aristotle is the potentiality for assuming form, is decidedly subordinate to the other three. This four-fold categorization of causes provides an indispensable framework for analyzing the fundamental structure of natural substances – as well as of metaphysical ("theological"), or eternal, substances and, indeed, artefacts.

3. Finally, what changes by a natural internal impulse is also contrasted with what happens spontaneously ("automatically," in Greek), by chance or coincidence, and by force. Aristotle thinks that just because natural substances have an internal principle of change and rest, their behavior is, to one degree or another, predictable and regular (*Physics* 198 b 35). What happens spontaneously or by coincidence does not conform to this pattern. Nor does what happens by external force, which makes a natural process deviate from its built-in pattern of motion. In other words, Aristotle denies that what happens spontaneously, coincidentally, or by force can be regular and lawlike. He also denies that what happens spontaneously, by chance, or by force can be the object of scientific knowledge. For scientific knowledge depends on logically necessitated demonstrations from secure first principles, as we have already noted, and Aristotle thinks that only non-incidental changes in the objects of a science can be necessitated in this way. For Aristotle, what is spontaneous, chancy, or forced cannot be scientifically known (*Physics* 199 a 1–6).

It is precisely in these areas that Aristotle's thought differs most fundamentally from modern science. Modern science is founded on the notion that regular behavior can be explained by equilibria arising among entities governed by external forces, or that emerge from the spontaneous statistical sorting of chancy events. This tenor in scientific thinking has been made possible by explicit denials of Aristotle's claim that what happens by force, spontaneously, or by coincidence cannot be studied scientifically. Aristotle certainly does not deny that the world is full of loose change, as it were, or of irregular, violent motions. He simply denies that appeals to what happens spontaneously, coincidentally, or by the exertion of force can figure in systematic, cognitively worthwhile explanations of natural processes. Spontaneity, chance, and force do not count for him as causes in the same way that the four causes do. When they are appealed to, it is as excuses for mere oddities, not as basic explanations.

Thus marked off from unmoved movers, externally moved artefacts, and irregular occurrences, Aristotle's physics includes, in the first instance, the study of the four elements, and especially of their natural process of conversion into one another in regularly necessitated cycles. These processes are explained in terms of the various inherent, natural tendencies of the elements that figure in them – fire goes up by nature, earth down. But Aristotle's notion of physics also includes the study of substances whose internal source of motion and rest is soul (psuchē). For Aristotle, soul means primarily "organizing principle." It is not a separate substance that ingresses into the body, as it is for Descartes and various Christian theologians.[2] Soul is instead a principle of life. It integrates beings composed of differentiated parts, or organs, into substantial unities – that is, organisms. This integration-by-differentiation enables ensouled substances to *do* various things – sometimes very clever things – rather than, like elemental cycles, merely undergoing predictable changes. Organized beings – beings with organs – have distinctive "works" or functions (erga) that make them capable of distinctive sorts of activities (On the Soul 412 a 27-b5). Plants, for example, have souls that initiate and guide reproductive, metabolic, and growth functions. Animals have, in addition, sensory and locomotive capacities, as well as affections. For, unlike plants, they must move over space to find food and mates, and so must have not only means of locomotion, but desires to drive them toward some things and away from others, as well as senses to guide them in doing so. Human animals, finally, have rational soul functions, as well as those characteristic of animals and plants (On the Soul 414 a 29-b1).

For Aristotle, organisms have a natural life-cycle; they are not only born and grow, but also age and die ("of natural causes"). From this fact, in conjunction with his view that organisms are paradigmatically

---

[2] The intellectual soul of human beings is, Aristotle concedes, separable from the body. The centuries' long effort by Christian theologians to give the intellectual soul a personal identity led to Descartes's understanding of the intellect as a separate *substance*, an inference that departs from the connection to Aristotle's conception of soul as the form of the body to which Thomas Aquinas, for example, still clung. Of great importance in this transformation was Descartes's denial that plants and animals have souls at all. For him, they are just machines (see Chapter 2).

natural substances, we can see why Aristotle says that what is natural has an internal principle of rest as well as of motion (*Physics* 192 b 14–15). We can see, too, why for Aristotle the study of organisms, and by extension of all natural substances, calls for the use of all four causes. There is an end-oriented temporal dimension in Aristotle's natural substances (growth), as well as an integration of matter and form at each point (metabolism). By contrast, modern physics, restricted as it is to the study of local motion under external forces, involves only material and efficient causation.

So far, then, it looks as though our modern concept of biology corresponds fairly well to the study of Aristotle's ensouled, or organ-ized, physical substances. However, we must be careful. For Aristotelian physics extends not only to living things in our sense, but also to the study of some substances that are "ungenerated, imperishable, and eternal" – namely, the stars and planets (*On the Heavens* 192 b 16–18). Admittedly, these immortal, and hence (by Greek usage) divine, beings do not reproduce, since they are free from the dependencies of plants and animals and from environmental wear and tear. As a result, they can maintain themselves in existence as numerically identical substances forever. In this respect, the heavenly bodies differ from plants, animals, and human beings, which are subject to "generation and decay," and so can live forever only in the sense that they regularly engage in the "highly natural" act of replacing themselves with offspring that have the same characteristics – "an animal [of a certain kind] producing an animal [of the same kind], a plant a plant," in an endless chain of species regeneration (*On the Soul* 415 a 28). (Aristotle says that in acting to replace themselves, mortal ensouled beings – organisms – strive to "partake in the eternal and divine" to the extent that is possible for them [*On the Soul* 415 a 30-b1; *Generation of Animals* 731 b 24–732 a 1][3]). None of this is to deny, however, that for Aristotle the heavenly bodies too are living beings. They are rationally ensouled natural substances whose internal principle of motion and rest is mental. Here we encounter an aspect of Aristotle's thought totally alien to our way of thinking. Although he is not as animistic or panpsychic as, say, the ancient Stoics, who maintained that the whole *kosmos*, as distinct from Aristotle's system of individuated substantial beings, is itself a single living substance,

---

[3] James Lennox argues that this desire or urge (*hormē*) is to be taken as applying to each individual in a species, and as aimed at the eternal persistence of one's own form (see Lennox 2001, p. 134).

Aristotle is willing to assign life and soul, even intellectual soul, to some beings that we have come to regard as decidedly inanimate.

At the end of *Parts of Animals I*, Aristotle admits that the study of "natural substances that are ungenerated, imperishable, and eternal" – that is, the stars and planets – is highly attractive (*Parts of Animals* 644 b 32). But he also says (in what reads like a pep-talk designed to induce reluctant students to study zoology) that we cannot know much about them, while "respecting perishable plants and animals we have abundant information, living as we do in their midst, and ample data may be collected concerning all their various kinds, if only we are willing to take the trouble" (*Parts of Animals* 644 b 28–32). Acknowledging that his prospective scholars may regard even thinking about "the humbler animals," let alone touching, manipulating, and even opening them up, as beneath their dignity, Aristotle suddenly waxes lyrical. He points out that "if some animals admittedly have no graces to charm the senses, nature, which fashioned them, still affords amazing pleasure" when we inspect them (*Parts of Animals* 645 a 7–11). For our attention as scholars is not to be on "blood, flesh, bones, vessels, and the like," but on the causes, particularly the formal and final causes, which reveal in perishable natural substances "absence of anything that is haphazard and conduciveness of everything to an end" (*Parts of Animals* 645 a 24). In this passage, we are afforded a rare glimpse into the motives that induced Aristotle to become the first true philosopher of biology.

Aristotle's Biological Works Surveyed

In the spirit of wonder evoked by the passage we have just summarized, Aristotle sets out to inquire systematically into a number of questions raised by the general picture of mortal ensouled substances we have sketched. The key word here is *systematically*. Each of Aristotle's treatises on natural philosophy, including what we call his biological works, marks off part of what amounts to a highly organized cycle of lecture courses. It is remarkable just how tidily related Aristotle's natural treatises actually are. This can be seen clearly at the outset of his treatise on "meteorology" (by which he means such things as comets, meteors, and the weather). Aristotle remarks here, speaking to his students, "When this inquiry has been concluded, we can consider what account we can give . . . of animals and plants . . . When this has been done, we may say that the whole of our original undertaking [into natural science] will have been carried out" (*Meteorology* 339 a 6–20).

While Aristotle's biological inquiries certainly embrace plants as well as animals, he seems personally to have concentrated almost exclusively on animals.[4] Indeed, he seems especially interested in marine animals, probably because he spent some time conducting first-hand research in the tide pools off the Asian coast.[5] He did this while he was living from 347 to 343 B.C.E. in self-imposed exile in Assos, a city-state on the coast of Asia Minor, and in nearby Mytilene, on the island of Lesbos. Aristotle had gone there at the invitation of Hermias, a former student in the Academy and fellow supporter of Philip of Macedon, when, upon the death of Plato, his own situation in the Academy and Athens had become difficult. It was on Lesbos that Aristotle began to collaborate with Theophrastus, a native of that island, who worked on plants while Aristotle worked on animals. Theophrastus eventually returned to Athens with Aristotle to become a founding member of the Lyceum. He is in this respect the father of botany, composing a *History of Plants* and a *Causes of Plants* modeled on Aristotle's treatises.

The longest of Aristotle's zoological treatises is *History of Animals*. It surveys, in the first instance, the entire range of traits that an animal kind can have: morphological (*morphē*), characterological (*ethē*), behavioral (*praxeis*), and ecological (*bioi*). Aristotle calls these descriptive terms "differences" (*diaphorai*) and, presumably when they are predicated of substances, "attributes" (*symbebēkota*).[6] *History of Animals* also considers how combinations of these traits can be attributed to groups of animals. Aristotle remarks that after "we have grasped the differences and attributes of animals, we must attempt to discover their causes" (*History of Animals* 491 a 10). Given this clean distinction between descriptive and explanatory biology – a distinction that appears in other places as well (*Parts of Animals* 646 a 8–11; *Progression of Animals* 704 b 9–10) – we are prepared to appreciate that Aristotle, consciously utilizing a relevant selection of the phenomena noted in *History of Animals*, explains the distribution of body parts to different animal

---

[4] There is a Peripatetic treatise *On Plants* in the corpus, but it is of doubtful provenance.

[5] On Aristotle's researches in marine biology, see Lee 1948.

[6] This interpretation of the relationship between differences (*diaphorai*) and attributes (*symbebēkota*) is perhaps controversial. Differences are often taken to mean essential differences, and attributes to mean accidental differences. We think it more likely that the terms refer to the same properties, but that the term *diaphorai* is used when Aristotle is referring to a trait that differentiates one kind of animal from another in a given context. The term attribute is used to predicate a property of an animal without necessarily referring to other kinds.

kinds by citing what we could call adaptive or ecological rationales for that distribution (*Parts of Animals* 246 a 8–11). This is the task of the treatise *Parts of Animals*.[7]

Body parts (morphology) are not the only traits that are distributed to animal kinds. What is needed in addition is reference to the actions that an animal must carry out by means of its body parts if it is to live in a specific environmental niche, as well as to the character traits that undergird and regulate these actions. The detailed vocabulary for identifying and comparing ethological and behavioral, as distinct from morphological, traits fills the later books of *History of Animals* (V–IX). There, Aristotle discriminates between character traits (*ethē*), such as tame and wild; actions (*praxeis*), such as crawling or walking or flying; and ways of life (*bioi*), by which he means ways in which animals integrate various actions to make a living in one or more of the three great environments – land, air, and water. A number of treatises then deal with ways in which these characterological and behavioral traits, grouped together under a set of soul or life functions, are embodied and carried out. Texts in this group include *Generation of Animals* (reproduction), *Movement of Animals* (kinesiology), *Progression of Animals* (locomotion), *On Length and Shortness of Life* (aging), and *On Youth, Old Age, Life and Death* (life-cycles). Treatises dealing with what we would call animal (and human) psychology should also be included in this group. These include, most importantly, *On the Soul* (*De Anima*) (sensation, memory, imagination, desire, knowledge), as well as certain "little" self-standing natural treatises ("*parva naturalia*") such as *On Sense and Sensibilia, On Memory, On Sleep, On Dreams*, and *On Divination in Sleep*.

As we have already said, and as the treatise *On the Soul* makes clear, soul is for Aristotle the form, or principle of integration and identity, of living substances: substances that are materially composed of differentiated, functional parts (*On the Soul* 412 a 20). Inanimate, or soul-less, beings do not have functioning parts or organs; they are merely aggregates or heaps of the same material, as a pile of sand is merely an aggregate of grains of sand. In *History of Animals*, Aristotle says that the characters and behaviors of each species are "footprints" of the distinctive life functions that make individual organisms into integrated substantial beings in the first place (*History of Animals* 588 a 20).

---

[7] Arguments in favor of the relationship between *History of Animals* and *Parts of Animals* expressed in this paragraph have been made by David Balme, Allan Gotthelf, and James Lennox.

Organisms are not, accordingly, mere loci where a variety of traits just happen loosely to be assembled wherever a specific ecological niche brings them together. Certainly, plants and animals do have just the set of traits that will enable them to survive in the environments that are natural to them. But it is more accurate to say that, for Aristotle, traits are distributed so that the organisms possessing them will have the *best* chance of developing and expressing the species-specific array of life-functions that makes each of them substantial individuals in the first place. There is an evaluative element at the heart of Aristotle's biology.

Speaking generally, the basic life-functions include "copulation, reproduction, eating, breathing, growing, waking, sleeping, and locomoting" (*Parts of Animals* 645 b 32–35). A plant, Aristotle says, "seems to have no work other than to make another like itself" (*History of Animals* 588 b 23–25, our translation). Accordingly, their life functions are restricted to reproduction, feeding, growing, and reproducing again (*On the Soul* 413 a 25–31; 414 a 34).[8] Precisely because they must move through space to acquire resources and avoid dangers, however, animal kinds also possess other classes of traits that we do not find among plants, which are stationary. In addition to means of locomoting, they have various sorts of sensory information-acquisition systems, ranging from the slightest responsiveness to touch to the most elaborate combinations of touch, vision, hearing, taste, and smell to tell them where to move. They also have complex patterns of desire and aversion in order to orient them to an appropriate set of objects. Animals also oscillate between a waking state of alertness, in which they carry out these higher life functions, and a sleeping state, in which Aristotle thinks they live a plant-like life (*Nicomachean Ethics* 1102 b 2–12).

Aristotle admits that the ways of life of animals require them to bend all of their abilities to feeding and reproducing themselves (*History of Animals* 589 a 3–5). This does not mean, however, that other capacities, especially those of sensation, play only an instrumental role in ensuring survival. Because it has an incipiently cognitive aspect, sensation is for Aristotle valuable in its own right (*On the Soul* 434 b 23–25). This insistence becomes clear in the case of human beings, whose intellectual abilities, which are themselves predicated on vivid imaginations and good memories, equip them not only to make a living for themselves

---

[8] Since Aristotle thinks of reproduction, as well as growth, as a result of nutrition – for reasons that will be made clear in the text – he assimilates the life of plants to the function of nutrition.

and their offspring, but to understand the world in which they are living. In this case, the orientation of life functions reverses itself. In proportion as things go well, metabolic and reproductive functions, as well as sensation, imagination, memory, and desire, can be directed toward ever more penetrating accounts of the world through the leisured cultivation of the cognitive soul for its own sake.

One can see, then, that, in the end, mere survival is not for Aristotle the governing principle of biological order, as it is, for example, from a Darwinian perspective. Instead, biological order is determined by a value-laden hierarchy of soul functions:

> Nature proceeds little by little from things lifeless to animal life in such a way that it is impossible to determine the exact line of demarcation, nor on which side thereof an intermediate form should lie ... In plants there is a continuous scale of ascent toward the animal. In the sea there are certain objects concerning which one would be at a loss to determine whether they are animal or plant ... In regard to sensibility, some animals give no indication whatsoever of it, while others indicate it, but indistinctly ... And so throughout the entire animal scale there is a graduated differentiation in the amount of life and the capacity for motion.
>
> *History of Animals* 588 b 4–22

This passage – the *locus classicus* for the long-lived notion of a "ladder of nature" (*scala naturae*) or "great chain of being" – shows that Aristotle's biological inquiries project and defend a certain metaphysical picture. No empiricist in the ordinary sense of the word, as he is often reputed to be, Aristotle appeals to his distinctive notion of soul as the form of organized beings not only to link the highest, divine aspects of the universe with the lowest, merely material dimensions, but also to cut a path between the reductionistic materialism of Democritus and Empedocles and Plato's tendency to overlook or minimize our embodiment. So prominent are these themes in Aristotle's writings that it is easy to believe (as we in fact do) that Aristotle's whole philosophy was stimulated, as well as confirmed, by his biological preoccupations. If so, one might go further, holding that Aristotle's entire conceptual arsenal – not simply the four causes, but more especially the way in which he applies these causes to the study of substances by means of such concepts as substrate (*hypokeimenon*), essence (*to ti en einai*), potentiality (*dynamis*), and actuality (*energeia*) – was designed explicitly to resolve the problems posed by research into living things (Grene 1963). That is to say, the "meta-vocabulary" that Aristotle devised for biological inquiry went into the making of his metaphysics. It is no less true, however, that Aristotle was insistent that all of physical science, including biology, must conform to the autonomous principles of first philosophy. Such, for example, are the principles that govern our determination of what really is and is not a substance, as well as the principles from which Aristotle derives his account of the overall structure of the cosmos as a *scala naturae*. This hierarchical metaphysical and cosmological framework was by and large retained by Descartes's predecessors and contemporaries, even though latter-day "Aristotelians" were by no means faithful to Aristotle's own detailed scientific thought. Signaled by Descartes's exclusive disjunction between extended matter and pure thought, the collapse of the *scala naturae* marks not only the dividing line between the modern and the premodern, but also the increasing unavailability of the mean that Aristotle vigorously sought to find between reductionist materialism and disembodied rationalism.

From Descriptive to Explanatory Biology

Given the general Aristotelian framework that we have sketched, let us now ask in more detail how his three major zoological works, *History of Animals*, *Parts of Animals*, and *Generation of Animals*, figure in his program for natural science. In this section, we consider the relationship between *History of Animals* and *Parts of Animals*; in the next section, Aristotle's account of reproduction in *Generation of Animals*.

History of Animals contains four kinds of information:

1. Observations that could be made into species-by-species "natural histories" of the sort later made familiar to Romans by Pliny and to Frenchmen by Buffon when they wrote about the morphology, life-cycles, and habits of various species. This dimension of the text is especially noticeable when Aristotle relies on reports from fishermen or beekeepers, for example, to inform himself about the habits of species with which they were familiar. On the whole, Aristotle and his collaborators were remarkably good observers, not just of the intricate structures that could be seen only through dissection, but of the sexual and other habits of animals, particularly the marine animals with which Aristotle was most directly familiar.

2. A division of animal kinds, first into the bloodless and the blooded; then into nine "great kinds" (*megista genē*) – five blooded, four bloodless; and finally into an indefinite number of "lesser kinds," or what we call species. Species share the general characteristics of one

or another of the great kinds, but vary from each other by "the more and the less," as when one sort of bird is reported to have a longer beak or a shorter leg than another, or a more pugnacious character (pigeons), or a more cooperative or "political" way of life (cranes) (*Parts of Animals* 692 b 4–7).[9] As for the great kinds themselves, the blooded ones are identified in terms of mode of locomotion combined with mode of generation. There are two-legged live-bearers (human beings), four-legged live-bearers, four-legged egg-layers (turtles, for example), two-legged egg-layers (birds), and legless egg-layers (fish). The bloodless kinds are mostly the sorts of animals Aristotle might have observed in shallow marine tidal pools. He appears to have invented names for these kinds by considering whether they are "softies," such as octopi[10]; "softshelled," such as crabs; "potsherd-skinned," among which he includes animals whose hard outsides protect markedly soft insides, such as sea snails and clams; and, finally, segmented or "cut-into" (*en-toma*, in-sects) animals, such as centipedes, which are hard all the way through.

3. *History of Animals* also contains a catalogue of traits. Indeed, the treatise as a whole seems to be organized in terms of names for body parts (Books I–IV), actions and ways of life (IV–VII), and, finally, character traits (VIII–IX). As David Balme has remarked of this text, "Animals are called in as witnesses to differences, not in order to be described as animals" (Balme 1987b, p. 88.) The most notable feature of the resulting "trait vocabulary" is that it makes use of the term *eidos* to refer not to species of animals within a larger genus – both great and lesser kinds are still kinds (*genē*) – but to quantitative differentiations or determinations that mark off lesser kinds within large ones. Sparrows differ from other animals within the great kind of birds because they have beaks, legs, and bodies of such and such a size. All of these are quantitatively varied forms (*eidē*) of bird (Lennox 2001, pp. 160–162). A trait vocabulary that is to be useful for the purposes of a scientific comparative anatomy and ethology must also include, however, terms for traits that perform analogous functions in different great kinds. Sometimes natural language already provides names for these these. "Gill," for example, names in

---

[9] Aristotle includes some things that we would call "qualitative" variation, such as different intensities of a character trait or a behavior, under the notion of "quantitative" variation. That is because quantity in ancient Greek mathematics includes both discrete and continuous quantity. That means it refers not only to numerical differences but to any feature whose different intensities can be compared by proportionality.

[10] The softies (*malakia*) are also called "cephalopods," because their head (*kephalos*) is down by their feet [*podes*]).

fish what is analogous to lungs in most blooded land-dwelling animals (*Parts of Animals* 645 b 8). Often, however, such terms have not been discriminated by speakers of natural languages. Thus, although snakes have something that corresponds to the neck and chest in humans and quadrupeds, these have no common names (*Parts of Animals* 691 b 29–692 a 10).[11] One can, of course – and Aristotle often does – continue to use terms such as "chest" to describe the analogous parts of snakes or birds rather than invent a term of art (*History of Animals* 503 b 30). Since most names in natural language refer to us, this leads to a certain anthropomorphism that is unintended in Aristotle's biology.

4. *History of Animals* also contains a survey of the co-extension or co-distribution of various traits. In this part of his treatise, Aristotle seems especially concerned to discover linked series of traits that are found always, or for the most part, in animals falling within the same great kind. We learn, for example, that "as many as are four-legged and live-bearing so many also have both an esophagus and a windpipe" (*History of Animals* 505 b 32–33). In addition to noting where "all X's have Y's" (or, more often, "as many as have X also have Y" [Lennox 2001]), Aristotle also takes pains to notice where only "some X's have Y's," and where "no X's have Y's," as when he remarks, "No fish has a neck, limbs, testicles, or breasts" (*History of Animals* 504 b 17–18). This effort to find levels at which traits co-vary, or cease to co-vary, is even more evident in *Parts of Animals*, where Aristotle focuses on cases where a trait that one would expect in a species that falls with a certain great kind happens to be missing.

Aristotle's survey of co-variant or co-extensive traits seems to us to be the most important of the four aspects of *History of Animals* we have distinguished. This is so because co-extended traits are the primary *explananda* of Aristotle's zoological project. Natural history in the medieval and modern senses – description of the life-cycle of each species – takes a back seat because, for Aristotle trait distributions are to be explained at their highest shared level, not species by species (Lennox 2001, p. 62). If, for example, all viviparous quadrupeds (a great kind)

---

[11] Something like this happens in Aristotle's moral psychology as well. Aristotle is impressed by the way in which names for virtues and vices that have culturally evolved fit nicely into his scientific scheme, according to which for every virtue there are two kinds of vice – an excess and a deficiency. Occasionally, however, names for one or another excess or deficiency, or even for a moral virtue, are missing, and must be supplied in the interests of a complete scientific account,.

have bladders and kidneys, then bears or horses have kidneys and bladders not because they are bears or horses, but because they are viviparous
quadrupeds (*Parts of Animals* 506 b 25–8). If anomalies and exceptions
occur, they deserve explanation, to be sure. But that is only because what
is to be explained in such cases is a deviation from an expected distribution of traits. A species of octopus with only one row of suckers, for
example, when all other kinds of octopi have two, must be explained as
an anomaly (*Parts of Animals* 685 b 12–16). So too must the apparently
sightless mole rat (*History of Animals* 533 a 3).

If natural history is less central than explanations of co-extensive
traits in Aristotle's zoological inquiries, systematic taxonomy of animal
kinds is more recessive still. Even Aristotle's most fervent modern admirers must admit that if his intention were to produce a system that
would assign an unambiguous place for every species within a system
of taxa that included families, orders, and classes, as well as genera and
species, he did not get very far. He did not even get far enough to distinguish between genera and species in our sense, for that is not what
genos and *eidos* mean, as we have seen.[12] Moreover, Aristotle recognizes that there are whole groups of animals that do not fit into his nine
great kinds. At one point, he suddenly seems to elevate the live-bearing
cetacea (whales, dolphins, sharks, and the like) into a great kind of its
own (*History of Animals* 490 b 8–9). Where, too, are we to place the
snakes? In addition, Aristotle admits that there are many "dualizers" –
species that "fall on both sides" of a large genus, such as apes (*Parts of
Animals* 689 b 31–34), seals (*Parts of Animals* 657 a 22), and hermit
crabs, the last of which, because they commandeer a shell to live in,
dualize between soft-shelled animals such as crabs and externally hard-
skinned, but internally soft ones such as clams (*History of Animals* 529

---

[12] David Balme has shown that Aristotle cannot be doing systematics in the modern sense
because the terms *genos* and *eidos* do not work in the way that would be required if
this were his aim. Among Balme's evidence is the fact that Aristotle speaks of a *genos*
of octopus that has one, rather than the usual two, rows of suckers, thereby turning
genos into a variety or sub-species (*History of Animals* 523 b 28). Conversely, we
are told that viviparous and oviparous animals form two *eidē*, even though they are
both *megista genē* (*Generation of Animals* 729 b 234). Finally, the same things can
be picked out as both *eidos* and *genos* in different contexts. At one point, two *genē*
of cicadas are mentioned, even though Aristotle says elsewhere that there are "many
eidē of cicadas" (*History of Animals*, 555b a 14; 532 b1 4). The general point is that
genos-eidos, at whatever level these terms work in a given context, refer to the plural
eidē ("looks") that spring from a common *genos* (Balme 1962).

b 20–25). In the famous passage on the *scala naturae* from which we have quoted, one can sense Aristotle's hesitation about pronouncing on borderline cases.[13]

In the eighteenth century, when biologists such as Linnaeus were intent on systematically classifying kinds into a hierarchical system that unambiguously afforded a unique address for every species, even species not yet identified, the medieval tendency to view Aristotle through Pliny's eyes, as a natural historian, gave way to thinking of Aristotle as a systematicist, in spite of his deficiencies. Aristotle's high-level differentiation between the blooded and the bloodless was taken, for example, to be a good first pass at the vertebrate-invertebrate distinction. Similarly, his supposed discrimination of a two-level taxonomic system was presumed to be a down payment on a more ramified system of taxonomic levels. Nineteenth-century comparative anatomists who were skeptical of Linnaeus's "artificial" approach to classification did nothing to undermine the notion of Aristotle as a systematicist. On the contrary, John Hunter, Georges Cuvier, and Richard Owen all admired Aristotle precisely because they assumed that he was, essentially, a systematic taxonomist who was working toward a "natural" system of classification which, like their own, respected the complexity of the phenomena.[14] This was not, however, Aristotle's aim, for at least two reasons. One reason follows from the fact that trait co-extensions are the object of Aristotle's explanatory biology. If this task can be accomplished without a systematic taxonomy in our modern sense, then so be it. The other reason is that Aristotle seems to have been aware of just how hard it

---

[13] Many contemporary Aristotle scholars believe that Aristotle eventually comes down on one side or another of a classificatory divide in the case of dualizers. We are inclined to think that one can admit more hesitation than that. Aristotle does not regard statements of the form "A is B with respect to (or in comparison with) one thing, but A is C with respect to something else" as flirting with relativism, but as objective statements about the proportional or relational structure of the world, which is mirrored in human knowledge. On this issue, we concur with Lloyd 1996, pp. 67–82.

[14] In his *Hunterian Lectures* of 1837, Richard Owen writes as follows: "The sketches which Mr. Hunter has left of his views of the arrangement of the animal kingdom more nearly correspond with that of Aristotle and nature than the more artificial system of Linnaeus: – but the attempts at zoological classification which the Hunterian manuscript [on which Owen was supposed to be commenting] contains cannot of course be compared to the bold and clear enunciations on this subject which the Stagyrite gave to the world; and the merits of which it required the profound researches of a Cuvier to enable us fully to appreciate" (Owen 1992, p. 95). See Chapters 3–5 for further discussion of these topics and figures.

would be to produce a jointly exhaustive and mutually exclusive set of taxonomic categories.

Aristotle's sense of how difficult it would be to produce a system of classification can be seen in his criticism of efforts that were underway in the Academy throughout the period of his residence there to "cut nature at the joints" (Plato, *Phaedrus* 265 e 1–3) by using successive dichotomous divisions (*diaireseis*) to define animal kinds by giving them a unique address in a larger system. Although the details of this method are unimportant here, the main idea is that starting from any arbitrary generality, the meaning of a more specific term can be marked off by a process of successive differentiation until a unique definition has been isolated at the end of the chain. A record of this research program is found in Plato's so-called "late dialogues," such as *Sophist*, *Statesman*, and *Philebus*. We will never know for sure whether Aristotle's biological researches were stimulated by his doubts about this program, or whether his concrete research into animals gave rise to his dissent. What we do know is that Aristotle's appreciation of the complexity and diversity of living things constituted a devastating criticism of classification by dichotomous division.

In *Parts of Animals I*, which constitutes a little treatise on method in biology in its own right, Aristotle claims that, when applied to animals, any procedure of successive dichotomous division will end up dividing through a series of characteristics that have no biologically realistic path from one to another:

> I mean the sort of thing that comes about if one divides off the featherless and the feathered, and among the feathered the tame and the wild, or the white and the black. For the tame or the white is not a differentiation of feathered, but begins another differentiation and is accidental here.
>
> *Parts of Animals* 643 b 19–24

The result of failing to understand that divisions must be realistically based on the pattern of real differentiation in a growing, developing organism may well be that we will take "proper accidents (*idia*)," such as "featherless biped" in the case of human beings, as defining characteristics; that we will treat non-existent, privatively defined classes, such as "featherless," as ontologically real (as if one could divide through nothing); and that we will turn definitions into long, unwieldy, overly inclusive collections of haphazardly collected sets of properties, which at most reveal the superficial meaning of names, but are not true definitions.[15] Aristotle admits that one can get by division (if not always by dichotomous division) from (generic) footed to (specific) two-footed, or four-footed, or many-footed; or from tame (a characterological trait) to gregarious (a behavioral trait); or from gregarious to political, or indeed from political to leaderless or governed (*History of Animals* 488 a 1–13). He does not want to do away with division altogether; he merely wants to reform it (Balme 1987b). Still, one cannot get from two-legged to tame or feathered. For only traits that have a common root in the developmental process of an organism, as it moves from less determinate or generic to more determinate or specific, can be divided. There is no hope, accordingly, of defining each species, or even large genera, by presuming that they will all be neatly arranged, one "specific difference" away from their nearest neighbor, at the bottom of a branching diagram.

How then *does* Aristotle find his great kinds? He says that the proper procedure is to "follow the many, who define the bird-*genos* and the fish-*genos* . . . by marking off many differences" – that is, by simultaneously taking a number of traits, both morphological and behavioral, and seeing whether or not they all appear in a large number of species, each of which differs from the others by quantitative variation in one or more of those generic aspects (*Parts of Animals* 643 b 10–12). This passage certainly expresses Aristotle's confidence in the ability of ordinary people to learn by adding the results of one generation's inquiries to those of earlier generations and embedding the results into language itself. Aristotle was quite serious when he claimed at the outset of his *Metaphysics* that "human beings by nature desire to know," and serious, too, in thinking that they are pretty good at it (*Metaphysics* 1.980a22). Still, the recommendation to "follow the many" is not meant to suggest that all theoretical biologists need to do is to add newly formed categories,

---

[15] The objections to definition by successive dichotomies listed in this paragraph summarize the discussion in *Parts of Animals* 642 b 5–644 a 12. It should be borne in mind that the term "definition" translates two different concepts in Aristotle's Greek. Verbs formed from *horizein*, and nouns such as *horismos*, are definitions in the sense of something, or a word, as marked off from others. Terms such as *logos ousias*, or "formula for the substance," are often called definitions, or essential definitions, in Latin (and hence English), since they refer to the what-it-is-that-makes-something-the-thing-it-is (*to ti ēn einai*). Balme 1987b is at least ambiguous when he says that "the purpose of collecting traits is to form definitions of animals." If this refers to markings off, it is false; if it refers to definitions in the substantive sense, it would be more correct to say that definitions *explain* why animals are marked off the way they are.

such as "softies" and "cut ups," to commonly recognized ones, such as "bird" or fish." For the process of systematic, theoretically driven inquiry, which leads to the recognition of new kinds, also requires that the concepts that are found in common language, such as bird and fish, be independently reaffirmed by analysis as valid. To qualify as a "great kind," it seems necessary that a genus should constitute a site where co-extensive traits vary quantitatively to give a large number of lesser kinds (Lennox 2001).

Viewed in this light, the function of the great kinds is to serve as a benchmark for explanations of traits. Sometimes, traits will be shared at a level above these kinds. Terms such as blooded and bloodless, for example, do not form incipient super-kinds, such as vertebrate and in-vertebrate, but instead identify traits that happen to be shared across five kinds in the case of the blooded and four in the case of the bloodless. Having grouped the blooded animals together, Aristotle notes, among other items, that there is a correlation between having lungs and being blooded. But this correlation is not universal, in the sense that would allow Aristotle to say, "As many as are lung-possessors so many are also blooded." For fish are blooded, but do not have lungs. It is anomalies like these that especially attract Aristotle's efforts at explanation. Fish would have lungs, he implies, were it not for the fact that they live in water. So they have something analogous to lungs – namely, gills. The same pattern can be found below the level of the great kinds. Some-times a trait that is found to one degree or another in most, or all, lesser kinds that are clustered around a given kind happens to be absent in one. The mole rat, for example, is unique among others of its kind in being blind. The reason is that, unlike other species in this genus, it lives entirely underground and cannot in any case use its eyes (which can be found by dissection beneath the skin) to see (*History of Animals* 533 a 2–7). From this perspective, to describe the hermit crab, the ape, or the seal as lying between two great kinds is not to fail in a classificatory task. On the contrary, it is to pick out precisely what is puzzling about these animals, and so to pick out what needs to be explained. Not only would it be useless to find unambiguous homes for these animals within a taxonomic system; it would obscure the status of their most salient characteristics as cases calling for explanation.

It is in *Parts of Animals* that Aristotle provides explanations for the distributions of morphological traits that he establishes in *History of Animals*. In general, these explanations are teleological. They work on the principle that "nature does nothing in vain, but in every case, given the possibilities, does what is best for the substantial being of each kind of animal" (*Progression of Animals* 704 b 12–18, translation Lennox). Ideally (*malista*), Aristotle says, the procedure should be one in which inquiry explains traits as necessitated by the essences of the kinds that bear them. Definition is not, in this respect, a question of being able merely to identify and distinguish species. Definitions are expressions of the essences (*ta ti en einai*) that make animals the substantial unities they are (*Metaphysics* 1045 a 12–19; *Parts of Animals* 643 b 18–19). To the extent that they are available, definitions in the sense of "formulas of the essence" afford inquirers a principle by reference to which they can ask and answer questions about why this or that kind has the precise array and hierarchical organization of traits that it does. Aristotle illustrates this in the case of human beings in the following way: "Because 'what it is to be human' is this, therefore [human beings] have these things" (*Parts of Animals* 640 a 34–35, our translation). If this kind of information is available, one might be able to deduce why each animal kind has the properties it does from formulas that pick out a set of life-functions, or "formal natures." Some parts will follow directly from the essence as part of the constitutive conditions of the kind; failure to mention them will mean failure to refer to the kind in question at all. Other traits will play a supportive, but necessary, role. Others still will be functionless concomitants of the material conditions for the emergence and existence of an animal of such and such a kind, such as eye color.

True, knowledge of essences is not easy to acquire. Happily, however, another approach is available, and in fact pervades the explanations actually given in *Parts of Animals*. Lacking knowledge of a defining essence, Aristotle says, "the nearest thing is to say that there cannot be a human being [for example] without these parts" (*Parts of Animals* 640 a 35–36). This approach, rather than working by demonstration from known essential definitions of formal natures, works by way of hypotheses from the "material nature" of an animal kind (*Parts of Animals* 641 a 25). If an animal is to live in its environment, the zoologist can identify the structural and dispositional properties of the matter from which the organism and its parts are made in order to see how these properties enable it to make a living, reproduce, and exercise its life-functions in the particular environment it is placed in. This is what Aristotle calls reasoning in terms of hypothetical, conditional, or *ex hypothesi* necessity. "If a piece of wood is to be split with an ax," he writes, "the ax must of necessity be hard; and if hard, it must be made of bronze or iron." "In exactly the same way," he continues, "the body, since it too is an instrument . . . must of necessity be of such and such a character, and made of such and such materials" (*Parts of Animals* 642 a 10–13).

Hypothetical necessity is necessary in two senses. It relies on the unconditional necessities of Aristotle's theory of elements and their cycles, together with his neo-Empedoclean theory of what *we* would call chemical and biochemical compounding (*mixis*), to identify what he calls the "material nature" of animal kinds. It also relies on a necessary bond between these materials, or mixtures of materials, and their inherent dispositional properties, which sustain life-functions. Since connections between materials, traits, and life-functions of this sort generally hold at the level of shared traits, we begin to see more clearly why Aristotle places traits and their co-extensions, rather than individual kinds, into the foreground of his zoological theory. They constitute the *explananda* of his zoological inquiries.

Nature (metaphorically understood) does nothing in vain, to be sure, but Aristotle's teleological principle also recognizes that since nature relies so heavily on hypothetical necessity, it must work under constraints that arise from the material natures with which it is working. It is not only that nature does nothing in vain; it does what is best, given what is available in a given set of environmental circumstances (*Progression of Animals* 704 b 12–18). That is why the absence of a trait that might be expected at the level of a great kind is often explained as an apt trade-off between the special ecological requirements of this or that animal, and the finite materials, both inorganic and organic, that nature has at its disposal. "Nature," Aristotle writes, "invariably gives to one part what it subtracts from another" (*Parts of Animals* 658 a 35).

The camel's dentition is an illustrative case. It follows directly from the basic requirements of growth and metabolism that all "animals must have . . . some receptacle for the food they take in," and another for storing the concocted food from which moisture has been removed prior to elimination (*Parts of Animals* 650 a 2–33; 674 a 12–16). Aristotle calls these "essential parts" stomachs and bowels. He goes on to discuss various different kinds of stomach, and their associations with other linked traits (*Parts of Animals* 674 a 20–b 35). He notes in this connection that "those of the blooded live-bearers that have two sets of front teeth also have one stomach." This co-extension holds for all blooded live-bearers with solid hooves, but applies to only some of the cloven-hooves. There are, it would seem, some cloven-hoofed live-bearers that have two stomachs, but only one row of front teeth. But then Aristotle notices something else. This reversal of the expected pattern of one stomach and two sets of front teeth occurs for the most part among horned animals. At this point, Aristotle attempts to provide an ecological rationale for this distribution of traits (at their highest level of co-extension). On the teleological principle that animals get only the traits they require, it would seem that the more teeth a species possesses, the fewer stomachs it will have, since mastication is already a first step in concoction. What separates the live bearers that have two stomachs but only one row of front teeth from those with two sets of front teeth but only one stomach is rumination. For the fact is that cloven-hoofed and horned animals that generally possess two stomachs and one set of front teeth are ruminants. Thus it would seem more important that, given their diet, they have two stomachs rather than two sets of front teeth. This explanation is confirmed by a case that at first glance seems to be an exception – the camel. Even though the camel is cloven-hoofed and ruminates, it has only one set of front teeth, and it lacks horns. It probably could make good use of two sets of front teeth. For it is a large animal, and so needs a good deal of food. But it lives on a diet of hard-to-digest thorny and woody materials. Why, then, has not nature used the material that might have gone into making horns to make an extra set of front teeth? Here, Aristotle reapplies the explanatory warrant that he has used in the case of other ruminants. Camels, he says, need multiple stomachs more than they need teeth. Having counted no less than four stomachs in the camel, he concludes, "Nature has made use of the earthy matter saved from the missing teeth [and the missing horns (*Parts of Animals* 663 b 29–36)] to make the roof of the mouth hard" in view of the camel's rugged, desert diet (*Parts of Animals* 674 b 4–5).

This example (analyzed in great detail in Gotthelf 1987) illustrates Aristotle's conviction that organic structure is necessarily hierarchical as well as differentiated into different kinds of parts. The order that must be followed in explaining the distribution of body parts follows the ontological composition of living things (*Parts of Animals* 640 a 33–7). First, the available material must be worked up into body parts that are "in the essence" – that is, are immediately called for by a kind's definition. (This appeal to definition serves to remind us again that, for Aristotle, "essence" is a notion that plays a role in explanations; it is not limited, as later tradition would have it, to finding necessary and sufficient conditions for placing an entity in a class.) Next, material must be assigned to parts that are derivatively required if the essential ones are to work properly under the particular set of environmental conditions that mark off a generic or specific way of life (Cooper 1982). Fish, for example, essentially need eyes if they are to see prey, predators, and mates. But they live in water. Hence their eyes must be protected by lids. Eyelids, then, are also necessitated. Next, if there is still material available, nature will work it up into "back-up" traits, such as two kidneys. Finally, there may be materials that have no function at all, but are merely concomitant products of the basic physical and chemical processes on which functional traits rely. The color of eyes is Aristotle's standard example. Bodily residues, such as urine and pus, are also concomitants.

As long as Aristotle's *History of Animals* was viewed either as a curious compendium of notes on natural history or as an incipient, but not very successful, taxonomic system, it was difficult to see any substantive relationship between this work, or any of Aristotle's other biological treatises, and his *Organon* – that is, his treatises on logic, such as *Prior Analytics*, and on scientific method, notably *Posterior Analytics*. However, when the focus of *History of Animals* is placed on establishing the distribution of co-varying sets of traits at the highest level at which they are shared, the stress that Aristotle puts on the quantifiers – all, some, and none – in his logic assumes a much more central role in his biological works. Moreover, the ecologically based explanations of these trait distributions that are offered in *Parts of Animals* approximate to what is required for genuine scientific explanation. Admittedly, the explanations of co-extended traits that we find in *Parts of Animals* are seldom traced all the way up to, or derived from, essential definitions. Nor are they derived and displayed in syllogistic form. Nonetheless, the midrange explanations we find in Aristotle's *Parts of Animals* do contain material that could in principle be set forth in the form of a finished, deductive, demonstrative science (Bolton 1987; Gotthelf 1997; Detel 1997; Lennox 2001).

From *Parts of Animals* to *Generation of Animals*

The purposiveness of nature is especially apparent, Aristotle argues, when we turn from the synchronic match between matter and form that is the focus of attention in the comparative morphology of *Parts of Animals* to the diachronic development of individual organisms in and through the process of generation, which binds efficient causes to final causes in the production of a mature member of its kind capable of beginning over again the cycle of conception, development, birth, growth, and reproduction. This is the subject of the third keystone work of Aristotle's zoological inquiries, *Generation of Animals*. When Aristotle speaks of hypothetical necessity, he refers not only to the fact that matter must be put to the uses of form in each species, but also to the fact that in order for an organism to develop and perform its defining functions through characterological and behavioral traits that are founded in material dispositions, its parts must develop in a specific order. "We must also show," he writes, "that this or that part of the process is necessitated by this or that earlier stage of it" (*Parts of Animals* 642 a 32–3; cf. 639 b 29–31; 640 b 1–4). Examination of this diachronic aspect of Aristotle's zoology makes clear that his teleology is not the external teleology of a craftsman, but the internal teleology of a complex system as it moves through many contingencies toward an end.

Aristotle's recognition of the spontaneous generation of organic compounds as occasionally resulting in functioning organisms shows just how much power he attributes to material natures (Lennox 2001, pp. 229–249). Nonetheless, organized substances come into existence more reliably, and hence naturally ("always or for the most part"), by way of a process in which the male parent, working up already highly processed material supplied by the female, transmits specific form – the what-it-is-to-be an X [*to ti einai anthropos*, or whatever other species is in question] – to an embryo. At some point, the nascent organism takes over the process of growth and differentiation for itself, thereby renewing the reproductive cycle, "like giving birth to like" in a chain that resembles the cycles of the heavenly bodies. The process is described in great detail in *Generation of Animals*. It is clearly end-directed.

The ontogenetic (or developmental) process depends on the ability of organisms – first parents, then offspring themselves – to transform food stuffs, specifically adapted to their natures, into organic traits by what Aristotle calls "concoction" (*pepsis*). This occurs in several stages. Uniform parts, such as sinew, fiber, bone, blood, cartilage, horn, hair, marrow, milk, and semen, are worked up into complex compounds from specific combinations of the four elements. These stuffs then become material for non-uniform parts, such as hearts, stomachs, bowels, lungs, eyes, ears. It is only with the formation of the heart, Aristotle believes, that a growing organism starts to produce uniform body parts on its own and to distribute them to various sites in the body where non-uniform parts are to be constructed. For the function of the heart is to further concoct food that goes into the formation of uniform stuffs, such as bone, blood, or marrow, and to push these between the emerging structures of the body, like purée through a food mill, so that when these products cool, they will become part of the structure itself. (Aristotle's picture of ontogeny is highly colored by analogies to the processes of cooking [Gill 1989; 1997].) Bone material congeals into a solid frame to support the weight of a developing and differentiating mass. After that, organs of sensation form at various appropriate sites, beginning with skin, the organ of touch. For Aristotle reasonably thinks the parts that support the most basic life functions must develop before those that support sensation, desire, and cognition. As long as organisms are able to perform these feats, they remain individual substances. Thus the passing away of an organism is tantamount to its inability to continue its activities. As an organism grows older and colder, matter gets the better of form.

Aristotle regards reproduction as an integral part of the ontogenetic process. This process is driven by heat. Thus he holds that the stuff that transmits form – a subtle fluid carried by the male sperm called *"pneuma"* – is the most highly concocted, and hence the hottest, of all bodily parts. Since, moreover, this process goes further in humans than in lower animals – humans have longer gestation times, more fully differentiated bodies, and a complete range of senses – Aristotle takes human *pneuma* to be hotter than that of other animals. Some interpreters have fancied that Aristotle's *pneuma* carries all the "information" needed to specify offspring (Delbrück 1971; Furth 1988; Kullmann 1998). This can be misleading. Aristotle's *pneuma* possesses merely the precise degree of heat needed to set in motion a predictable cascade of events, in which what happens at any point is caused by what has immediately preceded it. We do not find Aristotle's biological teleology in a predetermined script. Nor do we find any gaps between a succession of causal transactions of a sort determined by physical and chemical necessities. Aristotle's stress on material natures and hypothetical necessitation shows that if there is no physical and chemical mechanism for developing a functional morphological or behavioral trait, then no such trait can possibly arise. At the same time, lower-level processes are themselves insufficient to explain the regularity of organic form and the purposive behavior of organisms. For development is not the work of matter, but of form, which reaches down and shapes material into functional parts that are fit to carry out the "soul functions" that define the life, and life-cycle, of an animal. Formal and material causes are fused within a temporal process in which efficient causes – first the parent, then the offspring itself – are oriented toward the generation of a fully functioning, mature, reproducing organism. This process takes place, as Aristotle says in several places, "both for an end and of necessity" (*Generation of Animals* 739 b 20–28; 743 a 36). At the same time, the role of heat and cold in these processes reminds us that, on the whole, Aristotle's theory rests on a balance of qualities.

Aristotle's theory of reproduction is in part a response to a Hippocratic belief that semen is drawn from the whole body, special seeds for each body part congregating in the semen and being reassembled in the next generation. This early version of what Darwin was to call "pangenesis" was designed to account for inheritance. Aristotle's underlying worry about the Hippocratic view is philosophical. With its openness to environmental influences and to different contributions from the sexes, pangenesis threatens the unity of form that is required to make organisms into truly unified substances (Furth 1988). Aristotle's own theory of inheritance, accordingly, stresses the unity of form in the male parent more for metaphysical than for empirical or ideological reasons, although it must also be admitted that the resulting doctrine embodies a good deal of sexism. When, for one contingent reason or another, spermatic heat does not fully "cook up" an embryo into a more or less good copy of its male parent, it becomes, first, a female, then a "throwback" to earlier generations, and finally a nondescript member of its kind (*Generation of Animals* 767 b 7–768b14). For Aristotle, accordingly, only the transmission of form from male to male can explain the descriptive continuity of species. For only males embody the form in a non-defective way. Aristotle admits that this process fails in virtually every case to make absolutely perfect replicas of paternal form. He seems to hold, however, that it is effective enough to result in recognizably unchanged lineages that have already been in existence for an infinite number of generations. If the past is any guide to the future, these kinds will continue to exist for another infinity of generations. Aristotle's philosophical concern with the unity of form, as well as his resistance to chance and force as providing scientific explanations, is quite obvious in his attack on Empedocles' and Democritus' rival accounts of ontogeny. For materialists like these, the only things that endure long enough to count as substance are the elements. Thus, Empedocles and Democritus, in different ways, think that organisms are the result of chance encounters between more basic substances, but that they are not themselves substantial at all. Empedocles has a sophisticated theory of chemical mixing, more or less adopted by Aristotle, according to which the uniform stuffs out of which living bodies are constructed are compounded out of specific numerical proportions of the four elements (*Parts of Animals* 646 a 13–23). Non-uniform parts are compounds of these (*Physics* 198 b 23–27). But the subsequent aggregation of such parts into whole organisms, Empedocles holds, is a chance occurrence:

> Whenever all the parts happened (*sumbainei*) together *as if* they had come to be for the sake of something, these animals were preserved (*esothe*), having been suitably (*epitedeios*) constituted spontaneously (*ap tou automatous*). But those of whom this is not the case perished and continue to perish, as Empedocles says of "human-faced oxen."
>
> *Physics* 198 b 27–33, our italics, our translation

Empedocles does not deny that the properties and parts in virtue of which organisms have come together have a preservative, adaptive effect. He simply denies that they come together *because* they have this effect. Democritus speaks of inheriting all manner of contingencies that affect the developmental processes, such as the apparently bits of broken bones that form the spine (*Part of Animals* 640 a 18–24). But if organisms had come into being in this way, Aristotle counters, they could not exhibit the ability they clearly do successfully to sustain themselves in existence through responding adaptively to contingencies. To do this, they must be true unified substances, and not mere aggregates. That is to say, they must be definite "thises," as Aristotle puts it, that do not depend for their existence on other things, and that are able to serve as subjects (*hupokeimena*) for a vast number of transient properties without losing their identity as individual and highly individuated substances (*Metaphysics* 1029 a 7–32). Given these criteria for substantiality, the form that makes organisms into substances must itself be unitary. It must consist of a coherent set of essential life, or soul, functions that require for their actualization a coherent set of parts and other traits. These traits are developed ontogenetically in the prioritized way we have outlined here. In Aristotle's explanatory hierarchy, material necessity as the primary cause is to be invoked only when appeals to the essence, to hypothetical necessity, and to utility fail, as they do in the case of eye color. By appealing exclusively to the lowest ranked item of this hierarchy, materialists such as Democritus and Empedocles are condemned to misreading organic processes from the outset.

Aristotle's conviction that biological form can come for the most part only from entities that have already actualized such forms, thereby constituting an endless chain that runs from parents to offspring, rests on his conviction that in most cases it cannot come from anywhere else. It cannot "always or for the most part" come from material processes below, as his argument with Empedocles and Democritus shows. But neither can it come from Platonic forms above. Aristotle's quarrel with Plato turns, in fact, on the same considerations as his quarrel with the materialists. The same arguments that call into question whether an organism could ever be a unitary substance if it had been assembled haphazardly from a series of chance events are also arguments against the possibility that Platonic forms, when pressed into physical materials, could ever produce anything of more substantial integrity than a mere artefact like a bed. In Plato's theory, organisms can at most be matter-form compounds that are too weakly unified to count as substances. Aristotle's conception of form, accordingly, is more dynamic than Plato's. The form of an organism is identical with its end-oriented process and pattern of formation, its movement from the potentiality of undifferentiated, generic matter to its actualization as a functionally differentiated, or ensouled, substance, which retains its substantial identity by processing matter through itself by way of metabolism, growth, and reproduction. When we consider the four causes in this context, we find final and formal cause closely associated. The efficient form, too, can be identified with form insofar as an actual, or fully formed, begetter is needed to initiate the reproductive process, and indeed to serve as the agent in its own growth and functioning.

In order to assert just how thoroughly form penetrates and quickens matter, Aristotle is sometimes inclined to say that form as substance, rather than matter-form compound, is itself substance (*Metaphysics* 1029 a 30–33). This has led some interpreters to assert that for Aristotle, each individual has its own individuating form (Frede 1985; Balme 1987a). This is true in the sense that organisms are their own formedness. It is not true, however, in the sense that there is something like a Platonic form for every individual. If there were, Aristotle's conviction that scientific knowledge depends on grasping universals – in this case the forms that are preserved from generation and generation – would conflict with his ontology.[16]

---

[16] Many contemporary adherents of the individual forms interpretation concede that this view is *prima facie* inconsistent with the claim that knowledge is of the universal. A variety of sometimes tortured interpretations have been put forward to make Aristotle's ontology consistent with his epistemology. Balme, for example, thinks of

The Crisis of Aristotelian Biology

Aristotle's philosophy of biology is a delicately balanced philosophical juggling act, holding in place a number of empirical observations and keeping others at bay. This delicate balance was disrupted when Alexandrian physicians discovered the nervous system, which disrupted Aristotle's image of a qualitative body centered in the heart. This was made possible by the fact that they worked in the second century B.C.E. in Alexandria, a city devoted (partly by Aristotelian influence) to pure intellectual inquiry unconstrained by cultural taboos against dissecting human corpses – or even vivisecting those of criminals and non-human primates. The discovery of nerves was in the first instance the work of the anatomist Herophilus, who also found the valves of the heart, the ventricles of the brain, and the cornea and retina of the eye, and who identified distinctive functions of veins and arteries (von Staden 1989). His work was advanced by his successor Erasistratus, who determined that the heart was a pump (though not a circulating pump, as it would later seem to be for Harvey), distinguished motor from sensory nerves, located the source of the nerves in the brain, and followed their route down the spinal column (von Staden 1997c). The effect of these discoveries was the emergence of the image of the body that we still recognize. This new image shifted the center of life, awareness, and emotion away from the heart – the traditional Greek folk-biological locus, which Aristotle still shared – toward the brain, which Aristotle had regarded merely as a sort of radiator or cooling device. The discovery of the nervous system also unsettled Aristotle's rather ambiguous concept of *pneuma* as the motor of life, replacing it with a system of pulls and pushes. In short, the discoveries of Alexandrian physicians led to a more mechanistic biology. Nor is this surprising, given that Herophilus and Erasistratus lived and worked in a city that was as technologically oriented as it was devoted to pure inquiry, a city in which automata

species universals as simply empirical summations of the total number of traits that get replicated when male parents replicate themselves. This cannot possibly be the case, as it undermines the necessity and universality that Aristotle ascribes to species essences, forcing Balme quite implausibly to claim that Aristotle was not an essentialist (Balme 1987a). Michael Frede's defense of individual forms takes the other horn of the dilemma we are posing (Frede 1985). Form, Frede correctly points out, is what makes an individual into this thing. But he infers from this, that two objects cannot have the same form. From this, Frede infers that Aristotle must deny that knowledge is of the universal. We reject this conclusion because we reject the Leibnizean criterion on which Frede's Scotistic epistemological claim is based.

such as water clocks achieved a degree of perfection not approximated again until the seventeenth century, when another bout of mechanism broke out (see Chapter 2).

Sometime during this period, Aristotle's biological research program began to evaporate. Many reasons can be cited for this turn of events. The texts of Aristotle's *History* and *Parts of Animals* were reworked into species-by-species compendia of natural history, thereby obscuring the design of his explanatory zoology (Lennox 2001, pp. 114–117). These histories were seen as prolegomena to a grand project of classification, another Hellenistic preoccupation that was taken up again in early modernity. The epistemological turn most vividly represented by skepticism must also have put on the defensive Aristotle's apparently naive confidence in human knowing, as well as his conception of thinking as a life-function (Lennox 2001, p. 123). Even if things had gone better, however, it would have been difficult, given the shift toward mechanism, and hence toward the primacy of efficient over formal and final causes, to have resumed Aristotle's biological program in the form in which its founder had originally envisioned it and in which his immediate and most faithful disciple, Theophrastus, practiced it.

To be sure, this is not because the new mechanistic biology was anti-teleological. Alexandria had been founded under the influence of Aristotelian scholars. Demetrios of Phaleron, for example, a former governor of Athens and later advisor to Ptolemy I on founding the Alexandrian Library, was influenced by the Peripatetics, as Aristotle's school was known. Like many Alexandrian intellectuals, Erasistratus himself, if not Herophilus, was a Peripatetic, perhaps a student of Theophrastus (von Staden 1997c, p. 185). As such, Erasistratus did not believe that providing an enhanced role for materially necessitated, automatic processes such as nerve stimulation entailed, in and of itself, a rejection of Aristotle's teleology. After all, Aristotle himself had acknowledged that not everything in an organism comes into existence for the sake of an end, and Theophrastus had moved even further in that direction (Lennox 2001, pp. 259–279). In some ways, it can be said that Hellenistic biologists even enhanced the role of teleology, as when Erasistratus claimed that nature is so artful (*technikos*) that it can make good use even of bodily residues, such as pus, which according to Aristotle himself did not come into existence for an end. Yet in this claim, the very meaning of teleology subtly shifts. According to Aristotle, art imitates nature insofar as it constructs the right parts in the right order for the right function. It is because they know and follow this natural sequence that artful physicians can cure (*Physics* 199 a 12–15; *Generation of Animals* 734 a16–735a20). A teleology based on putting materially necessitated entities to good *use*, however, puts the matter the other way around. It makes *nature imitate art*. Taken to an extreme, it makes organisms into artefacts, not self-moving substances, and reconstructs Aristotle's Nature as a puzzling external agent.

Accordingly, when several centuries later Galen (129–199 C.E.) mounted a highly rhetorical attack on Erasistratus, accusing him (falsely) of not being a teleologist at all, but a blind materialist, he converted Aristotle's maxim, "Nature does nothing in vain," into an a priori conviction that a purpose can be found for virtually every trait. Galen's teleology is purely intentionalistic. There is for Galen no difference between a trait that comes into existence for a purpose and one that comes into existence through purely material causes, but is subsequently put to good use. Good use is all there is, and good use can include the utility of one species for another as among its reasons for existing, as horses exist, in his view, in order to serve human beings. But Galen happened to be the most influential medical writer for the next millennium and a half. Accordingly, when he wrapped himself in Aristotle's mantle, Aristotle's teleology came to be interpreted in such a way that its emphasis shifted from internal, developmental teleology to external use-oriented teleology. In this way, the *scala naturae* passage from *History of Animals* was turned into a universal teleology in which every thing that exists at one level of the cosmic hierarchy exists to serve the purposes of beings at higher levels. This is the view that was inherited by the Scholastic philosophies of Islam, Judaism, and Christianity. When Aristotle's texts were republished in critical editions beginning in the Renaissance, readers of his biological works began to find something quite different, and much more interesting, there. Five centuries later, they are still doing so.

✦

CHAPTER 2

Descartes, Harvey, and the Emergence of Modern Mechanism

Introduction

Aristotle is the one philosopher in our history who is also a great biologist, and, indeed, whose metaphysic, as we have come to call it, serves as grounds for his biological interests. But even in ancient times, as we have seen, his life-centered philosophy was modified in favor of a more mechanistic perspective. Here we will be looking at the seventeenth-century confrontation of the Aristotelian tradition with more mechanistic views, especially in Descartes. In England at least, Gassendi, one of the most outspokenly critical Objectors to the *Meditations*, was perhaps as influential as Descartes. However, we are not attempting a survey here; and Descartes can certainly be taken as one of the chief proponents of the new mechanism in biology.

Like many such labels, "mechanism" is a term imposed by critics and historians. Moreover, it is an ambiguous term. In connection with the sense of "Mechanics" introduced apologetically by Robert Boyle in the late seventeenth century, mechanism suggests billiard-ball causality, just one thing after another (see O.E.D. entry under "Mechanics"). In an earlier meaning, which still resonates in the notion of mechanistic biology, mechanism is concerned with machines. Thus, for example, when Huygens asked Descartes for some examples of mechanics, Descartes sent him accounts of several "engines by means of which one can lift a very heavy weight with a small lever" – we might say, several mechanisms (Descartes 1637; AT I, pp. 395, 435).[1] In the first sense,

---

[1] This "practical" sense of "mechanics" occurs in almost all Descartes's frequent uses of the term, although sometimes the more "mechanistic" sense is also present. See, e.g., Descartes's exchange with Froimont (mediated by Plempius) in 1637. When Fromondus

mechanistic biology enumerates a series of movements, each of which evokes its sequel, all necessitated *a tergo*, and all conceived as movements of matter. It is as if Aristotelian efficient and material causes were to function without the correlates of end and form. In the second sense, as we have already seen happening in Galen's reading of Aristotle, living things become machines, designed for an end externally imposed. When we speak in this chapter in the rise and spread of mechanism in the early modern period, we have both these senses in mind.

We may take Descartes as the great initiator of modern mechanistic biology. In his thinking about living things, both senses of mechanism play a role. First, consider the Boyle or post-Boyle sense. According to the Aristotelian tradition, there are four kinds of change: substantial, qualitative, quantitative, and local. Substances (under the moon) come into being and pass away. Remaining the same substances, they also change their accidental characters. The uneducated man becomes educated, the pale man sunburned. Things grow. And they move from place to place. For Descartes, in contrast, there is only the last kind of alteration – that is, local motion. There is an indefinite stretch of extended substance, within which parts change place. Except for minds, no new substances come into being: At a material level, birth and death are just rearrangements of bodies, or of parts of body. (It is impossible to tell whether, in the last analysis, Cartesian *res extensa* is a count term or a mass term.[2]) Thus, except for minds (and of course God), matter changing place is all there is. For Aristotle, physics included the

---

objects that his explanation of the parts of bodies is "too mechanical," Descartes replies: "I don't understand what he is objecting to . . . For if my philosophy seems to him too crude, because it considers only sizes, shapes and motions, as Mechanics does, he is condemning in it what I consider one ought to praise above all things, that I philosophize in such a manner that I use no argument that is not mathematical and evident, and such that its conclusions are confirmed by true experiments, so that whatever I have concluded can be done on the basis of my principles, really can be done, as long as actives are appropriately applied to passives. I am surprised he himself has not noticed that what has been practised till now, Mechanics, is nothing but a part of the true Physics, which, because it could find no place with the vulgar philosophy, withdrew to the Mathematicians. Thus this part of Philosophy remains truer and less corrupt than others, since, because it relates to use and practice, those who make mistakes in it usually suffer by the loss of their expenses, so that if he condemns my style of philosophizing for its similarity to Mechanics, this seems to me the same thing, as if he were to condemn it because it is true." (AT I, pp. 420–1.)

[2] The ambiguous status of Cartesian body is confirmed by Helen Hattab's argument about secondary causes in Descartes. She points out that for the Scholastics' substances,

disciplines *we* call biology (see Chapter 1). Paradoxically, this holds for Descartes as well, except that biology dwindles to an offshoot of the science of local motion. The difference is that biology is conceptually paradigmatic for physics in Aristotle, whereas for Descartes, the physics of local motion swallows biology whole.

In denying any uniqueness to living things, Descartes is also invoking "mechanical" considerations in the engineering sense. Notoriously, animals for him are mere machines, automata. He likes to compare them to the artefacts on display in the Royal Gardens: a bathing Diana, who if you approach her coyly hides in the reeds, or a Neptune, who if you try to pursue her threatens you with his trident (AT XI, pp. 130–132).[3] However, engineered as they are by an infinite God, animals are machines infinitely more ingenious than those of our making. But machines they are nonetheless, whose purposes, presumably, only God knows. We had better not ask, but just follow their mechanical operations one by one.

As a matter of fact, as we shall see when we look at the doctrine of the *bête-machine*, Descartes himself sometimes assimilates the engineering sense to the billiard-ball causality conception. Although when he speaks of "mechanics" as a discipline he is usually referring to something we might call a kind of technology, in a letter to his friend Mersenne in 1639, boasting about how much he has learned about animals during eleven years of dissection, he remarks:

> The number and the orderly arrangement of the nerves, veins, bones and other parts of an animal do not show that nature is insufficient to form them, provided you suppose that in everything nature acts in accordance with the laws of mechanics, and that these laws have been imposed by God.
>
> 1639, AT III, p. 525; CSMK III, p. 134

The "laws of mechanics," however, are identical, Descartes himself tells us in the *Discourse*, with the laws of nature (1637, AT VI, p. 54; CSMK I, p. 139). But the laws of nature, which will be stated in the *Principles* (Part II, art. 37–40; AT VIIIA, pp. 62–5; CSMK I, pp. 240–242), are plainly laws of mechanical motion, of what we would call mechanics,

---

Descartes has substituted the laws of nature as secondary causes of particular events in the material world God has created (Hattab 1998, pp. 274–302).

[3] The quotation is from the undated *Treatise on Man*, written in the same period as the (also unfinished) *World* – that is, in the early 1630s.

in part even anticipating Newton's three laws. Thus, although the discipline of Mechanics is still, for Descartes, a practical or engineering discipline, it is closely related to, and sometimes nearly identical with, what we would call a "mechanistic" conception of causality.

In this chapter, we consider Descartes, first, as a thinker schooled in the late Aristotelian tradition, who turned against it from within it, and then as the thinker who abolished life from nature – the inventor (or discoverer?) of the *bête-machine*. And finally, we will look at his account of the circulation of the blood, especially of the motion of the heart, in opposition to the peculiarly Aristotelian thought of Harvey. Although Descartes acknowledged Harvey's discovery, he differed from him on the question of cardiac movement, and in fact in many cases his own more mechanistic reading of that process facilitated the acceptance of the fact of circulation.

Descartes and the Late Scholastics

Descartes is taken to be the founder of modern philosophy, as well as, more particularly, the founder of modern mechanism, the definitive conqueror of the Scholastic or Aristotelian tradition. Like most historical generalizations, this one contains a core of truth, but needs modification in respect to the nature of late Scholasticism as well as of Descartes's relation to it.

It is true that official philosophy in France in the seventeenth century – in the universities and colleges – was dominated by what both its friends and its enemies considered the heritage of Aristotle. The curriculum was organized in terms of "the Philosopher's" works. Even innovators, if they were academicians – as Descartes never was – had to present their lectures within this framework. Their discourse was still cast in a recognizably "Aristotelian" mold.[4] There was form-matter talk, particularly in terms of the post-Aristotelian concept of substantial form: the form that is imposed on matter when a new substance comes into being and that disappears when it perishes. There was talk of the four causes – material and formal, efficient and final. There was reference to potency and act, or possibility and actuality. (Note, when we look at Descartes's arguments, that since it is form "informing" matter that makes a substance really a substance, "formal" means actual or real!)

---

[4] For the Scholastic background, see, e.g., Ariew 1999; Garber 1992. On the Scholastic format of innovators' lectures, see Grene 1993.

In addition to substantial change, the coming to be and dying away of a substance, there was use of the three kinds of accidental change: qualitative, quantitative, and local motion.

At the same time, in the seventeenth century, we are 2,000 years, and, intellectually, eons away from Aristotle himself. By the standards presented in the last chapter, we can now only very remotely be said to be living in an Aristotelian world. This story, too, is complicated. In the present context, we can touch on it only hastily and superficially.

We should note in particular that by the early modern period, form and matter have changed the roles they had in Aristotelian explanations. Granted, Descartes was trained in what seems at first sight to be a fairly faithful reading of the Philosopher. He attended the Jesuit college of La Flèche, and the Jesuit order was committed to following Aristotle in philosophy wherever possible. True, Aristotle had been modified somewhat to fit Christian theology. As Joseph Owens put it, Aristotle, like every catechumen, had to be instructed (Owens 1957, p. 300). Yet, given a certain Augustinian framework in theology, Jesuit Thomism appears relatively close to Aristotle. At the same time, the Jesuits of La Flèche were clearly sophisticated teachers, who at least reported, if only to reply to, recent controversial issues. Despite the admonition to teach nothing novel, they may well have even adopted some relatively controversial positions. At least their students would have heard about these developments. So what had happened to the basic concepts of matter and form in the four centuries since St. Thomas – let alone the twenty since Aristotle?

In Aristotle's thought, as we pointed out in the last chapter, form is emphatically prior to matter: "Matter is potentially, so that it may go into the form, and when it is actual it is in the form" (*Metaphysics* 1050 a 15–16). However, partly because of the problems involved in describing, if not "explaining," the mystery of transubstantiation, whereby bread and wine allegedly become the body and blood of Christ, and especially through the influence of Duns Scotus and his school, matter had come to acquire a quasi-independence.[5] (The bread and wine still look and taste like bread and wine, so their matter must be somehow independent of the Divine form that they now presumably express.) In Aristotle's system, prime matter certainly did not exist independently. Indeed, it probably did not exist at all.[6] For Thomists, it existed only in

---

[5] See, e.g., Duns Scotus, *Opus Oxoniense*, II, dist. 12, quaest. 1 and 2.
[6] See Charlton 1992, Appendix, pp. 129–143.

principle, so to speak, as the "matter" of the elements, air, earth, fire and water. Each of these was thought to be the product of two of the basic qualities – hot, cold, moist, and dry – organizing an underlying matter that never in fact existed apart from those primitive organizing principles. Throughout the hierarchy of matter/form relations, moreover, the substances we find around us were all this-suches, potentialities organized by a form that made them the kinds of things they were. In the view of Thomists, matter is the principle of individuation. It is relatively accidental. It is form that provides the specific character that marks the substance as being what it essentially is: a being of a given kind. Socrates is a little snub-nosed human being with beautiful discourse to be elicited from within his odd exterior. Those are all accidents predicated of this human being. Form gives a nature; matter singles out one instance of a nature from another.[7]

By the early seventeenth century, however, in many writers at least, this relationship is oddly reversed. Forms are formalities that contract a chunk of matter, first to a species, then a genus, then an individual. Thus, matter now exists as everywhere the same, while form marks off one individual from another. This holds, for example, for the Paris professor Eustace of St. Paul, whom Descartes in the early 1640s considered taking as the best example he could find of Scholastic doctrine, and also of the more popularizing Scholastic writer, Scipion Dupleix.[8] So, in a sense, matter has shed its intimate dependence on form, and is already on its way to becoming Cartesian *res extensa* – even though when Descartes actually took the step of calling extension matter's only principal attribute, he roused angry reaction, and indeed official condemnation from a number of sources.[9] Form, on the other hand, is now moving close to becoming simply the shape of pieces of matter. In fact, in Descartes's case, it nearly drops out altogether as an explanatory concept.[10] Substantial form does survive in the human case, but as an anomaly in an otherwise non-hylomorphic world.[11] From a distance, it looks as if Descartes's Scholastic predecessors and contemporaries had

---

[7] See Thomas Aquinas, *Summa Theologiae*, I. quaest. 7, art. 2; quaest. 66, art. 1.

[8] Eustacius a Sancto Paolo 1629, Part 4, disp. 2, quaest. 4.; Dupleix 1992, pp. 131–2. Texts from these authors are included in Ariew, Cottingham, Sorell 1998, pp. 68–135; the new meaning of form is clear from Dupleix's statement on p. 122 of Dupleix 1992.

[9] See R. Ariew 1998, in Ariew, Cottingham, and Sorell 1998, pp. 252–260.

[10] For an account of Descartes's use of "form," see Ariew and Grene, 1997, pp. 317–320.

[11] See Grene 1991.

been almost ready to take the additional step he took, shocked though they were when he in fact took it.

Let us look at one example of what has happened to the form-matter relation in late Scholastic commentators. The Coimbrans were a group of Jesuit teachers from the University of Coimbra in Portugal, whose subtle work was much cited. We can assume it was discussed at La Flêche. Consider the Coimbrans' account of Aristotle's treatment of matter in some crucial passages in the *Physics*. In *Physics* I.9, Aristotle introduces into his argument a parenthetical definition of matter. "By matter," he says, "I mean the first subject for each thing, from which it arises without qualification and not accidentally" (*Physics* 192 a 31).

Does "first subject" here mean prime (first, unqualified) matter, or does it mean proximate matter (the matter/form closest to the thing in question)? A cat, for example, is organized by its feline form. Its proximate matter is the whole of its organ systems. These in turn provide organizing principles for the tissues, which, as we move downward, inform the elements of which they are composed. Only remotely do we approach prime matter, which is (abstractly, though not really) the vague potentiality actualized in one or the other element. The Coimbrans, however, take Aristotle's definition of matter to be referring directly to prime matter, on which they comment at length (Conimbricenses 1592, pp. 250 *ff*). When, in *Physics* II, Aristotle discusses matter at greater length, he speaks of the wood of a bedstead or the bronze of a statue (193 a 10), or of the silver of a bowl (194 b 25). Now these are clearly instances of proximate, not prime, matter. Remarkably, however, the Coimbrans pass them over in silence. It seems as if, in general, we are to understand generation, the coming to be of a substance, as the imposition of substantial form on prime matter. Never mind the levels in between. They drop out altogether. The complex form-matter hierarchy of Aristotelian nature here seems already oddly impoverished. Some Scholastic comments on generation confirm this impression. Thus, for example, we are told that when an animal dies, a *forma cadaveris* has to be summoned to organize the prime matter formerly subject to that animal's substantial form (Conimbricenses 1592, p. 62).[12] This is a far cry from Aristotle's account of reproduction or of the many-leveled organization of a living being at a given moment of its life history.

---

[12] See also Suarez, O*pera*, 25, 478, cited in Des Chene 1996, p. 147. Des Chene remarks: "The *forma cadaveris* with its scars and dwindling heat would occupy a useful though morbid chapter in the history of Aristotelianism" (p. 147).

In the light of such examples, it seems that we need to modify what appears the most obvious contrast between an "Aristotelian" and a Cartesian nature.[13] In Aristotle – and, one would suppose, in any Aristotelian view of nature – we find hylomorphism, the intimate connection of form and matter, virtually everywhere, except in the odd case of "active reason" or "agent intellect," or, in Christian terms, the rational soul.[14] In the new Cartesian nature, on the contrary, we have pure matter – no form except in the residual sense of shape, and this time it is only in our case that we have a form, the human mind, informing a body. It is our hylomorphism, during this life, that constitutes the exception in an otherwise lifeless, formless nature. Is this contrast valid?

If we think of ordinary people in a pre-industrial society, it seems reasonable to suppose that their view of things begins with the plants and animals they see around them, which are born and die and grow and change qualities and, in the case of animals, move about. But perhaps, considering the late Scholastic view of generation, this contrast is too simple. Perhaps for a learned friar sequestered in a monastery or a learned professor in a university setting, the immediacy of living things was no more striking than it is for modern city-dwellers. In a monastery, it was the lay brothers, not the learned, who tended the livestock and worked the fields. Nor could men of learning compensate for this rather disembodied way of life by consulting Aristotle's biological works, which were not widely known. They were introduced late in the revival of ancient learning, and even then they were far from central to the curriculum.[15] As we suggested in the last chapter, it was a Christianized Aristotelian metaphysic with its attendant physical and logical works that caught the attention of Renaissance and early modern, as well as medieval, thinkers. The biological focus of Aristotle's own thought had all but disappeared. Thus the "Aristotelian" tradition of Descartes's own day offered a severely modified version of Aristotle's own life-centered vision. The stage was set for Descartes's entrance.

What, then, of Descartes's dramatic opposition to the Aristotelians, as he himself often liked to proclaim it? For any question Aristotle and his followers claimed to have answered, he could show the solution to

---

[13] See Gilson 1930; Grene 1991.

[14] The original source of this controversial concept is Aristotle's discussion in *De Anima* III.5. It would take us too far afield to deal with the debates about it, especially in Padua, in the early modern period.

[15] See Toletus 1589. *Prolegomenon*, cap. 3, fol. 6b.

be "invalid and false" (AT VII, pp. 579–580; CSMK II, p. 391). Reasonable people should put school learning behind them and turn to "good sense," which we all share; that is a major theme in the *Discourse*.[16] Descartes states the contrast, as he sees it, emphatically in the open letter he issued in reply to a savage attack by the Dutch (Protestant) scholastic, Voetius:

> The philosophy which I and all its other devotees are engaged in pursuing is none other than the knowledge of those truths which can be perceived by the natural light and can provide practical benefits to mankind; so there is no study that can be more honourable, or more worthy of mankind, or more beneficial in this life. The ordinary philosophy which is taught in the schools and universities is by contrast merely a collection of opinions that are for the most part doubtful, as is shown by the continual debates in which they are thrown back and forth. They are quite useless, moreover, as long experience has shown to us; for no one have ever succeeded in deriving any practical benefit from 'prime matter,' 'substantial forms,' 'occult qualities' and the like.
>
> AT VIII b, p. 26; CSMK III, pp. 220–221

Nevertheless, as Etienne Gilson argued long ago, it is impossible to understand Descartes without paying attention to the scholasticism he professed to disdain, but within which his thought had developed (Gilson 1930). Although, as Descartes tells us in the *Discourse*, he soon turned to read the great book of the world rather than the books reported by his teachers, Descartes clearly retained a good deal from their instruction, and later, in replying to the Objections to the *Meditations* that he had asked for, he made an effort to recall and supplement his knowledge of the School.[17]

To get a better view of Descartes's relation to the School, let us ask two questions. What was Descartes after in the *Meditations*, particularly in the initial move of hyperbolic doubt? And, in the course of that famous doubting procedure, what did the Meditator fail to doubt?

In answer to the first question, Descartes's aim, he tells us, was to lead the mind away from the senses (AT VII, p. 12; CSMK II, p. 9). For any adherent of any even remotely Aristotelian view of our cognitive powers, this is indeed heretical. We must distinguish between what is first in being and what is first for us, and what is first for us

---

[16] See also the unfinished *Recherche de la Verité*, AT X, pp. 495–527; CSMK II, pp. 399–420.

[17] The *Principles* were the resulting presentation in relatively scholastic form.

comes from the senses, from our perception of the things around us. Why does Descartes want to make that unsettling move? The answer is straightforward enough. His return to intellectual problems had come, in 1618, during a visit in Holland to a young mathematician-physicist, Isaac Beeckman (Rodis-Lewis 1971, I, pp. 25–32). Together they had embarked on some problems in mathematical physics not unlike the problems Galileo was wrestling with in Italy. Descartes, who was, as it turned out, a highly original mathematician, wanted to apply mathematical methods to physical problems. But this was something that, on Aristotelian principles, could not – or should not – be done. True, there were many areas in applied mathematics that look, in hindsight, like parts of physics.[18] Nevertheless, in Aristotelian methodology, there was supposed to be a sharp separation between physics, or natural philosophy, which dealt with the full, qualitied nature of real existent substances, and mathematics, which dealt abstractly with the merely quantitative aspect of such substances, whether discrete or continuous.[19] So to produce a mathematical physics, it was necessary to break loose from the Aristotelian foundation of our knowledge of nature in order to be able to use mathematical principles and methods directly in the understanding of the natural world. To this end, a good deal of Scholastic baggage had to be discarded, particularly the deeply empirical basis of Aristotle's, or the Aristotelians', method of gaining knowledge. And indeed, the cleansing process of methodological doubt was in fact deeply disturbing to many of Descartes's readers, for this very reason.[20]

At the same time, there was a good deal Descartes did not doubt. He retained much of the terminology and of the accompanying conceptions of his teachers and his Scholastic contemporaries. If he wanted, as he claimed, to overthrow all his former opinions, there was a good deal of apparatus that was, in his view, not opinion at all, but a collection of clear and distinct ideas evident to natural light. For every opinion there is a contrary opinion; it was that contrariety of opinion in much of what he had learned that convinced the young Descartes he had gained no knowledge at all in his school years. But when something was clear by natural light, Descartes believed, no one would contradict it. Thus if we look in the *Meditations* for things Descartes failed to doubt – and believed no

---

[18] See Dear 1995.

[19] See, e.g., Grene 1963.

[20] See especially the Seventh Objection to the *Meditations*, AT VII, p. 451 ff.; CSMK, II, p. 302ff. See also Ariew in Ariew, Cottingham, Sorell, 1998, pp. 257, 258.

one could doubt – we find an impressive list. To embark on this intellectual adventure in the first place he had to have a firm grasp of a number of concepts. He had to know what thought is, what extension is, and so on (*Principles* I, 10, AT VIIIB, p. 8; CSMK I, pp. 195–6). He also claimed to know – and apparently believed we all knew – indubitably maxims such as the causal principle: that a cause must contain, either formally or eminently, at least as much reality as its effect (AT VII, p. 40; CSMK II, p. 28). Understanding causal connections in terms of degrees of reality is a very traditional process, which the "new mechanical philosophy" that Descartes himself was initiating would soon abandon. (Indeed, it would be completely abandoned in Descartes's own lifetime in the more radical modernism of Thomas Hobbes (Hobbes 1655, II, 9, 19). Descartes also accepted a more controversial – non-Thomist – Scholastic distinction between the formal and objective reality of ideas – that is, between an idea as an actual – "formal" – occurrent in my mind and an idea as representative of its object in this sense, objective.[21]

What about God? The first reference to an all-powerful maker occurs in the *First Meditation*, where, in the order of reasoning from everyday confused ideas to the clear and distinct ones we hold with greater evidence, we hear a rather muddled tale of a possible maker who might deceive us even about simple arithmetic (AT VII, p. 21; CSMK II, p. 14). (Note that this is not the malignant demon who appears toward the close of *Meditation* I and who may deceive us about external existences, but not about arithmetic and geometry.) However, when in due course we clear a path to our proper and wholly convincing idea of our Infinite maker, that knowledge turns out to be innate; we have always had it, ready to be grasped with the greatest certainty once we clear our minds of their sense-based confusion (AT VII, p. 51; CSMK II, p. 35).

Finally, despite his dualism, and like the anti-Aristotelians condemned in 1624,[22] Descartes retained the conception of the human mind as the substantial form of the body while the mind and body are united in this life.[23] In short, although in reducing matter to extension and soul to thinking mind he was initiating what was indeed to be a radical reform,

---

[21] Remember that in the Aristotelian tradition, matter was long thought of as pure potency, whereas it was form that gave actuality to that mere possibility. "Form" thus comes to mean actuality, and "formal" means "actual." This usage is exemplified in Descartes's contrast of formal and objective, or formally and eminently.

[22] See Ariew and Grene 1997, p. 313.

[23] See Grene 1991, pp. 18–19, for a list of passages in which Descartes asserts the substantial unity of mind and body.

it was a reform conceived from within, not beyond, the Scholastic tradition. It was, after all, Aristotelians he wanted to cure of their Aristotelianism; but one can convince only those whose language one speaks and whose beliefs one at least partly shares.

Cartesian Biology

Descartes's primary interest was in physics, and especially in the development of mathematical, or specifically, geometrical physics. He told Mersenne, "All my physics is geometry" (AT II, p. 268; p. CSMK III, 119). In 1629–30, when he had moved to the Netherlands, he discovered that he needed a metaphysical foundation for his physics. If he wrote any of this project then, we do not have it (Rodis-Lewis 1971, I, pp. 112–120). But eleven or twelve years later, the *Meditations* fulfilled that mission. At the same time, the philosopher-physicist developed an interest in biology: he performed dissections and experiments. He reported on his biological work briefly in Part V of the *Discourse*, summarizing what he had done in the chapter "On Man" that would have been part of the discarded *World* (AT VI, pp. 45–56; CSMK I, pp. 134–139; see "Treatise on Man," AT XI, pp 119–202; CSMK I, pp. 99–108 [partial]). And much later he composed a *Description of the Human Body*, which reports some experiments as well as observations – and speculations – about our bodily machine (AT XI, pp. 223–286; CSMK I, pp. 313–324 [partial]). In all these places, Descartes makes it quite plain, first, that he considers our bodies, as well as those of animals, in themselves inanimate machines. As he puts it in the chapter "On Man," "I consider the body to be nothing but a statue or machine made of earth . . ." (AT XI, p. 120; CSMK, I, p. 99). To this piece of matter, our minds have been added. Animals, however, having no minds, are *nothing but* machines. This is stressed in particular in the text of the *Discourse* and in the reply to Antoine Arnauld in the Fourth *Objections*.[24]

This doctrine of the *bête-machine* was as notorious among opponents of Descartes as it was celebrated by his adherents. For example, it was

[24] There is a problem with a remark in Descartes's letter to Henry More in February 1649. While Descartes has made it very clear elsewhere that animals lack sensation as well as thought, since sensation is a function of mind (e.g., *Principles*, I, p. 66, AT VIIIA, p. 32; CSMK I, p. 216), to More he says: "I do not deny life to animals, since I regard it as consisting simply in the heat of the heart, and I do not even deny sensation, insofar as it depends on a bodily organ" (AT V, p. 278; CSMK III, p. 366). At the same time, he emphatically repeats the position he has held all along.

one of the items condemned by the Jesuits in 1706, and its condemnation was implicit in the objection to taking extension as the principal (or sole) essential attribute of matter.[25] There are no vegetative or sensitive souls, only the rational souls that inform our bodies; everything else is just stuff that takes up space. Indeed, the celebrated Father Daniel, in the sequel to his *Cartesian Voyage*, called the conception of the *bête-machine* "the very spirit and sap of Cartesianism."[26]

How did Descartes come to hold this view? As we have seen, he wanted to alter the method and therewith the doctrines of Scholastic physics. The School, he held, had never solved a single problem, whereas with his method one could go on, in the right order, to answer all the questions that would ever confront the human mind. Both parties to this dispute agree, of course, that God made the world as He made it. The question is: how to describe the regular, yet subtle, changes that occur in God's creation. Both parties agree, also, that what nature consists of is substances and their modes or accidents. The problem is to explain satisfactorily the changes that take place in these naturally existing entities. Now the position of the School, according to Descartes, is roughly that of common sense; it has elaborated with pompous verbiage, and so solidified, the prejudices of infancy and childhood. On this view, which, in an abstract and impoverished way, the Scholastics share with Father Aristotle, the most conspicuous and paradigmatic example of substance on this earth is living things. And what is most conspicuous about living things is that they come into being and perish: That is the first and most fundamental kind of change in Aristotle – and, as Gilson has argued, it is in connection with this phenomenon that the concept of "substantial form" was needed (Gilson 1930, ch. I, esp. pp. 143–176). Every time a new living entity comes into being, a new substantial form comes into being with it: For this is a substance, uniquely characterized by its form, this instance of the form of its kind. Again, there are of course also changes of quality: An ignorant man becomes educated, a hairy man goes bald, a ripening apple changes from green to red, and so on. And there is growth: The child grows taller and becomes a man. Finally, also, substances move about: Hares and even tortoises go from place to place, as do less clearly substantial entities such as water or fire. In four of the ten categories – substance, quality, quantity, and place – there

---

[25] See Ariew, Cottingham, and Sorell 1998, p. 259.

[26] In a Latin version: *spiritus est et succus, si ita loqui, diceat puri Cartesianismi* (Daniel 1694, p. 3).

is change, and it is the task of the physicist to study all of these. Most of this, Descartes believes, is empty talk, both childish (we have all thought in our pre-Cartesian youth that the world was so full of a number things, or kinds or qualities of things) and pretentious (in Descartes's time one was taught, even in the best schools, such as the one he had attended, that substantial forms, real qualities, intentional species, and the like could support in our philosophical maturity those infantile delusions). In fact, Descartes has discovered, there is only one kind of physical change: local motion. Birth and death, qualitative alteration, growth, are all reducible to this one, measurable, manageable type: the change in position of bodies relative to one another. Granted, if growth is change in quantity, that should be all right, from the Cartesian perspective, since it is measurable and hence quantifiable. But if we think of growth as one of the characteristics of animate existence, as what happens typically to little plants and animals, then it too must go. The very phenomenon of life, archetypical for ordinary people, is and must be overlooked and abolished. The reduction of animal existence to a congeries of movements is a necessary consequence of this fundamental shift in the concept of natural change. Thus the thesis of the *bête-machine* forms a coherent part of the style of explanation initiated by Descartes, first, of course, in *Le Monde* and in the chapter on man, publicly, first in the essays of 1637.

The best-known text expounding the *bête-machine* is the account in Part Five of the *Discourse*, where Descartes enumerates the two criteria by which we distinguish ourselves as beings endowed with minds from animals, who have no such source of thought or sentience (AT VI, pp. 57–59; CSMK I, pp. 140–141). We have speech – genuine speech – and we have, or can develop, a variety of skills. Language, Descartes tells us, takes very little intelligence; so animals, having no language (like ours), have no intelligence at all. As Descartes notes elsewhere, if you ask an animal a question, it cannot answer it (AT II, p. 40: CSMK III, p. 49). Second, if a given animal can do one thing, like building a nest, skillfully, it does so "by nature," in a fixed way, and cannot learn some other technique, as a person, endowed with mind, can do. In the Fourth Objections, Arnauld wonders about this view. Could the light reflected from a wolf onto the eyes of a sheep really provoke its flight with no interference from soul? (AT VII, p. 205; CSMK II, p. 144). Descartes replies by pointing out that even many of our seeming "actions" are mechanically induced by the body without interference from the mind. The animal spirits and organs of the body do lots of things without mental direction, and in the case of the sheep, and the wolf pursuing it, that is obviously how it works (AT VII, pp. 229–30; CSMK II, p. 161).

The way Descartes reasons to arrive at this conclusion is made especially clear by some passages in his correspondence. In 1638, for example, one Alphonse Pollot had sent to Descartes, via an intermediary, some comments on the *Discourse* and its accompanying essays, published the previous year (AT I, pp. 511–519). On Descartes's view of animal automatism, he wrote,

> Experience makes us see that beasts make their affections and passions understood by their sort of language, and that by several signs they show their anger, their fear, their love, their grief (pain? "douleur"), their regret for having done wrong . . . it is clear that animals function by a principle more excellent than the necessity stemming from the disposition of their organs, that is, by an instinct, which will never be found in a machine, or in a clock, which have neither passion nor affection as animals have.
>
> AT I, p. 514

Descartes replied in terms that illustrate his affection for automata as well as for a kind of science-fiction reasoning. We have, he admits, a habit of judging, on the basis of the resemblance between the behavior of animals and many actions of our own, that animals have some sort of interior principle similar to our minds. But this judgment is one rashly formed in infancy. And of course "those who want to discover truth must above all distrust opinions rashly acquired in childhood" (AT II, p. 39; CSMK III, p. 99). How shall we reinforce this distrust, and so move to Cartesian intellectual maturity? Descartes suggests the following:

> Suppose that a man had been brought up all his life in some place where he had never seen any animals except men; and suppose he was very devoted to the study of mechanics, and had made, or helped to make, various automata shaped like a man, a horse, a dog, a bird, and so on, which walked and ate, and breathed, and so far as possible imitated all the other actions of the animals they resembled, including the signs we use to express our passions, like crying when struck and running away when subjected to a loud noise. Suppose that sometimes he found it impossible to tell the difference between the real men and those which had only the shape of men, and had learnt by experience that there are only two ways of telling them apart.
>
> AT II, pp. 39–40; CSMK III, p. 99

Descartes proceeds to summarize the two criteria described in the *Discourse*. He asks:

> [W]hat would be the judgment of such a man when he saw the animals we have; especially if he was filled with the knowledge of God, or at least had noticed how inferior is the best skill shown by our artefacts when compared with that shown by nature in the composition of plants . . .
>
> AT II, p. 40; CSMK III, pp. 99–100

Plants apparently offer no problem for the automatist; vegetative souls are not considered; it is only animals that seem to present a difficulty. Be that as it may; our science-fictional man has noticed plant mechanisms,

> . . . and so believed firmly that, if there were automata made by God or nature to imitate our actions, they would imitate them most perfectly, and be incomparably more skillfully constructed than any which could be invented by men
>
> AT II, p. 41; CSMK III, p. 100

In this reflective frame of mind, let our science-fiction character observe the animals that we in fact see in our current world. He will notice between them and us the same two differences he had observed between his automata and ourselves. What will his judgment be now?

> There is no doubt that he would not come to the conclusion that there was any real feeling or emotion in them, but would think they were automata, which, being made by nature, were incomparably more accomplished than any of those he had previously made himself.
>
> AT II, p. 41; CSMK III, p. 100

Now, which judgment should we trust? The one we made as children, and retained only through unthinking habit, or the one our fictional character makes in his peculiar circumstances, "unprejudiced by any false opinion"? (AT II, p. 41; CSMK, III, p. 100). To Descartes, the answer is clear. Ingenious reasoning, guided by clear and distinct ideas, must prevail over common sense.

It is worth noting that Descartes had his adherents as well as his critics. So before we leave the topic of animal automatism, let us look at one of his defenders. Jacques Rohault's *Treatise on Physics* (1671) was widely read, both in France and England. In the same year, Rohault also published two "conversations on philosophy" in which he took up the two Cartesian doctrines to which devout, or official, people had chiefly taken exception: in the first dialogue, the alleged explanation of transubstantiation, and in the second, the thesis of animal automatism (Rohault 1978/1671). One Monsieur N. has come to inquire of the author about the alleged dangers associated with this novel doctrine, especially with respect to the miracle of the Eucharist. Satisfied with the author's explanation of this (in the first dialogue), he returns the very next day to report that he has heard tell of another strange development. There are some people, he has been told, who believe that animals do not act from knowledge, but are pure machines (Rohault 1978, p. 38). (In contrast to Pollot's concern for the emotions of animals, it is their knowledge that seems to be the chief point at issue here and in many such debates.[27]) This, says Monsieur N., seems absurd. However, the author persuades him that it is a probable view, if we consider probability as attested by the phenomena, rather than by the majority opinion. For what we have before us, in appearance, is simply a variety of movements.

N. agrees, but raises several objections, all easily answered. First, there are so many movements. Answer: A clock with only ten chief parts can tell the time beautifully; think how much better God could do with the parts at His disposal.

N: Clocks must be rewound.

R: Beasts need food.

N: Animals behave purposively.

R: So do magnets.

N: Beasts flee from danger.

R: Magnets also move to the opposite pole.

N: What of their cries of pain?

R: That's like organ music. [There is always some mechanical analogy.]

---

[27] In this and some other passages, Rohault is clearly alluding to Montaigne's famous "Apologie de Raimond Sebond." Montaigne 1965, pp. 415–589.

N: complex behavior, like nest building or insect metamorphosis, demands knowledge; R.: stones move toward the center of the earth, and flowers, too, which certainly don't know anything, exhibit complex development. So this must all be a question of the movements of the parts of matter.

Besides, R. continues, "this argument, drawn from the exactitude with which beasts perform their various actions, proves too much: for one would have to conclude that beasts have more perfect knowledge than men" (p. 143) – and that would indeed be a scandal!

N. now considers that it is probable that beasts are machines; but he still asks if it will be necessary to say they have a soul.

The author explains how confused the scholastic view of soul has been; so that's all right.

But N. has heard, he says, some arguments to the effect that this opinion about animals may lead to atheism.

No, says his host, it is the view that beasts think that is more dangerous, since it threatens the principal proof of immortality, and he proceeds to recapitulate, via question and answer, the proof implicit (though of course not explicit) in the *Meditations*.

N. is convinced, but he still worries that almost everybody does think beasts act with knowledge, and so libertines will argue that since their souls are mortal, ours are too.

"You see, then," says the author, "how important it is, not to attribute to beasts something of the certainty of which one has not the least proof, and of which the contrary is very probable" (p. 147).

Fine, says N., but surely those who hold to the existence of animal souls also have sure-fire proofs of our souls' immortality.

The author has of course answers to these alleged proofs: First, we have knowledge of universals.

R: So do animals, if they have knowledge at all.

Second, perhaps beasts just have less knowledge than we do – it's a question of degree.

R: In that case, we are a kind of beast! Irrefutable answer!

Third try: We can reason; answer:

So do beasts, if we permit them any knowledge.

Fourth: We can reflect on our thoughts.

Answer: Reflection is but a return of thought to itself, no harder than a return to the past, which any being capable of thought can manage.

All this shows, finally, says the author, that you can't get good consequences from a bad principle. And one last fling, anticipating a later century: Some may say that if beasts are machines, men are machines as well. But to say so, they must speak, and so contradict their assertion of their own mere-machine nature.

All is well. The thesis of the *bête-machine*, far from endangering orthodoxy, secures the mind, in its indestructibility, against the challenge of our lower, bodily nature.

The Movement of the Heart and Blood

If the concept of animal automatism is characteristic of Descartes's attitude to biological questions in general, the thesis he most often emphasized – in the chapter "On Man" in the unpublished *World*, in the *Discourse*, and again in the *Description of the Human Body* – was the circulation of the blood, discovered by William Harvey and announced in his *De Motu Cordis* in 1628 (Harvey [W]). Descartes acknowledged Harvey's discovery (although he said he had dealt with the circulation before he read Harvey's book). However, his interpretation of the movement of the heart was strikingly different from Harvey's, and allegedly more "mechanical."[28]

---

[28] Further technical details of Descartes's view can be gleaned from his correspondence, via Plempius, with Fromondus, AT I, pp. 399–401, pp. 496–499; AT II, pp. 62–69,

Both Descartes and Harvey, we should note in passing, first describe the motion of the heart and then explain it by reference to the circulation. It even appears that Harvey first discovered how the heart moves, and later added the account of the circulation, with the independent, and partly quantitative, arguments by which he supports it (Frank 1980, p. 11). We are inclined to ask who accepted the circulation, and then to insert the account of cardiac movement as a subordinate matter. Yet in both the discoverer of the circulation and his most illustrious Continental supporter, it was the question of the heart's movement that came first.

Before we look at the two accounts, and ask how they differ, let us imagine, if we can, how the heart and its movement appeared to physicians of the early seventeenth century. To begin with, the heart was one of three chief centers in the body – it sent blood and vital spirits out through the great aorta. The venous system, with its natural spirits, was based in the liver, while the subtler animal spirits had their seat in the brain (Frank 1980, pp. 2–9).[29] Putting the heart so centrally in control would be one of the novelties forced on reluctant physicians by the doctrine of the circulation. However, we are looking here primarily at the heartbeat itself. How did it look to the contemporaries of Harvey and Descartes? On the one hand, as one writer put it, "the nature and cause of this perpetual motion is so full of obscurity and entangled with so many difficulties, that the very learned Dr. Fracastor thought there were only God and Nature who had the true knowledge of it."[30] On the other hand, difficult though it was to explain through what strange powers this organ keeps pulsing from birth to death, there was some definite medical lore, and language, involved in its description. When it beats, the heart hardens and strikes the chest. This movement was called "diastole" or stretching. (Forget our current terminology if you know it; it was Harvey who was to introduce it, and we are trying to imagine the thinking of his predecessors, or his more conservative contemporaries.) To use a crude analogy, when you lift the arm hanging limply at your side, and move it upward, you are stretching it. And after the beat, the heart relaxes and seems to grow smaller: This is its

pp. 343–345. Parts of Descartes's letter are translated in CSMK III, pp. 60–66, pp. 76–77, pp. 79–85, pp. 92–96. For a discussion of this exchange, see Grene 1993.

[29] See also Fuchs 1992, 2001.

[30] du Lauren 1610, p. 1068. Quoted in Bitbol-Hespériès 1998, p. 30. Harvey refers to the same remark in Harvey 1847. p. 19. 32.

"systole," or coming together that is – it would seem, its contraction. In the same way, when you stretch upward, you're taller, your arm seems longer, than when you let it hang limply at your side. So diastole was the name for the condition of the hardened, beating heart, and systole the name for its relaxed condition. It was also agreed by all that for life, the body needed some kind of vital heat, and the heart, which felt warm in the living (or recently dead?) animal, was the source of that necessary heat. Obviously, also, there was some connection between heart and lungs; but try to forget that. Although there was a good deal of experimentation and speculation about the function of the lungs later in the century, in Harvey and Descartes's time and later – in fact, until Lavoisier – no one could do much with this problem (see Frank, 1980). We shall see, in passing, what our two investigators thought about it.

Now let us look at Descartes on cardiac movement. In Part Five of the *Discourse*, he recounts the chief points he had made in his unpublished *World*, and in particular his account of the movement of the heart and arteries, "the first and most widespread movement that we observe in animals" (AT VI, p. 46; CSMK, I, p. 134). He advises his reader to look for himself: Get someone to dissect before him the heart of a large animal, observing the chambers with the vessels attached to them and the valves ("little doors") that open and close between the chambers and the arteries. Observe also, as any one can feel (!), that the heart is hotter than the rest of the body (AT VI, pp. 48–9: CSMK I, p. 135). Now what happens is that a drop of blood entering the heart is rarified by the ambient heat, and when enough blood has entered and expanded to stretch the heart, so that it becomes hard and strikes the chest, the blood is thrust out into the arteries – the great aorta on the left, the pulmonary artery (or the arterial vein, as it was called) on the right. Rarefaction appears to be a straightforward response to heating; so that's all there is to it. The heart, when beating, is in diastole, as all physicians agree, but we now have a good mechanical explanation for that otherwise mysterious event.

Where, then, does the blood go when it is forced out of the heart? Round in a circle – that is clear from the position of the valves, which Descartes has adjured his reader to observe. And here he gives credit to "an English physician, who must be praised for having broken the ice on this subject" (AT VI, p. 50; CSMK I, p. 136). Harvey, however, as Descartes will argue in detail in the *Description of the Human Body*, had been mistaken in his explanation of cardiac movement. Let us look very briefly at Harvey's account in the *De Motu Cordis*, and then compare the two.

Harvey's work opens with an introduction to the whole text, including the later chapters on the circulation (Harvey [W], pp. 9–19).[31] He then sets out his view of the motion of the heart and blood in Chapters I to V, treating the pulmonary circulation in chapters V and VI. He has come to his conception, he tells us, "by using greater and daily diligence, having frequent recourse to vivisection, employing a variety of animals for the purpose and collating numerous observations" (p. 19). What he has found is that "the motion which is generally recognized as the diastole of the heart is indeed its systole" (p. 22). This is a very radical suggestion, indeed. Remember, neither the nature of muscular movement nor that of involuntary muscles was well understood. The heart turns out to be a muscle that hardens when contracting, and pushes the blood out into the arteries, causing their diastole: a most uncomfortable reversal of terminology and of conception.

A muscle that works all one's life, whether one knows it or not? Surely Descartes's view was much easier to accept in traditional terms. The heart stretches when it hardens – in medical terminology, it is in diastole, which means stretching, and when it collapses, and becomes limp (and smaller?), it is in systole. Harvey wants his medical colleagues simply to reverse their usual conceptions. Moreover, as Descartes points out, all the physicians know the heart is hotter than the rest of the body, and that is all that is needed to produce the effect in question. The heart is a kind of furnace, made of balloon-like material. It contains a hidden fire, like that in badly cured hay; blood coming in is rarefied by the heat and makes the heart swell up, or stretch out. Harvey, on the contrary, not only wanted to switch systole and diastole, but to locate the vital heat in the blood itself rather than in the heart, so denying the obvious fact of the heart's greater heat. Besides, Descartes held, the blood changes color in the heart: How does Harvey account for that? Well, Harvey did see the change in color in the lungs, but thought it an artefact of the circumstances incident to dissection. Besides, since he could not well call the relaxation of the cardiac muscle "diastole," he had to introduce it as part of a threefold rather than a twofold movement: the stretching (diastole) of the auricles followed by the systole of the ventricles and then their relaxation. Descartes thought his own explanation better because simpler, and, in his view, more machine-like, but it seems it would also

---

[31] All references to Harvey are to Harvey 1847.

have appeared less radical and unsettling to traditional physicians. If you have to admit the circulation instead of the traditional separation of the venous and arterial systems, which stay right and left rather than moving around and replacing one another, at least you can keep your habitual placement of diastole and systole and the "fact" of the greater heat of the heart.

How did these two ingenious investigators reach such different conclusions? It has sometimes been argued that Descartes was an a priorist, while Harvey relied on observation and experiment (Passmore 1958). That is scarcely fair to Descartes, who did after all adjure his reader to look, and even to feel, what his senses told him to be the case. Moreover, he himself performed experiments, on which he reported in the *Description of the Human Body*; we shall return to these shortly. On the other hand, Harvey was not just a simple-minded observer; he certainly recognized the inference to causes (see, for instance, *On Generation*, "Introduction," in the translation of Willis (Harvey 1847, p. 163). So we need at least to modify this very crude contrast. Let us look at the style of each one's observations and experiments, and the style of each one's explanations. We find, in fact, impressive differences on both sides.

Harvey was a physician. He had begun his studies at Cambridge, where the curriculum was more humanistic in nature than the stricter scholasticism of Descartes's Jesuit teachers. In his college, Gonville and Caius, there was even some interest in anatomy. He had gone on to the greatest center of medical training of his time, the University of Padua, where his teacher was Fabricius of Aquapendente, successor to Realdus Columbus, himself successor to the great Vesalius. Moreover, by the time he published the *De Motu Cordis*, Harvey had added varied medical experience and extensive researches on animals of all varieties: deer, fish, snakes, amphibia – you name it, he had studied it. His manuscripts, sadly lost during the English Civil Wars, even included a *De Insectis*. Perhaps there had never been a more skillful and industrious comparative zoologist than Harvey (see Keynes 1966). In these pursuits, he is following the "way of the anatomists," for whom inspection is fundamental and doctrine secondary (Wear 1983). The latter may indeed be nobler, but the former is more certain. As Harvey put it in the "Second Disquisition to Riolan,"

> This is what I have striven, by my observation and experiments, to illustrate and make known. I have not endeavoured from causes and probable

principles to demonstrate my propositions, but, as of higher authority, to establish them by appeals to sense and experiment, after the manner of anatomists.

W, p. 134

Now Descartes, too, anatomized. In the *Description*, he in fact described three experiments that, he tells us, would fit either Harvey's or his own account of cardiac motion, and three that would fit only his (AT XI, pp. 241–245; CSMK I, pp. 317–319).[32] The fact that the heart hardens, changes color, and emits blood at the moment of striking could follow equally well from Harveyan contraction or Cartesian dilation. But then there are three observations that fit only Descartes's reading: The ventricle is enlarged (a little) while stretching, the cut tip of a rabbit's heart stretches when hardening, and the blood changes color in the heart. By the experiment on the rabbit, Descartes remarks in an earlier correspondence on this subject, "the opinion of Harvey about the motion of the heart has its throat cut" (*quo experimento Harvaei sententia de motus cordis jugulatur*) (AT I, pp. 526–7; CSMK III, pp. 81–2). He is supremely confident of this, remarking to his correspondent: "I have added this in passing, so that you may see that no opinion different from mine can be imagined, which some experiments of the greatest certainty do not dispute" (AT I, p. 527: CSMK III, p. 82). It happens that all his observations were mistaken; yet his account seems to follow good clean scientific, even Popperian, methodology: You eliminate the hypotheses that your experiments contradict, and accept what's left – only as certain, not merely probable. Harvey, on the other side, also wants certainty, not probability. Since, however, as we have already pointed out, he is following the "manner of the anatomists," he holds that what makes his conclusions "true and necessary," if his premises are true, is "ocular inspection, not any process of the mind" (W, p. 133). To us, with our maxim "all observation is theory-laden," this seems naïve compared to Descartes's reasoning.

Yet Harvey turned out to be right and Descartes wrong. After all, Descartes was a mathematician with an interest in physics (and metaphysics), who turned relatively late (at age thirty) to anatomy and vivisection. His reasoning may have been sound, but his practice left a good

---

[32] See also the exchange with Plempius, AT I, pp. 496–499, AT II, pp. 52–54, 62–69. Descartes's replies are translated in part in CSMK III, pp. 79–85, 92–96.

deal to be desired. So although Descartes did make observations, and although he seems to have had a good sense of experimental method, he could not in fact compete with Harvey as an observer and comparative zoologist. At the same time, as we noted earlier, Descartes's more conventional "observations" may have made it easier for ordinary physicians to accept his account than the more accurate, systematic, and original inspection conducted by Harvey.

What about the two investigators' styles of explanation? Even though Harvey puts inspection before doctrine, he does want to explain what he has seen. But like their styles of observation, the two writers' styles of explanation are very different. Descartes is guided by his belief that there is nothing unique about life that makes it different from machinery. The body is a machine, with no touch of soul or mind, and thus its every movement, including the heart beat, should be explained in terms of the rules of the mechanical arts, just as the movement of a clock, once made, is explained purely through reference to a series of local motions. In this, as we have seen, Descartes is opposing the Scholastic invocation of substantial and qualitative kinds of change. Nor do we need any but what an Aristotelian would call material and efficient causes in explaining the processes we observe. Again, for Descartes, the father of the *bête-machine*, there really is nothing alive.

Harvey, in contrast, is himself an Aristotelian of a sort. As we have noticed, he is in a way more radical than Descartes in his interpretation of the heartbeat, switching systole and diastole, denying the greater heat of the heart (since for him it is the blood itself that is the seat of vital heat). But, philosophically, he is much more conservative. In particular, he relies heavily on teleological reasoning. "Nature always does that which is best" (W, p. 39).[33] Note, this is Aristotelian, internal teleology, not the external variety invoked either by Galen or, for that matter, by Descartes, who has to take it that God has made the ingenious machines whose operation he is analyzing. Harvey, too, does indeed sometimes invoke machinery, likening the heart beat to the firing of a gun (W, pp. 31–2). But his machinery is always in the service of purposes within the organism, not imposed from without.

Harvey's Aristotelianism, however, it must be stressed, is very different from that of Descartes's contemporaries. What Harvey chiefly does is to quote chapter and verse of Aristotle's biological works, those

---

[33] A more global or universal teleology is apparent in the *De Generatione*, but what is fundamental, it seems to us, is the harmony discovered within the phenomena of life.

treatises so thoroughly neglected by Descartes's teachers and critics, let alone by Descartes himself. True, in the introduction to his *De Generatione*, published in 1651 through the agency of his young friend George Ent, Harvey invokes the *Posterior Analytics*, and praises Aristotle's double course of inference, from the senses to principles and back again (W, pp. 151–166). Even there, however, he stresses the perceptual beginning and end of the process. But chiefly in the "exercises" that constitute the *De Generatione*, it is the biological works he refers to, as well as the writings of his teacher Fabricius and other investigators. These include Galen, among traditional teachers, and also, among more recent works the *Teatrum Anatomicae* of Caspar de Bauhin (in Bitbol 1990), which it seems Descartes had also studied.[34] Harvey is of course also quick to disagree with any or all of his sources, even with his master Aristotle, when his observational and experimental results contradict their views.

Nor, as we noted, is Harvey the kind of "Aristotelian" characteristic of late French Scholasticism, who invokes "substantial forms" imposing themselves on mere matter to produce the entities we see around us. It is a very immediate sense of the ensouled character of living things that drives him. In this, it seems to us, Harvey is genuinely a follower of Aristotle, even though he explicitly differs from Aristotle in finding the primary seat of life in the blood as such rather than in the heart. Following the development of the hen's egg, he sees a speck of blood pulsating even before the formation of the heart: that is the "first to live, the last to die" (W, p. 29). Thus the blood is the locus of innate heat and hence of soul. In general, it is awareness of the uniqueness of the living, of its deep difference from the non-living, that Harvey shares with Aristotle. Harvey even seems to have believed that there is non-life only as a falling off from life – a very different conception from Descartes's reduction of animals to beast-machines (W, p. 517).

Harvey's devotion to the living – as well as to the closely perceptible – is evident, we may add, in another respect in which he differs from both Aristotle and Descartes: his indifference to, and even dislike of, astronomy. True, Aristotle stressed the need to separate one science from another in terms of its unique subject-matter and unique principles, while Descartes sought to develop a unified science, founded principally on mathematics. At the same time, both philosophers, in very different ways, developed an overall cosmology that would embrace a range of disciplines. Aristotle, in the treatise *On Generation and Corruption*,

---

[34] See Bitbol-Hespériès 1990.

explained how the circle of reproduction follows, as best it can, the circular course of the celestial spheres (*On Generation and Corruption* 336 b 35–337 a 8). And Descartes certainly thought his vortices would account for the order of the solar system as well as, at a farther remove, more mundane physical events. Harvey, however, has no interest whatsoever in a general cosmology beyond the harmony he finds in the phenomena of life. If he ever does refer to the heavens, it is in terms of a traditional geocentric picture (he obviously took no interest in the recent development of astronomy). And in fact he even pours scorn on astronomy as a science. No one can palpate the planets or the stars; no one can dissect, let alone vivisect, them. So why pretend to know anything about them? ("Second Disquisition to Riolan," Harvey 1947, W, p. 124).

True, it has sometimes been argued that Harvey's discovery of the circulation sprang from his reliance on Aristotle, with his circling celestial spheres.[35] We see little evidence of this. If at the beginning of the dedication of the *De Motu Cordis* he calls the heart "the sun of the microcosm" (W, p. 4), that is the kind of overblown language one used in dedications; it tells us nothing about the writer's own beliefs. (Indeed, in Harvey's own view, it should have been the blood, not the heart, that was the center of our vital cosmos).[36] Moreover, when Harvey introduces the circulation in Chapter VIII , he tells us how he came to it. Having discovered the true motion of the heart and arteries, he wondered where all that blood was going, and conjectured: Could it be going in a circle? (W, p. 46). True, he then proceeds to invoke Aristotle's analogizing of circular motion beneath and beyond the moon. However, this seems to us partly a polite bow to Aristotle, whom he did admire, and a pretty piece of rhetoric, but less than a statement of deep or central motivation. Then, in the succeeding chapters, as he has promised to do, he adduces evidence to show that the blood does in fact circulate. There is nothing here of a fanatical love of circles as such. Harvey does indeed find in living things a rhythm and a harmony that contrast sharply with the mechanism of Descartes and his successors. But he is no

---

[35] This view was put forward in particular by Walter Pagel in Pagel 1967.

[36] See, e.g., *On Generation*, W., p. 249: "I am of opinion that the blood exists before any particle of the body appears; that it is the first-born of all the parts of the embryo; that from it both the matter out of which the foetus is embodied, and the nutriment by which it grows are derived, that it is in fine, if such thing there be, the primary generative principle."

more an Aristotelian than he is a Cartesian cosmologist. Such remote, unmanipulable matters lie beyond the reach of his ruling passion: the structures and functions of life.

That was by the way. We have been trying to show how, even when they both explain what they have seen, Harvey and Descartes part company in their styles of explanation. It remains to ask what became of their theories of cardiac motion in the succeeding generations. This history, like every history, is complicated; we shall certainly not attempt to follow it in any detail (see Fuchs 1992/2001). It is safe to say, in general, however, that as Harvey's doctrine of the circulation came to be accepted, especially in the Netherlands and in Britain, his Aristotelian approach to living phenomena was largely replaced by a more mechanistic and often Cartesian view. His friend and supporter, George Ent, for example, the disciple who urged Harvey to publish his *De Generatione*, and in fact saw it through the press, held that the master had confused systole and diastole.[37] On the Continent, Dutch investigators tried – in vain, at last – to defend Descartes's furnace conception of the heart. But even when that collapsed in the light of advances in anatomy and physiology – when it finally became clear, for example, that the heart is not hotter than other parts of the body – they retained a mechanical view of cardiac activity more Cartesian than Harveyan in spirit.

For example, Regius, though Descartes quarreled with him, echoed Descartes in his medical writings. In 1676, Cornelis van Hoghelande declared:

> ...we are of the opinion that all bodies, however they behave, are to be viewed as machines and that their actions and effects...have to be explained only according to mechanical laws.
>
> von Hoghelande 1676, pp. 124, 137

And in 1666, Nicholas Steno had announced:

> No one but [Descartes] has explained all human functions, and above all those of the brain, in mechanical fashion. Others describe for us man himself; Descartes speaks only of a machine, which at one and the same time shows us the inadequacy of the doctrines of others and points out a method of investigating the function of the parts of the body just as insightfully as he describes the parts of his mechanical man. We should

---

[37] Ent 1641, p. 35: "*systolen pro diastolen acceperit.*"

not blame Descartes if his system . . . does not correspond entirely with experience. His soaring spirit . . . makes up for the errors of his hypotheses.

Steno, 1666, II, p. 8[38]

Even in England, where Harvey was certainly revered, interpreters of the heartbeat and circulation showed the heavy influence of the "new mechanical philosophy," an approach, as we have seen, alien to the great initiator's thought. How Cartesian this new mechanism is in its English versions it is hard to say. It is at least indirectly Cartesian, but also strongly influenced by Gassendist atomistic arguments as well as by new chemical speculations and experiments. In particular, it was respiration that was studied by a number of English workers in the circle around Robert Boyle (see Frank 1980). Neither Descartes nor Harvey had made much headway in this area, Descartes – as we noted – holding that the blood changed color in the heart, and Harvey considering that, if something happened to it in the lungs, the brighter observed color was perhaps a consequence of a straining action (Frank 1980, pp. 15–16).

Whatever the subtleties of this history, what interests us as philosophical students of biology is the paradox that the reduction of animals to machines clearly facilitated the acceptance of a doctrine that had been first put forward in what was philosophically a much more conservative, fundamentally Aristotelian spirit. Similarly, at a later time and in a different context, Albrecht von Haller, who considered himself a mechanist, would introduce a concept of "irritability" that helped support an anti-mechanistic tendency among late eighteenth-century biologists.[39] These two contrasting emphases in the attitude of life scientists to their disciplines have of course recurred conspicuously a number of times in the twentieth century, along with new controversies about them. We shall return to them also in later chapters.

---

[38] This and the two preceding quotations are taken from Fuchs 1992/2001.

[39] On von Haller, see Fuchs 2001, pp. 187–190. Because of the heavy ideological connotations of the term, we avoid using "vitalism" whenever possible.

CHAPTER 3

✦

The Eighteenth Century I

Buffon

Introduction

In looking at episodes in our philosophical tradition that bear in various ways on what we now call biological questions – and anticipating what we have come to call the philosophy of biology – we have been dealing so far with figures at the opening and the close of the Aristotelian tradition, a tradition lasting more than 2,000 years. If Descartes wanted to overturn Aristotelian science, he wanted to do so from within a Scholastic environment, and he was speaking to Scholastic readers. However, when we come to Georges-Louis Leclerc, famous in his time as the Comte de Buffon, we enter a different world. Before we look at some of the major features of Buffon's work, we need to specify briefly the most striking novelty in the intellectual climate of his time: that from the last days of Scholasticism, we have moved to a post-Newtonian era. Looming over every area of scientific work, there falls the shadow of "the great Newton." There were still thinkers in the Aristotelian tradition, such as Buffon's critic Malesherbes (Malesherbes 1798).[1] But, on the whole, Newton was the authority figure to be followed, or perhaps in some ways challenged. In cosmology, Newton, rather than Descartes or Aristotle, commonly served as the starting point for new speculations. When, for example, in his account of the origin of the planets, Buffon invoked an "impulsive force" in addition to gravitation, he was reflecting against a Newtonian background (Buffon, "Proofs of the Theory of the Earth," OP, p. 1). Methodologically and epistemologically, Newton's seeming inductivism, reinforced by the methodology of Locke's *Essay* (1690), supported attention to the value of observation

---

[1] See Malesherbes, in Lyon and Sloan 1981, pp. 329–345.

and experiment, as distinct from system-building. Thus it is no surprise that the young Buffon, already a member of the French Academy of Sciences, prefaced his translation of Alexander Hales' *Vegetable Staticks* (Hales 1727/1969) with an impassioned eulogy in defense of experiment and observation as against any liking for systems:

> Works founded on experience (or experiment) have more value than others; I can even say, in the matter of physics, that we must seek experience as much as we must fear systems. I admit that nothing would be so lovely as to establish a single principle to explain the universe.... But sensible people see well enough how vain and chimeric that idea is. Perhaps the system of nature depends on several principles. These principles are unknown to us. How do we dare to flatter ourselves that we can unveil these mysteries with no guide but our imagination? And what do we do to forget that the effect is the only way to know the cause? It is by fine experiments, reasoned and carried through, that we force nature to uncover her secret. All other methods have never succeeded, and true physicists cannot help seeing the older systems as ancient dreams. And they are reduced to reading most of the new ones as they read novels. Compilations of experiments and observations are thus the only books that can increase our knowledge.... So let us always amass experiments, and distance ourselves, if possible, from all spirit of system.... It is this method that my author has followed; it is that of the great Newton; it is the method that Verulam, Galileo, Boyle, Stahl have recommended and embraced; it is the method that the Academy of Sciences has made it a law to follow.
>
> Buffon 1735; OP, pp. 5–6

In the thirty-one volumes of natural history that he was to publish over nearly forty years (from 1749 to 1788, though some pieces date back as far as 1744), Buffon himself often produced strange enough theories. Nor, as we shall see, was he a simple-minded Lockean. Yet, however global and far-fetched his theories may seem to be, he never abandoned his stress on detail, on facts and more facts, or his distrust of what appeared to be claims to have grasped ultimate causes. His model here was indeed Newtonian: Gravitation as a general phenomenon specifies, in a way, the cause of particular phenomena, but the cause of gravitation itself is and will remain unknown. As we go from particular to general effects, we call the latter "causes." As we shall see, this is not skepticism, or even phenomenalism. All it means is that we want to stick close to what we see with our eyes – and what we really see. More of this shortly. Our point here is simply that it was Newtonian cosmology as well as a version, or a vision, of Newtonian methodology that dominated the thought of Buffon and of many of his contemporaries from the start. Buffon wrote admiringly of Aristotle's *History of Animals*. But Aristotelia*nism*, however altered from the original, had emphatically been overcome – though not quite in the form in which Descartes would have wished it!

In this chapter, we will look at Buffon's contributions in several biological areas: taxonomy, and, fundamental to that discipline, the concept of species; reproduction; the relationship between mechanism and vitalism in his approach to the phenomena of life, and finally, very briefly, his cosmogony. But before we deal with those special questions, we need to assess, as best we can, what we may call Buffon's philosophy, or at least his notion of the method he is following in his study of nature.

Buffon's Methodology

Buffon started his professional life as a mathematician. In fact, he translated Newton's work on fluxions as well as Hales's book (Buffon 1735; 1777). As a landowner in Burgundy – he always spent at least half the year at his estate in Montbard – Buffon had become interested in the problem of improving wood for naval and other practical purposes. In March 1739, in part perhaps as a result of these and similar interests, he left the Mechanics section of the Academy for the section on Botany. Later that year, he was appointed head of the Royal Botanical Garden, an appointment that triggered his life work: the elaboration of a most ambitious and epoch-making succession of volumes on natural history, a genre that goes back to Aristotle and Pliny, but that had not yet been put into a modern scientific context.

As to the method he used in this enterprise, Buffon is often, almost routinely, counted among the many French followers of Locke. His most authoritative biographer, the great historian of science Jacques Roger, labels him in this way, and refers a number of times to his subject's "sensualist philosophy."[2] And there are indeed passages that support such a reading. For example, in a criticism of Plato's ideas, Buffon writes:

> Do we not see, then, that abstractions can never become principles either of existence or of real knowledge; that on the contrary such knowledge

---

[2] See, e.g., Roger 1989/1997, p. 135. However, Roger also notes Buffon's sense of concrete reality, which appears to us to contradict a Lockean reading (pp. 89–90, 426).

can never come from anything but the results of our sensations, ordered and followed through; that these results are what is called experience, unique source of all real science; that the use of any other principle is an abuse, that every edifice built on abstract ideas is a temple raised to error?

OP, pp. 257–8

This certainly appears to be a strong "empiricist" statement. But the lesson of Locke's epistemology is that, except for "sensitive" knowledge, which is purely particular and local, we are limited to intuitive relations among our ideas or demonstrative relations in mathematics, morality, or theology. Everything else rises only to the level of *belief*, not *knowledge* (Locke, 1690, Part IV, ch. 4, para. 3). For Buffon, on the contrary, the evidence of our senses, when regularly repeated, does indeed offer us "real science," by which he means certain knowledge, not only of our sensations, but of how things actually are. We cannot push through to ultimate causes, but we can know effects with sufficient scope and reliability to amount to certainty. That is not by any means a Lockean attitude. The difference is clear from the paragraph that precedes the passage just quoted, and it is even clearer from Buffon's exposition of two kinds of truth in the "Premier discours," which he set at the head of the first volume of the *Histoire naturelle* (Buffon 1749b). Consideration of both these passages should give us some idea of the peculiar style of Buffon's empirical realism.

Attacking Platonic ideas, Buffon observes in the passage just quoted that while "in speculation they seem to start from noble and sublime principles; in application they can arrive only at false and puerile consequences." He continues:

Is it indeed so difficult to see that our ideas come only from the senses; that the things we consider real and existent are those of which our senses have always given us the same evidence on all occasions; that those we take as certain are those that always occur and present themselves in the same way; that this way, in which they present themselves, does not depend on us, any more than does the form under which they present themselves; that consequently our ideas, far from being able to be the cause of things, are only their effects, and very particular effects, effects so much the less like the particular thing, the more we generalize them; that finally our mental abstractions are nothing but negative beings, which do not exist even intellectually, except through the excision that we make of sensible qualities from real things?

OP, p. 257

That all our ideas come from the senses sounds Lockean enough. That the way they present themselves does not depend on us is a commonplace, which might justify Locke's "sensitive knowledge": the awareness that something non-me is here before me now (Locke 1690, IV, X, 3). For Locke, that is as far as our knowledge of bodies goes. As his argument proceeds, the new corpuscular philosophy of the illustrious Boyle and the incomparable Newton falls off into the realm of probability, which warrants only faith or opinion, not knowledge (*Essay*, IV, XV, 2). Buffon's attitude, by contrast, radically undercuts such incipient skepticism. What occurs with total regularity, we can count as real. Of course, we are dealing with effects, not hidden causes to which our limited intellects cannot reach. But when such effects are regular, constant, utterly unvaried, we have good reason to be certain it is existences we are dealing with, not just our own sensations as subjective happenings. When things happen sometimes, we have to rely on analogies, or use a calculus of probabilities. But a very high probability amounts to certainty, and hence to knowledge.

Indeed, Buffon himself – an adept of contemporary probability theory – applied just such calculations to the very example Hume used to skeptical effect: Would the sun rise tomorrow? (Buffon 1777, in Lyon and Sloan, pp. 56–7).[3] Hume, of course, is more radical than Locke, but since he was clearly taking Locke's "new way of ideas" to its logical conclusion, the comparison seems a fair one. For the Scottish philosopher, a prediction of the sun's rise tomorrow would be guesswork. For Buffon, it would be the *reasoned* result of calculation. Our prediction has a very high probability indeed; so we could be certain that the sun would rise tomorrow (and indeed, it did). Admittedly, even for the skeptic Hume, there are "proofs" as distinct from "probabilities" (Hume 1739, Book One, Part III, Section 11). But even his proofs are just the effect of the constant conjunction of impressions and ideas combined with the force of habit. Only the propensity to feign induces us to believe there is a sun at all, let alone an ordered planetary system and everything else. In contrast, "constant conjunction" for Buffon produces certainty of existence, not just certainty of subjective expectation. This is "empiricism" all right, if the term means anything, but empiricism of a very concrete and realistic sort. What Buffon is supporting here is what in the "Premier discours" he calls "physical truth" (OP, p. 24).

---

[3] For Buffon's participation in the new work on probability, see Daston 1998.

Buffon set the "Premier discours" at the head of his essays on the history of the earth, his essay on the origin of the planets, and his general history of animals, perhaps in parallel to Descartes's placement of his *Discourse on Method* at the head of his *Dioptrics, Meteors, and Geometry*. At the close of this text, Buffon sets out explicitly his conception of truth and the way it is to be obtained (OP, pp. 23–26). He distinguishes two kinds of truth – mathematical and physical. Mathematical truth consists in the manipulation of definitions of our own devising. Thus it produces *evidence*, in the sense of perspicuous clarity, but at the cost of any relation to reality. Buffon's discussion of the problem of infinitesimals in the introduction to his translation of Newton's work on fluxions foreshadows this skepticism about the basis of mathematical conceptions, and his *Essai d'arithmétique morale*, a compilation of materials from various periods, confirms it (Buffon 1777).[4] In sharp contrast to the abstract nature of mathematical truth, Buffon postulates what he calls "physical truth," which, he claims, produces not evidence, but certitude. Here we begin with what our senses tell us, concretely, and then if we follow what they tell us with regularity and constancy, we arrive at generalizations worthy of our total trust. Mathematical concepts can indeed be applied to reality in certain limited areas – clearly, those treated by Newtonian natural philosophy. But even there, as we have seen, Buffon insists that what Newton achieved was a knowledge of general effects, such as the fact of gravitation, not an ultimate explanation (in this case an explanation of why the inverse-square law should obtain). When it comes to the kind of subjects Buffon himself was dealing with, however – geology, cosmogony, and what is now called biology – even such limited mathematization is beyond our scope. Nevertheless, if we rely on observation of constant and regular events and objects, and on concrete results of experimentation with what we really find around us, we can and should achieve, on a goodly number of empirical questions, not just probable conclusions, but certainty. Buffon changed his views on particular questions in the decades to come. Yet he never abandoned this fundamental distinction, with its dislike of abstraction and its confidence in the positive issue of reliable induction. Given the state of knowledge at the time, the history of the earth, for example, would seem to be a fairly speculative subject. But,

---

[4] The texts in question are Newton's *La méthode des fluxions et les suites infinies*, translated by Buffon in 1740 and incorporated into Buffon 1777. See discussion in Roger 1997, pp. 384–5.

in contrast to the speculations of others, Buffon was confident that if he stuck to the processes we now observe around us – in other words, if he practiced a kind of uniformitarianism – he could draw quite a few safe general conclusions.[5] Thus, even in what appear unlikely areas, there was a good supply of "physical certitude" to be obtained.

Beyond that, explanations of particular geological or cosmological events would be speculative, or merely hypothetical, and a similar situation obtains in the biological realm. Despite his apparent strict inductivism, Buffon does use hypotheses, some very central to his position on important questions. He seems to be happier when they are justified by experiment, as in the case of organic molecules, to which we shall return later. Our point here, however, is simply that in all his work, it is not just relationships among our ideas, but realities he is in search of.

Thus the distinction of the "Premier discours" and the emphasis that recurs throughout his works on the secure knowledge to be obtained by careful induction separate Buffon quite plainly, it appears to us, from more faithful followers of Locke. Granted that we start from our sensations, we can build knowledge from them in ways inaccessible to the author of the *Essay* and those who followed in his wake.[6]

Can we place Buffon better, then, by looking for other philosophical sources than Locke for his views? It is clear, for one thing, that he knew well, and applied in his own work, recent developments in the theory of probability. We have seen one example of this in the case of the sun's presumptive rise tomorrow. Phillip Sloan has argued in a number of articles that a further important influence was that of Leibniz, or Leibniz as expounded (and modified!) by Christian Wolff (Sloan 1979; 1992). We respectfully disagree. Admittedly (as Sloan stresses), Buffon, with his vivid sense of concrete reality and his dislike of abstractions, would have rejected Newton's concepts of absolute space and time and would have preferred the Leibnizian approach, which treated space and time as relative to the actual ordering of real events. For Leibniz, however, these would have been merely "phenomena." They were indeed well-founded phenomena, which we in our circumstances could take as real. But beyond or behind them lies the metaphysically real, which, at least in terms of the later Leibniz, bears little resemblance to the thick, perceptible realities grasped by Buffon's vaunted "physical certitude."

---

[5] See Buffon 1749b, p. 149.

[6] Our view is supported by the argument of Phillip Sloan in his authoritative paper on "The Buffon–Linnaeus Controversy" (Sloan 1976).

Further, Sloan includes in his otherwise wholly convincing analysis of Buffon's philosophy of science a reliance on the Law of Sufficient Reason in the formulation of hypotheses (Sloan 1992, p. 213). Yet Buffon ridiculed not only the use of final causes in general, but the Law of Sufficient Reason itself in its Leibnizian form (OP, pp. 258, 360). Or if it was indeed Wolffians like Madame du Chatelet that Buffon was following,[7] he would be just as unlikely to accredit their *version* of the Law; whether it was intuitive or based on contradiction, this was, again, a kind of reasoning that Buffon would be unlikely to accept.[8]

Perhaps we may add, as a conjecture, that Buffon's peculiar kind of realism, far from needing any particular philosophical sponsorship, might well simply express the attitude of a landowner passionately interested in what he could achieve and what he could learn from the cultivation of, and experimentation with, his land and woods and livestock. Much as he clearly enjoyed the prestige of his position in Paris, and much as he clearly enjoyed setting forth his views on every conceivable subject, often with only too much eloquence, he did after all spend as much time at Montbard as he possibly could, and he did carry out (or supervise) experiments in the growth of trees under various conditions; the possible, or impossible, crossing of animals of seemingly related species; and so on. His is the realism, we suspect, not simply of a philosopher – a philosopher with no trace in his makeup either of traditional teleological thought or of a more recently popular systematic idealism – but of a zealous and experimentally ingenious agriculturalist, a student of living things as deeply interested in the practical as in the theoretical, but always in the *real* implications of his investigations, in whatever field.

Given this very crude sketch of the "philosophy" of a very complex thinker, let us consider his contributions to the philosophical aspects of biology, or to the philosophy of biology, in a number of areas.

Taxonomy

The two centuries preceding Buffon's had seen the rise of systematic taxonomy, especially of plants, starting with the work of Andrea Cesalpino. Here we do need to return to the Aristotelian tradition, albeit in its late Scholastic form. Aristotle himself, as we noted in Chapter 1, was

---

[7] Sloan's detailed historical account is certainly trustworthy, but that does not entail a deep philosophical influence.

[8] For Wolff on the Law of Sufficient Reason see, e.g., Wolff 1730, ¶¶66, 70.

not a taxonomist. He did indeed believe that the mind, in scientific knowledge, could identify itself, so to speak, with the essences of its objects – provided, of course, that its possessor followed the canons of Aristotelian science. But he did not believe that the essential nature of any organism could be tagged by the name of a single genus and single specific character. Different organisms might need to be differentiated by very dissimilar criteria. Indeed, as David Balme has shown, and as we have reiterated, Aristotle did not use his own terms "genus" and "species" in anything like the univocal way in which modern classi-fiers were to use them (see Chapter 1 and Balme 1962). However, a late Scholastic thinker such as Cesalpino (1519–1603), armed with a more standardized use of these concepts, and confident in the knower's ability to apply them successfully, had, it appeared, an easier row to hoe. Plants are characterized by their vegetative souls – that is, by their powers of nutrition and reproduction. Cesalpino divided them all, by the first criterion, into two groups, depending on the hardness or soft-ness of the material at their "heart." All further divisions followed from differences in the other essential feature, reproduction – that is, fructi-fication. This criterion was later followed by other systematizers, like Rivinus (or Bachmann, 1652–1725) in Germany and Joseph Pitton de Tournefort (1656–1708) in France.[9] But it was the system of the great Swedish taxonomist Carolus Linnaeus (1707–1778) that brought this tradition to its most authoritative development and soon came to dom-inate European thought (see Mayr 1982, pp. 171–180; Larson 1971).

Although Linnaeus was born the same year as Buffon, his *Systema Naturae* was already sweeping Europe when Buffon began publishing his *Histoire naturelle* (Buffon 1749–1767). In Revolutionary France, it became, for some reason, the only approach at all permissible, but even in Buffon's lifetime it was widely accepted. Although it was not Linnaeus who first introduced the binomial system, as in *Felis leo* or *Felis tigris*, for example, it was his work that firmly established this nomenclature, as well as the series species, genus, . . . order, class that systematists still follow. Admittedly, Linnaeus could not apply the "fructification" crite-rion to animals; these he had to sort out by various marks in various cases. Thus he distinguished mammals from one another by their teeth, birds by their bills, fishes by fins, and insects by wings. But clearly, the whole procedure, applying the binomial nomenclature consistently and

---

[9] For a summary of these developments, see Sloan 1972, especially pp. 9–13, 27. Cf. Mayr 1982, pp. 154–166.

arranging a vast number of species in a series of widening categories, had a tidiness that was immensely appealing.

However, there was more to Linnaeus's method than appeared on the surface. Linnaeus was indeed a passionate classifier, who wanted things sorted out neatly. In this pursuit, he followed the canons of Scholastic logic. At the same time, he was a devout believer in God's creation. It was the essences of natural kinds as God had made them that he wanted to discover. In this search, it was the genus that was the principal object. So he was using relatively artificial methods of division to achieve a knowledge of certain essential realities. But essences can be hard to get at. And along with his reliance on traditional logic and his faith in the order of God's creation, Linnaeus clearly possessed very great gifts as a field naturalist, who was just astonishingly perceptive in sorting out one organism (especially one plant) from another. But that meant noticing all the perceptible characters of an item one was inspecting: the so-called *habitus*. Is that what a person does who is searching for God-given essences? You had to do it, Linnaeus said, "secretly, under the table, so to speak, in order to avoid the formation of incorrect genera" (Linnaeus 1751, §168). From our point of view, and even, it appears, sometimes for himself, Linnaeus's three great intellectual passions fit uneasily together: his love of logic, his faith in God's essential ordering of His creatures, and his devotion to detailed observation in the field. Even in the very definition of the genus, its "character," he distinguishes three aspects: "the factitious, the essential, and the natural" (Linnaeus 1751, §186).[10] It is in search of the natural that factitious – that is, artificial – criteria are used. The search for a natural system is what taxonomy in almost any version is all about (§163). But for Linnaeus at least, the natural is, or should be, expressed through the stable, God-given essence. The tension between those three criteria is what makes Linnaeus's work both fascinating and obscure.

At the same time, his achievements in classification were so striking, that, as we noted, by the time Buffon was developing his own contribution to natural history, the name and method of Linnaeus had come to dominate the field. True, there were taxonomists, such as John Ray (1627–1705) in his later period, or Michel Adanson (1727–1896), a younger contemporary of Buffon's, who insisted that a classification based on one character, or even just a few characters, must be inadequate. "Without doubt," Adanson remarked, "the natural method in

---

[10] Paragraphs 186–209 deal in some detail with Linnaeus's method.

botany can only be attained by consideration of the collection of all other structures" (Adanson 1763, p. clv). In the case of Ray, as Sloan has convincingly argued, this position seems to be linked with the skepticism inherent in Boyle's defense of the "new corpuscular philosophy" and in Locke's epistemological arguments (Sloan 1972). Indeed, Locke's insistence that we can know only "nominal" essences may well have had its origin in work he himself had done in botanical taxonomy, as well as in his early medical training.[11] There is no evidence, however, that it was Locke's skepticism that turned Buffon against Linnaeus. Our point here is that objections such as those of Adanson, or the earlier debates between Ray and Tournefort or Rivinus, concerned particular methods of classification: Would one character do, or were more needed, and if so, of what sort? Despite skepticism such as that of the older Ray, these were in the main debates within a growing taxonomic science, to which Ray himself had made important contributions. The aim was to find natural divisions among plants or animals, and so to establish a "natural" rather than a merely artificial system. As we have seen, that was an aim cherished by Linnaeus himself.

The case is quite different with Buffon. In the "Premier discours," he savagely attacks Linnaeus, not only for using single characters as a basis for classification, but, it seems, for attempting to raise up a taxonomic hierarchy at all (Buffon 1749b). Let the moderns boast of their new methods, he tells us; the ancients understood that "the true science is the knowledge of facts." "Aristotle, Theophrastus and Pliny," he continues, "who were the first naturalists, were also the greatest in some respects. Aristotle's history of animals is today still perhaps the best we have in this genre" (OP, p. 20). The reason is that Aristotle starts out by describing carefully the "history" of each animal before trying to collect them into classes of some sort. It's the concrete detail that matters. Indeed, since the accounts of each organism given by Aristotle or Pliny were relatively brief, Buffon admits that further description is necessary. Beyond that, he adds, we need to rise to generalizations from our detailed knowledge – but the method for that is not yet available. What is clear is that the arbitrary system-building of modern taxonomists is antithetical to proper procedure in the study of living things.

---

[11] On Locke's connection with botanical taxonomy, see Sloan 1972, pp. 21–2. For his views on medicine, closely associated with the phenomenalist philosophy of his teacher Thomas Sydenham – in publications only recently attributed to Locke himself – see Ducheneau 1973; 1998, pp. 239–164.

In opposing classification through single characters, Buffon is in agreement with Adanson, but in an even more radical fashion. What he wants is not simply a complete list of morphological characters, but a description that takes account also of the organism's behavior, habitat, and so on. Thus in natural history, "the precise description and the accurate history of each thing is . . . the sole end which ought to be proposed initially" (OP, p. 16). As to the description – the necessary starting point: ". . . one ought to show form, size, weight, colors, positions of rest and of movement, location of organs, their connections, their shape, their action, and all external functions" (OP, p. 10). If we can add an account of internal organs, so much the better – but not in too much detail! (Although, or perhaps because?, the anatomist Jean-Marie-Louis Daubenton was his collaborator for many years, Buffon seems to have wished to minimize the contributions of anatomy.) Then, once we have a well-rounded and fairly detailed description of the organism in question, we proceed to its history (so far at least, in the traditional sense of history – an account of whatever are the salient and interesting facts about an entity's style of existence): "The history ought to follow the description, and it ought to treat only relations which the things of nature have among themselves and with us (OP, p. 16)." We start with particular facts, carefully described, and proceed to investigate the relationships among them. It is in fact, in Buffon's view, only such relationships that we *can* investigate. But because of the constancies we find in the events and things that strike us, we can, it appears, go beyond just the account of particular cases. Thus "[t]he history of an animal ought to be not only the history of the individual, but that of the entire species." This will include:

> . . . their conception, the time of gestation, their birth, the number of young, the care shown by the parents, their sort of education, their instinct, the places where they live, their nourishment and their manner of procuring it, their customs, their instinctual cleverness, their hunting, and, finally, the services which they can render to us and all the uses which we can make of them.
>
> OP, p. 16

Again, we may add some anatomy, but not too much. This is a far cry from the arbitrary characters for the sorting of animals – teeth and mammary glands for mammals, for example. Anything less than full description would fail to do justice to the demands of physical truth, substituting the inventions of our imaginations for careful observation and experiment. True, we have to arrange our many objects somehow, in a kind of classification, but that is not our ultimate aim.

Moreover, the usual methods, unlike Buffon's, lay us open to the snares of linguistic innovation. Once we have imagined some distinction, we give it elaborate names and definitions and suppose that we are building a science, whereas we are in fact only heaping error upon error. Buffon's scorn for name-giving is positively Hobbesian. Thus he writes:

> Even in this century, when the sciences seem to be cultivated with care, I believe it is easy to perceive that philosophy is neglected, and perhaps more than in any other century. The arts that people would like to call scientific have taken its place. The methods of calculus and geometry, of botany and natural history – formulae, in a word, and dictionaries – occupy almost everybody. We think that we know more because we have increased the number of symbolic expressions and learned phrases, and we pay hardly any attention to the fact that all these arts are only scaffoldings to get to science, and not science itself.
>
> OP, p. 23

All this reflects Buffon's distrust of abstractions and his insistence that we stick as closely as possible to the concrete realities presented to us by our senses. We want to make generalizations, but we must do so with caution. Sometimes Buffon puts this warning in Lockean, or almost Berkeleyan, terms: ". . . our general ideas are only composed of particular ideas" (OP, p. 18). But in the overall argument of the "Premier discours," and in Buffon's work in general, we need to understand his warning about the risk of hasty generalization in more realistic terms. As he puts it at the close of this discourse. "The most delicate and the most important point in the study of the sciences" is "to know what is real in a subject from what we arbitrarily put there in considering it" (OP, p. 26). In taxonomy as practiced in his time, Buffon believed, there was much too much of the latter; abstraction ruled rather than the careful and detailed study of individual organisms considered in (and, it appears, even *as*) the species to which they uniformly and regularly belong. (That much generalization appears to be safe; we shall return shortly to the question of the nature of species in Buffon's account.)

If the criteria used for classification are abstract and arbitrary, clearly there is little hope that taxonomic hierarchies built on such foundations will meet Buffon's demands for concreteness and realistic detail.

He is confident of the reality of species, but building beyond that is
dangerous:

> For, in general, the more one augments the number of divisions of the
> productions of nature, the more one approaches what is true, since in
> nature only individuals exist, and genera, orders and classes exist only in
> our imagination

OP, p. 19

We can take this statement as programmatic for Buffonian natural history. Not even Linnaeus himself, of course, insisted on the reality of
higher taxa. But in his view, the genus is a fundamental reality. For
Buffon, it was the species he wanted to stick to if he could. (As we shall
see, there were some, but less than fundamental, modifications of this
program in his later years.)

Still, no writer can just jumble all his material together with no order.
What Buffon does do when he is forced to classify species one way
or another is to order his subjects according to their relationship to
us. If this appears arbitrary, and "anthropocentric," Buffon's answer
is that we know things only in some relation or other, and to take
them in relation to ourselves is the most ordinary and therefore the
most intelligible way to set about our task. So, for example, he treats
domesticated animals on the one hand and on the other savage beasts, as
Aristotle had done. More than that: if the dog in fact habitually follows
the horse, he wonders, why shouldn't he do so in our account of both
these familiar species? Why place the dog with the wolf or the horse
with the ass – let alone with the zebra – when we do not habitually see
them together? Except via Lockean skepticism about any classification
whatsoever, a revolt against taxonomic science could hardly go further!

At the same time, we must answer some obvious questions about
Buffon's program. To begin with, if there are only individuals, how can
we so confidently study the species? The answer will be clearer if we
look at Buffon's concept of species. These primary and really existing
units turn out to be, in a sense, individuals – or even, as he argues in the
"Second View," "the only unities in nature" (OP, p. 35). Before we come
to that notion, however, there are other difficulties to be addressed.

First, in his discussion of other systems, Buffon has spoken of genera, and even classes, as if they did exist (OP, p. 13). He talks of how
species are established through noting very close resemblances – genera
through somewhat less close resemblances, and classes still less. Admittedly, he may here be simply recounting the practice of others without

for the moment evaluating it. Beyond that, however, is a second question. What has appeared to many students to be a more serious difficulty is that some years later, he does admit that he must deal with birds in terms of genera, not only species (Buffon 1770, cited in Sloan 1976). He also speaks of "genera" among domesticated animals (Buffon 1764). It has been generally supposed, therefore, that Buffon finally gave in and admitted the validity of the Linnaean hierarchy (see, e.g. Daudin 1926, p. 143). We see no good reason, however, to read the history of Buffon's *Histoire* in this way. To be sure, the fact that Buffon allows for genera in the classification of birds is a partial concession. But his use of the term "genus" in the discussion of domestic animals belies any such intention. The various kinds of sheep, for example, turn out to be different genera within a given species! In the chapter on "The Wild Sheep and Other Sheep" (1764), he writes:

> In nature there exist only individuals and successions of individuals, i.e. species . . . We have altered, modified and changed the species of domestic animals. We have thereby made physical and real genera, very different from those metaphysical and arbitrary genera existing only as ideas. These physical genera are in reality composed of all the species, altered in different ways by the hand of man, which have, however, a common and unitary origin in nature.
>
> Buffon 1764, in Sloan 1976, p. 374

Apparently, like Aristotle in his biological works, Buffon is using "genus" to mean "breed" or "race," no matter at what taxonomic level. This is as far as one could get from what has come to be considered orthodox taxonomy.

There is a third, and less easily solvable, difficulty that should be mentioned, also reminiscent of Aristotle. One of the reasons Buffon gives for the unreality of the higher taxonomic divisions is the continuity we observe in nature. Nature, he tells us,

> proceeds by unknown gradations, and, consequently, it is impossible to rely entirely on those divisions, since she passes from one species to another species, and often from one genus to another genus, by imperceptible nuances.
>
> OP, p. 10[12]

---

[12] Sloan 1976 (p. 262, n. 32) traces this passage to Locke, and Piveteau to Leibniz; a *locus classicus* is Aristotle, *Historia Animalium* 588b4–12; cf. Aristotle, *De Partibus Animalium* 681a12, 686b26; also Plato, *Timaeus* 91D.

Buffon proceeds to describe how experience in botany shows this to be the case. Such a statement appears to be not so much a report of experience as the expression of a belief in the traditional *scala naturae*. "Imperceptible nuances" seem to be, by definition, outside the range of experience. Yet such global generalizations recur often enough in Buffon's many volumes, and, indeed, in very different forms. In a careful and illuminating study, Guido Barsani has shown that Buffon moved between three "images of nature": a scale, a network, and a tree (Barsani 1992). Thus, Barsani argues, Buffon was both mirroring and anticipating the major approaches characteristic of the naturalists of his century. Although it would take us too far afield to follow the details of his exposition, we should keep in mind that even those who pride themselves most in looking first at facts and more facts do so in the light of some more sweeping presuppositions, some guiding schema with which their cherished facts appear to harmonize.

Species

Whatever "images" guided Buffon, however, it was the description of a great variety of species with which he was primarily concerned. We have been told that only individuals exist, and yet that, in studying the individual, we are studying the entire species (OP, p. 16). In the *First Discourse*, Buffon does little to resolve this paradox. He simply tells us that the species consists of individuals closely resembling one another (OP, p. 13). Presumably, if the resemblances are very close, studying one is the same as studying another, and so in any case one is studying the species. Dissecting a frog tells us (unless we happen on a "deformed" specimen) about the anatomy of the frog.

But that is by no means the end of the story. A few years later, in his account of the ass, among domestic animals, Buffon proposes a more original, indeed radical, concept of species:

> It is neither the number nor the collection of similar individuals that constitutes a species; it is the constant succession and uninterrupted renewal of these individuals that constitutes it.

> OP, p. 255

To be sure, the ability to produce fertile offspring had often been taken as a criterion for separating species. But to make reproductive sequence the very *nature* of species shows a new stress on change over time as a fundamental characteristic of living things. Aristotle and Harvey had both admitted the significance of generation in the renewal of the form that distinguishes one kind of organism from another. But Buffon, with his distrust of abstractions, was able to go further, and to dismiss the timeless *eidos* altogether in favor of the continuously repeated generative process as the definitive characteristic of species. He continues:

> A being that lasted forever would not be a species, nor would a million similar beings that would also last forever. For species is an abstract and general word, whose object exists only in the succession of times, and in the constant destruction and equally constant renewal of beings. It is in comparing nature today with that of other times, and present individuals with past individuals, that we have obtained a clear idea of what we call species; and the comparison of number or of the similarity of individuals is merely an accessory idea, often independent of the first. For the ass resembles the horse more than the spaniel resembles the greyhound, and yet the spaniel and the greyhound are but one species, since together they produce individuals who can themselves produce others, while the horse and the ass are certainly different species, since together they produce only barren and infertile individuals.
>
> OP, p. 355–6

Thus a species, in Buffon's view, is not *either* an eternal form *or* a historical sequence; it is only the latter. It is a historical entity. We could even call it an individual in the sense of a single linked chain of organisms connected by genealogy. And it is as just such a link in a continuous chain that the individual, when we study it, can be taken as representative of the species.

To us, this sounds very modern. It is sometimes taken, indeed, as an anticipation of evolution, since only change over time can make a species (Gayon 1992b, p. 475). We think so, too, because for us life is history from start to finish. But it does not appear this way to Buffon. He does indeed see the reality of each species in historical terms, in terms of real, continuous temporal relations between successive individuals. In fact, as we have noticed, time is for him the real relationship between real events; it is no more abstract than are the real entities whose interrelationship it expresses. Yet Buffon firmly believed in the fixity of species. Though they are historical entities through and through, species are constant; they exist in time, or as temporal, but are basically unchanged through the passage of centuries. Some species, Buffon admitted, had gone extinct. And there is *degeneration of*, or better, *within* species, as in the case of the domestic animals deformed by man. But one species never turns into another or (in cladistic terms) produces a sister group.

As Jean Gayon has argued, the concept of species as historical entities is independent of the concept of evolution (Gayon 1992). Buffon accepts the notion of "descent with modification," if you like, but within the bounds of species – his "genera" of domesticated species, like the spaniel and the greyhound, are what we would call varieties. There is no descent of one species from others, only, as we have noted, *degeneration* within a species, a slow failure to retain descriptive identity under changing environmental conditions or through human interference, or possibly, in terms of his later works, some hybridization, but always within limits. We must accept Buffon's definition, then, as what it is: a significant innovation within Enlightenment thinking about species, and not an anachronistic anticipation of another time. There were indeed contemporaries of his who, in a speculative way, anticipated to some extent the more generalized view of descent that we now take for granted – notably Maupertuis, Diderot, and Bonnet.[13] But Buffon was not among them.

Buffon's position, in its limited historicity, seems in fact to have become more stable as he proceeded with his studies. In his "Second View" (Buffon 1764), which was meant to contrast with the "Premier discours," there is no longer a paradoxical stress on individuals that might be seemingly contrasted with the reality of species. Individuals have now given way to species as the primary units of nature. Buffon writes:

> An individual, to whatever species it belongs, is nothing in the universe. A hundred individuals, a thousand, are still nothing. Species are the only beings of nature: perpetual beings, as ancient, as permanent as nature herself. To judge this better we no longer consider it as a collection or a series of similar individuals, but as a whole independent of number, independent of time; a whole always living, always the same; a whole that has been counted as one in the works of creation, and that consequently constitutes but one unity in nature.
>
> OP, p. 35

Thus we "put the species in the place of the individual" (OP, p. 35). We still place man at the head of all species. But at the same time,

> however different in form, in substance and even in life, each one holds its place, subsists on its own, defends itself against others, and all together compose and represent living nature, which maintains itself and will maintain itself as it is maintained: a day, a century, an age, all the portions of

---

[13] See the discussion in Burkhardt 1977, pp. 82–4. Our account of Buffon on species owes a great deal to the illuminating Sloan 1987.

time do not make up part of its duration. Time itself is only relative to the individuals, to the beings whose existence is fugitive. But that of species being constant, their permanence constitutes duration, and their difference constitutes number. Let us count species as we have been doing, let us give to each one an equal right to the revenue of nature. They are all equally dear to her, since she has given each one the means of existing, and of lasting as long as nature herself.

OP, p. 35

Species exist as units of reproduction, each one, through its individual constituents, making more like itself – with variations, indeed, but always within the permanent contours of that one species rather than another. The reality of temporal succession, the realities of birth and death, are safely confined within the permanent reality of nature herself.

Admittedly, in his late *Epoques de la nature* (Buffon 1778), Buffon subjects nature, at the geological and even cosmogonic level, to a process of gradual cooling, which will lead eventually to the extinction of life. Yet this very history, as Hans-Jörg Rheinberger has argued, actually prevents a basically transformist conception of species (Rheinberger 1990). There is a general, physical law that predicts successive stages in earth's history, but no such law for the relationship of species to one another. Instead, once a given configuration of organic molecules has settled into a particular internal mold in accordance with Buffon's theory of generation, "the type of every species has in no way altered; the internal mold has maintained its form and has in no way altered" (OP, p. 125; for Buffon's account of generation, see the section on Generation, following). Thus, although each particular species consists in a temporal succession from generation to generation, as this species it is both temporal and unchanging.[14] There is degeneration, and even occasional hybridization, but no regular transmutation from species to species.

To see how this strange phenomenon of succession in permanence is maintained, we must look at Buffon's view of generation: of the way in which, as he believed, individual organisms keep making more of the same kind.

Generation

In the last chapter, we considered the rise of mechanistic thinking in the seventeenth century. One of the most conspicuous obstacles to the

---

[14] For the meaning of "organic molecules" and "internal molds," see the next section, on generation.

definitive triumph of mechanism was the problem of explaining reproduction and development in its terms. On that view, animals are machines, like watches. As Fontenelle put it in a much quoted passage,
"Put a male and female dog-machine side by side, and eventually a
third little machine will be the result, whereas two watches will lie side
by side all their lives without ever producing a third watch" (Fontenelle
1683, p. 312). An increasingly widespread answer to this problem was
the doctrine of preformation. Another name for it was "evolution," not
in the post-Darwinian sense, but in the literal meaning of "unfolding."
Somehow, all the plants or animals that ever would be were contained
in little in the first of their kind when God created them. As one English
writer put it "... it seems most probable that the Stamina of all the Plants
and Animals that have been, or ever shall be in the World, have been
formed *ab origine mundi* by the Almighty Creator within the first of
each respective kind" (Garden 1691, p. 476–7; and see Roe 1981, p. 1).

There were two competing versions of this theory: for animalculists,
or "spermists," the infinite series of little individuals resided in the
sperm; for "ovists," in the egg. Preformation in either version, however,
was opposed by those who accepted, in one way or another, Harvey's
conception of epigenesis: the notion that the organism developed from a
fairly simple, not yet organized egg. This view was adopted by a number
of naturalists in response to the discovery by Abraham Trembley of regeneration in the polyp. If you bisect a simple little hydra, the separated
section will regenerate a whole organism. This seemed to suggest that
the organism itself had some power to develop on its own, not needing
to have a miniature of itself contained in itself, like a Russian doll – let
alone an infinite series of such miniatures from creation to the end of
time. The eminent German savant, Albrecht von Haller, who had been
an animalculist, was for a time converted by "Trembley's polyp" to
the acceptance of epigenesis. However, he was reconverted, this time to
ovism, partly through his own experiments, but partly through reaction
against what he took to be Buffon's epigeneticist account.[15] Through the
opposition of others also, such as the Swiss naturalist Charles Bonnet,
Buffon was understood to be a typical epigeneticist. He was certainly
deeply engaged in this long-standing quarrel. But his view was more
complicated than his enemies made out. Here we can look only very
briefly at the major features of his account.

---

[15] Von Haller translated the first two volumes of the *Natural History* and wrote careful
critical introductions to both volumes. In our discussion here, we have chiefly followed
the account of Roe 1981.

From early in his decades as a natural historian, Buffon had announced his doctrine of "organic molecules." All the living world, he believed, was composed of living, though not yet organized, units, whose assemblage in various patterns produced the vast variety of kinds of living things that we in fact find in the world around us. According to Roger, Buffon had borrowed this concept from one Louis Bourguet (who, however, used it in a preformationist context) (Roger 1997, p. 129). He clearly worked it out in conversation with his friend Maupertuis, who published a similar account in his *Vénus physique* in 1745 (Maupertuis 1756, II, pp. 137–168).[16] Whatever his sources, Buffon adopted the concept of organic molecules as his very own. For most of his career, he took them simply as given. In his last years he raised, vaguely, some questions about their origin (Buffon 1778/1962). But for the most part, they were taken as simply available for the formation and reproduction of organisms.

It was a notion common to many naturalists of Buffon's time that the parts of living things are already like the wholes in which they were contained. This was a first principle of preformationist doctrines. At the outset of his chapter on reproduction in the second volume of the *Histoire naturelle*, Buffon seems to adopt it with enthusiasm. "It . . . appears very probable," he writes,

> that there really exists in nature an infinity of small organized beings, similar in every respect to the large organized bodies that appear in the world; that these small organized beings are composed of living organic parts which are common to animals and vegetables; that these organic parts are primitive and incorruptible; that the assemblage of these parts forms what in our eyes are organized beings; and consequently that reproduction, or generation is only a change of form made and operating through the mere addition of these resembling parts alone, as the destruction of the organized being by death or dissolution is produced by the division of these same parts.
>
> OP, p. 240[17]

Buffon is confident that in the succeeding chapters, he will place beyond doubt this conception of the processes underlying the existence of organisms, of nutrition and growth as well as of reproduction.

---

[16] The *Venus physique* was first published anonymously in 1745 and republished as *Système de la Nature* 1751, in *Oeuvres* 1756, II, pp. 137–168.

[17] Apparently this text was written in 1746; see Sloan 1992, p. 416.

We have, he says "no other rule to judge by but experience":

> We see that a cube of sea-salt is composed of other cubes, and that an elm
> consists of other small elms, since if we take an end of a branch or an end of
> a root, or a piece of wood separated from the trunk, or a seed, an elm tree
> will result equally from any of them. It is the same with polyps and some
> other species of animals, which we can cut and separate into different
> parts in all directions in order to multiply them. And since our rule
> for judging [in both cases] is the same, why should we judge differently?
>
> OP, 240

Here, Trembley's polyp seems to provide evidence for preformation
rather than epigenesis, or at least for the preexistence of basic units
at the level of transmission from one generation to the next. What is
different about Buffon's account is his vehement denial that organisms
had been created all at once, with all future individuals of each kind
boxed inside one another to infinity. Although he had himself spoken
of an "infinity" of organic particles, he returned to his favorite theme
of the unreality of abstractions, and insisted on the impossibility of our
conceiving of an actual infinite. So when it comes to the development
of the new organism, the preformationist theme of "emboîtement," or
encasement, was to be rejected.[18] If Buffon opposed as abstract the
insistence that composites are always to be reduced to simples – so that
all living matter can be already somehow complex – he also opposed the
"evolutionist" idea that the full organism was already there, and had
only to be enlarged. This is the epigeneticist element in Buffon's thought.
When we think of nutrition, it seems clear that the organic molecules an
animal assimilates in ingestion are seldom, if ever, tiny instances of its
own kind. The text we have been looking at is puzzling on this point; but
in general it appears that organic molecules are found in all plants and
animals, yet are confined within the bounds of one species by further
agencies that we have yet to consider.

Before we mention those other factors, we should note that beyond
the general rule of following experience, or analogies based on experience, Buffon did accumulate some experimental support for his theory.
Indeed, although he had stated his position about organic molecules in
the mid-1740s, by the time he published the first volume of the *Histoire
naturelle* he had found what he considered strong experimental justification for it. In 1748, together with the English priest John Turberville

---

[18] "Encasement" is Shirley Roe's rendering of "emboîtement." See Roe 1981, p. 3.

Needham, he carried out a number of microscopic observations that supported his view that organisms are composed of the very small motile parts that he called organic molecules. These experiments have been very carefully studied (Sloan 1992). It seems probable that what Buffon and Needham were observing were in some cases bacteria, in others organelles or other parts of cells (for instance, from the Graafian follicle) that are capable of what would later be called Brownian motion.

The Buffon–Needham experiments were faulted by preformationists both on the ground of alleged technical inadequacy and for more theoretical, or even theological, reasons. Although to us the notion of some special living stuff may seem anti-materialistic, to Buffon's critics it appeared just the opposite. Give power to the bits of living things to develop and reproduce themselves? Are we to have spontaneous generation, without God's creative power? Materialism and even atheism are waiting eagerly in the wings – or so it seemed to people like Haller and Bonnet (Roger 1997, pp. 193–4, 342–4).

However, if we take organic molecules in that generalized sense – as minute bodies of which all plants and animals are composed – the confinement of their aggregation to a given species presents a puzzle. Why do the organic molecules of dogs produce dogs, of horses, horses, or of elm trees, elm trees? If organic molecules can all mix with one another, how do we get reproduction confined to a single species? For this purpose, Buffon had to introduce another notion, that of the "internal mold" (*moule intérieur*), which, on the analogy of crystal formation, presumably guides and limits reproduction. In 1749, in the second volume of the *Histoire naturelle* (Buffon 1749–1767, Vol. II) he put forward this hypothesis in the following terms:

> As we can make molds, by which we give to the exterior of bodies any shape we please, let us suppose that nature can make molds by which she not only gives bodies their external shape, but also the internal form. Would not this be a way in which reproduction could be performed?
>
> OP, p. 243

Buffon was well aware that here he was leaving the firm ground of physical certitude, of reliance on what our senses invariably tell us. The concept of organic molecules itself is introduced only as "very probable."[19] The chapter "Of Reproduction in General" is filled with

---

[19] The passage quoted here begins: "Il me paroit donc très vrai-semblable par les raisonnements que nous venons de faire, qu'il existe réellement . . ." "Thus it seems to be very

elaborate defenses of the use of hypotheses in science, and of counsels about how good hypotheses are to be distinguished from more arbitrary ones (Buffon 1749–1767, Vol. II; OP, pp. 238–246). There are, Buffon argues, three kinds of questions we might ask. First, there is the question why, but that is useless, since we never can know ultimate causes. "Why do plants and animals reproduce?" That's a foolish question. Answers either just repeat the fact or try to produce final causes, which is ludicrous. Then there are questions about how something happens. These are permissible. So are quantitative questions in the right places, but in general, the question of reproduction is not one of them. So we can only ask, "How does it happen?" And here we draw on analogies from experience, on the basis of which we frame hypotheses, such as those concerning organic molecules (well-founded experimentally, in Buffon's view) and internal molds.

Even internal molds will not quite do the job, however. Something more is needed, especially in sexual reproduction, where, in Buffon's view, we have two sources that have to be brought together in the proper order. So, in addition to organic molecules throughout the body, and the internal mold that organizes them appropriately for a given species, we need "penetrating forces" to mediate the proper arrangement. Presumably some kind of force is always acceptable in a Newtonian cosmos, in analogy to gravity. We have certainly come a long way from the "facts and more facts" attitude that had marked Buffon's initial account of method. Still, he is confident that his hypothesis is more probable than any of the others that have been put forward, and he will certainly maintain it with confidence in the decades to come. Is it an epigenetic account, as his contemporaries believed? It is, in its rejection of *emboîtement*; yet it also appears, in the persistence of each species' internal mold, and even in the initial account of the nature of organic molecules, to include at least some features of the more popular doctrine.

Vitalism

If Buffon's opponents attacked him as a pure epigeneticist, and consequently, as they believed, a pure materialist, they had no reason to object to mechanistic explanations as such. Mechanism was widely accepted by naturalists of the period, but usually with Divine Providence prominent

---

probable from the arguments we have been constructing, that there really exists . . ." (OP, p. 240).

off, or sometimes on, stage. Buffon, however, although he occasionally paid lip-service to the Deity, was plainly trying to give to nature itself wider powers than were permissible to the religious sensibilities of such writers as Haller or Bonnet. At the same time, if his explanations were mechanistic, they could be so only in an expanded sense of that term – that is, in the sense that no teleological reasons were invoked. What is conspicuous in his natural history is his emphasis on the uniqueness of living matter and its operation, even its priority over the non-living. The units that make up living things are already alive. The non-living, for Buffon, is the dead – the detritus of life, as exemplified in the case of fossil shells (OP, p. 244). This stress on the uniqueness of the living, which will characterize the growing movement of vitalism in a variety of forms, is evident in all three of the basic concepts we have mentioned in Buffon's account of generation: organic molecules, the internal mold, and penetrating forces.

Organic molecules, as the parts of which all plants and animals are composed, were very widely discussed and even widely accepted. Roseleyne Rey, in her illuminating article on "Buffon et le vitalisme," points out that in 1753, Louis de La Caze referred to

> all that M. de Buffon, also ingenious and profound in all his physical researches, has succeeded in discovering and establishing firmly on the subject of the existence and of some of the principal properties of those elementary parts that he has made known under the name of living organic molecules.
>
> Rey 1992, p. 408

Moreover, the *Encyclopedia* article on "animal economy" speaks of the first elements of plants and animals as "living atoms or organic molecules" (Diderot 1751–1757, Vol. X, p. 361 a–b).[20] It should be noted, however, that Buffon is not here reviving some kind of animism or taking refuge in some sort of "entelechy" as coming in to save life from reduction to mechanism. Indeed, putting the organic molecules together with the internal mold, we may conjecture that Buffon was trying to develop something like the "organic mechanisms" that Leibniz had

---

[20] Buffon's relationship to the *philosophes* was complex; to treat it properly would demand a separate chapter. He was a friend of Diderot's, but had his differences with him as well as with others of the group, even though they borrowed his expositions in a number of articles. See Roger's account in his 1989/1997.

suggested in response to Stahl's animistic viewpoint (Rey 1992, p. 403). Even from this perspective, the internal mold must have seemed diffi-cult to accept. The third component of the Buffonian model, however, the "penetrating forces," may have appeared plausible enough to those who recalled Newton's remark in "Query 31" of the *Opticks*, where he suggested that just as the ordinary course of nature is controlled by the attractions of gravitation, magnetism, and electricity, there may be other kinds of attractions in other areas (Newton 1718, Query 31). Finding forces – including what would soon come to be called vital forces, which dominated late eighteenth- and early nineteenth-century biology – is a good, safe Newtonian thing to do.

What was lacking in Buffon's original program was any reference to organization itself, any kind of hierarchical ordering of structures, or interaction among parts within the developed organism. This would, of course, be a conspicuous feature of later versions of vitalism, including those influenced by Kant. Buffon did get around to it himself, however, in a 1758 chapter on carnivores. There he writes:

> In order that feeling ("sentiment") should exist in the highest degree in an animal body, the body must form a whole that is not only sensi-tive in all its parts, but again composed in such a manner that all its parts have an intimate correspondence, so that one of them cannot be disturbed without communicating a part of that disturbance to each of the others.
>
> OP, p. 368

This statement, confined as it is to sentient organisms, is less general than the Kantian conception of organic unity, in which all parts of all organisms are both cause and effect of all the others. It also continues by referring to the need, in such organisms, for a central presiding principle. Whatever its implicit scope, psychological passages like this, according to Roger, were often quoted with approval by the vitalistic physicians of Montpellier (Roger 1997, p. 167). At the same time, Rey argues that Buffon's view at this juncture may itself have been influenced by his reading of La Caze (Rey 1992, p. 412). Thus, although Buffon founded no school, and although many of his arguments were misunderstood by his successors and even his contemporaries, he stands squarely at the center of the complex interrelationships between biological and philo-sophical interests throughout more than half the eighteenth century, the century of the Enlightenment.

The Epochs of Nature

If we have stressed Buffon's preference for the physically real, for what we grasp through our senses before and around us, as against mathematical abstractions, we have noted also, at least in passing, some of the more comprehensive and speculative ventures included in his forty years' pronouncements in natural history. We mentioned his culminating cosmogonic work, *Les époques de la nature*, which was published in Buffon (1778/1962). Yet even in the first volume of the *Histoire naturelle*, in 1749, Buffon had included a "Theory of the Earth," together with "Proofs of the Theory of the Earth."[21] In this early work, indeed, he rejects what he considers fanciful theories of the earth's nature and origin, like those of Whiston or Burnet, claiming that his own approach is closely based on observation and cautious induction.[22] Yet, thirty years later, he has developed an account of the origin of our planetary system, along with a probable spontaneous generation of living things – not only monera and such, but organisms of every degree of complexity – on other planets as well as ours. (Giraffes on Jupiter or monkeys on Mars? This seems a long way from a sober love of "facts.")[23] Influenced in part, it seems, by Jean-Jacques Dortous de Mairan's "Dissertation on Ice," Buffon argued that the earth (and other planets, too) had been steadily cooling – and if organic molecules and internal molds were thrown up in this cooling process, the forms of life we know would be necessitated everywhere (Buffon 1778/1962). Life appeared as soon as the earth was cool enough. It appeared in the north first; animals and plants migrated from there in successive waves. Even stranger (but clearly related to the contemporary fascination with mastodons and other such beasts) is the claim that the first animals were huge. Cooling produced smaller and

---

[21] Translations of part of these documents are included in Lyon and Sloan 1981.

[22] "As historians," he writes, "we reject these vain speculations; they are mere possibilities which suppose the destruction of the universe, in which our globe, like a particle of forsaken matter, escapes our observation and is no longer an object worthy [of] regard; but to preserve consistency, we must take the earth as it is, closely observing every part and by inductions judge of the future from what exists at present; in other respects we ought not to be affected by causes which seldom happen, and whose effects are always sudden and violent; they do not occur in the common course of nature; but effects which are daily repeated, motions which succeed each other without interruption, and operations that are constant, ought alone to be the ground of our reasoning" ("Second Discourse," in Lyon and Sloan 1981, p. 149).

[23] A table of the occurrence of life on the planets is included in the "Second Supplement" to the *Natural History*. We are grateful to Peter McLaughlin for pointing this out to us.

smaller kinds, which moved south as increasing temperature killed off their larger predecessors. Allegedly, no secret causes are being invoked here, but just the impulsive forces that we see everywhere at work.

Such overall theorizing takes us beyond biology to a theory of cosmological or cosmogonic scope. However, it does set Buffon's work into the context of the popular concern with "theories of the earth," and thus with the pre-history of geology, a developing science that, especially through Lyell's magisterial work, would guide the young Charles Darwin toward his new and transforming vision.[24]

[24] For an account of Buffon's cosmogonic theories, see Hodge 1992.

✦

CHAPTER 4

The Eighteenth Century II
Kant and the Development of German Biology

Introduction

Although Immanuel Kant produced an important body of occasional writing about biological topics to which we will turn more explicitly at the end of this chapter, the bulk of his work on the philosophical aspects of biology can be found in the section of the *Critique of Judgment* entitled "Critique of Teleological Judgment" (Kant 1790; 2nd ed. 1793). Kant reflects there on the key philosophical issues in the life sciences, especially teleology and reductionism. It must be conceded that these issues were not central to Kant's lifework, and that they play only a supporting role in the overall line of argument of the *Critique of Judgment* itself. Nonetheless, Kant's ideas about the central questions of biological inquiry were influenced by, and in turn influenced, practicing biologists in early nineteenth-century Germany, as well as later philosophers. Indeed, some of those with whom Kant interacted, especially Johann Friedrich Blumenbach, were at that very time laying the foundations of the modern science of biology. For these reasons, Kant deserves a place in a history of the philosophy of biology.

Organisms as Natural Purposes

On its theoretical side, Kant's critical philosophy was an effort to demonstrate that the logical forms inherent in the very act of thinking (like the subject-predicate, if-then, and other such relationships) correspond to categories into which our objective experience – our experience of objects – falls. Our experience can be objective only because this conjunction of sensible matter and categorical form occurs. Kant's effort to

show how and why this must be so was aimed at securing not only metaphysics, but also scientific knowledge against skeptical attack. Hence, in
working out his critical philosophy, Kant was at the same time seeking
to demonstrate that categorically constituted material objects moving
through space and time are necessarily governed by the "pure principles" that undergird modern physics – the three Newtonian laws of
motion, and the dynamical forces, both attractive (gravitational) and re-
pulsive (elastic), that these laws govern. (Forces were multiplying apace
in the late eighteenth century; magnetic, electrical, and perhaps chemical
forces were becoming conjoined to the gravitational core.) These laws
certainly had to be discovered, verified, and applied empirically. But the
objectivity, certainty, and necessity of modern physics, which permitted
it to join the ranks of fully mature forms of systematic knowledge such as
logic and mathematics, was to be guaranteed by the a priori connections
Kant sought to establish between the categories, our pure intuitions of
space and time, the meaning of the concept of matter (which for Kant
referred to the inherently lifeless stuff that is pushed or pulled around
by measurable forces), and, finally, Kant's slightly reconstructed version
of the basic laws of mechanics.

An integral part of this analysis – and something of its motive as well,
given Kant's self-confessed attempt to refute Hume's skepticism about
the objective nature of the causal bond – was the ingenious suggestion that the formal notion of the hypothetical, or if-then, relationship
corresponds to the pure concept of cause and effect. When referred
to objects in time, this category reappears (in the Second Analogy) as
the principle that causes must precede their effects. "If I lay a ball on a
cushion, a hollow follows upon the previously flat smooth shape," Kant
wrote in the *Critique of Pure Reason*. "But if for any reason there previously exists a hollow in the cushion, a leaden ball does not follow upon
it" (Kant 1781 [A], p. 203; 1787 [B], pp. 248–9) (trans. Kant 1929,
p. 226).[1] Even in cases of the simultaneous interaction or "community"
that obtains between bodies governed by Newton's third law, Kant concluded that an exertion of force by an affecting body (not necessarily by
crude impact) must precede any change in the motion or rest of an affected body, even if only by an infinitesimally small interval (Kant 1781

---

[1] Citations of Kant's Critique *of Pure Reason* refer to the page numbers of the first (A)
edition and the second, or B, edition. All translations from the *Critique of Pure Reason*
are from Kant 1929. All other citations of Kant's work refer to the so-called "Akademie
[hence Ak] Ausgabe" of the *Gesammelte Schriften* (Kant 1908–13).

[A], p. 203). At bottom, there is a single direction of causality. That is a law of nature.

It is not our task to interpret this line of argument, far less to judge its success or failure. Our concern is with grasping the implications of Kant's theory of causality for the possibility of scientific knowledge of living things – or rather with grasping what Kant took these implications to be, and why. To put the problem simply, Kant thought that the doctrine that causes must precede their effects is in considerable tension with our experience of organisms. For in an organism – a "natural purpose," as he calls it, for reasons we will explore later, – each part is reciprocally means and end to every other. This involves a mutual dependence and simultaneity that is difficult to reconcile with ordinary causality. In fact, so concerned was Kant with the conflict between the efficient causes that uniquely explain natural phenomena and the final causes that seem required in the case of organisms – and with the related question as to whether we might someday find a point at which the laws governing matter also determine and explain the formation of organic stuffs and organs – that he was willing to assert quite confidently:

> It is quite certain that in terms of merely mechanical principles of nature we cannot even adequately become familiar with (*kennen lernen*) much less explain (*erklären*) organized beings and how they are internally possible. So certain is this that we may boldly state that it is absurd for human beings even to attempt it, or to hope that perhaps some day another Newton might arise who would explain to us, in terms of natural laws unordered by any intention (*Absicht*), how even a mere blade of grass is produced. Rather we must absolutely deny that human beings have such insight.
>
> Kant 1793, ¶75. Ak. V, p. 400; cf. ¶77. Ak. V, pp. 409–410[2]

Kant's concern with potential conflicts between efficient and final causes arises in at least three areas: the mutually supportive web of relationships between various species of plants and animals in what we call an ecological community; the fact, mentioned earlier, that the parts of a single organism are mutually dependent on one another in a radical way; and the process of ontogeny, in which a future end-state seems mysteriously to guide the successive stages of the development of an organism. Although we will deal with all three of these topics, we will be

---

[2] All translations of the *Critique of Judgment* are from Kant 1987.

especially concerned with the last: how the patterned growth of a plant or animal seems to be influenced by the very end-state toward which the organism is oriented. Kant deals with this question late in the argument of the "Critique of Teleological Judgment" (Kant 1793, § 81). It seems reasonable to consider his substantive, scientific views about ontogeny in connection with the general argument presented there. For on the face of it, there would be less conflict between the Second Analogy, on the one hand, and the reproduction, differentiation, and growth of organisms, on the other, if Kant had subscribed to a straightforward, "encasement" version of preformationism. The fact is, however, that he did not hold to this view, but was influenced instead by a decisive movement toward the epigenetic alternative by biologists in his own time.

On the preformationist family of views, the cause of an organism's coming-to-be is presumed to lie entirely in a pre-existing germ (whether sperm or egg) that already contains the whole differentiated organism in miniature, which merely unfolds or rolls out (the original meaning of the term evolution [from e-volvere]) under a set of perfectly mechanical, when-then conditions. For his part, Leibniz had no trouble with preformationism in this sense. He was not bothered by the implication that each germ cell must not only contain a whole organism in miniature, but must also encase within itself the germ cells of descendants down to the last generation. On the contrary, Leibniz showed marked enthusiasm for the notion of an infinity of infinitely small systems organized into functionally differentiated parts, his preoccupation with theodicy being supported by the contemporary fascination with microscopy.[3] For Kant, on the other hand, the notion of an infinity of already articulated and differentiated parts runs up against the requirement that organisms, as natural beings, must at some point be composed of inert, homogeneous, material stuffs that obey purely physical laws. In the *Critique of Pure Reason*, Kant rejects as "unthinkable" Leibniz's view that organization can go on until infinity (Kant 1781 [A], ¶526; 1787 [B], ¶554). But in addition to this philosophical objection, Kant was also affected by

---

[3] Preformationism can easily appear mechanistic, but not particularly naturalistic. Epigenesis, by contrast, can appear naturalistic, but not mechanistic. Clark Zumbach infers from this perception that Kant insists on design in order to qualify the presumptive naturalism of epigenesis (Zumback 1984, p. 92). We would suggest that design is invoked by Kant as an idea not to qualify the naturalism of epigenesis, but to keep naturalism from leading to complete mechanism.

empirical considerations. In his day, embryologists were no longer able (or inclined) to sweep under the rug the problems that fertile hybrids posed for preformationism. If either the egg (ovist preformationism) or the sperm (vermist preformationism) contains the whole organism, how can fertile hybrids possibly arise? And if a combination of factors from both parents is required, what can preformation even mean? Moreover, how could broken-off pieces of the polyp regenerate a whole organism, as they clearly did? In the face of these difficulties, Caspar Friedrich Wolff at the University of Halle and Albrecht von Haller at Göttingen were influential in giving new salience to the notion that each organism is a result of a *de novo* process of fertilization and development. This is, in a general sense, the epigenetic approach that goes back to Harvey's reworking of Aristotle's *Generation of Animals*. But epigenesis brought with it its own problems. It seemed to require some sort of teleology to guide the formation of an organism. For an Aristotelian like Harvey, of course, this was not a problem; he believed that final causes are truly causes. For Kant, however, the fact that a fully functional, living being is reliably the result of a series of material causal episodes seemed mysterious indeed: potentially a violation of the notion that causes must precede their effects, and inconsistent with any sound definition of laws of nature.

In fact, the situation was somewhat more complicated than that. Kant agreed with Buffon, von Haller, and Charles Bonnet that there *are* pre-existent germs for primordial parts. But he did not believe (any more than Haller did) that these germs are floating free in the world, as Buffon had, or that they are formed into a living whole by something as mechanical as an interior mold (*moule intérieur*) (Kant 1763. Ak II, ¶68; see Chapter 3 on Buffon). Kant's contribution to the scientific question (proposed as early as 1763) was to postulate that, in addition to germs, certain predispositions (*Anlagen*) are also heritable. These predispositions make themselves felt only under the specific conditions to which the organism-to-be is to be adapted, thereby guiding its formation toward its fully functional end. Germs are species-specific and constant; dispositions are conditional and variable. Kant called his approach "generic preformationism" rather than something like "germ-grounded epigenesis." In this phrase, the adjective takes back some of the noun; Kant's preformationism is "generic" because he recognizes that only "the form of the *species* is preformed *virtualiter*" or potentially (Kant 1793, ¶81. Ak V, 423, our italics). When it comes to individuals, Kant's position is fundamentally a version of epigenesis. The individual has no prior identity *in ovo* or *in utero*; it is a product of a complex interaction between germs, predispositions, and environmental contingencies. "Nature," Kant writes, "produces organisms rather than merely developing them" (Kant 1793, ¶81. Ak V, p. 424). He even admits explicitly that this position is substantially epigenetic. "The system that considers individuals as products is called the system of epigenesis," he writes. "It is what we call generic preformationism" (Kant 1793, ¶81. Ak V, p. 423).

Whether it was philosophical or empirical considerations that predominated in his thinking, however, what is important in the present context is that Kant's subscription to epigenesis, qualified though it may have been by his acceptance of pre-existent germs and rendered more plausible by his postulate of inherited predispositions, was still in considerable tension with the views of causal explanation countenanced in the *Critique of Pure Reason*, as well as with the sort of science whose foundations are grounded in the *Metaphysical Principles of Natural Science* (Kant 1786b). One part can certainly trigger off the development of another in a causal sequence under particular conditions within the womb of the mother as well as under external environmental conditions. But in a living thing, the existence and balanced functioning of each part still seems to depend on the prior or concomitant existence of all the other parts, as the leaves of a tree, for example, depend for their existence on its branches, but the branches in turn depend on the leaves. In the absence of an already preformed whole, how could such a system get going? More mysteriously still, if the whole is not already present *in nuce*, how can *exactly* those co-adapted and co-creative parts successively emerge that will later be needed for the successful life of the developed organism?

Kant might have relieved the tension arising from the goal-orientation of ontogeny and the strict when-then causality to which he was committed by the doctrine of the first *Critique* if he had been willing to believe that organisms are artefacts: productions, presumably, of a divine creator. The idea of a house, for example, is the "ground of the possibility" of the production of a particular house, as is the idea of any other technical product in which functionally different parts are subordinated to an overall intention. Such a product, Kant says, *is* a purpose, goal, or end (*Zweck*): "A purpose is the object of a concept insofar as we regard this concept as the object's cause . . . We think of a purpose if we think . . . of the object itself . . . as an effect that is possible only through a concept of that effect" (Kant 1793, ¶10. Ak. V, p. 220). There is no violation of the Second Analogy in regarding organisms as objects brought into being in this way. For a guiding idea occurs in the mind of the architect before the house comes into being, and initiates a sequence of actions that results in and guides its production. If organisms were artefacts, they would, in this sense, *be* purposes. (Note that, for Kant, the concept of a purpose [*Zweck*] is applied first to the things or states of affairs brought into existence by means of an intentional idea that individuates these objects (*Gegenstände*), and not to the guiding idea itself. A purpose, he writes, is the "object of a concept," not the concept of an object. This does not coincide with English usage.) For several reasons, however, Kant was unwilling to say that organisms are, or indeed are crucially like, artefacts.

For one thing, the inference to a superhuman artificer depends on the validity of the "physico-theological proof" of God's existence, commonly known as the argument from design. Unquestionably, Kant had a great deal of respect for this argument. In the *Critique of Pure Reason*, he calls it "the oldest, clearest and most accordant [of all such arguments] with the common reason of mankind" (Kant 1781 [A], ¶623; 1787 [B], ¶565, trans. Kant 1929, p. 520). Indeed, Kant was always eager to assign the argument from design a certain heuristic function in our thinking about living things; even in the first *Critique*, where arguments for the existence of God are pilloried, he remarked that the argument from design "enlivens the study of nature, just as it derives its existence and gains ever new vigor from that source" (Kant 1781[A], ¶623; 1787 [B], ¶651, trans. Kant 1929, p. 520). How precisely to construe the heuristic role of the design argument is, in fact, one of the central issues of the "Critique of Teleological Judgment." Still, given the limitations of our cognitive apparatus, which grants genuine knowledge only when we succeed in organizing sensory data under determinate concepts, Kant thought it presumptuous (*vermessen*) to "posit a different, intelligent being above nature as its architect" (Kant 1793, ¶68. Ak. V, p. 383). Moreover, as a fascinated reader of Hume's *Dialogues Concerning Natural Religion*, he was keenly aware that a vicious circle results "if we introduce the concept of God into the context of natural science in order to make the purposiveness in nature explicable and then turn around and use this purposiveness to prove that there is a God" (Kant 1793, ¶68. Ak. V, p. 38).

Perhaps more significant in the present context is the fact that Kant was unwilling to let the artefact line of reasoning even get off the ground, because he was also aware of certain disanalogies between organisms and artefacts. These render the inference to a divine maker questionable from the outset. There cannot be an argument *from* design to a designer unless there is an argument *to* design in the first place. But Kant was alert to the consideration that the parts of an organism are so mutually dependent and so tightly connected with the whole that it is difficult to say what, if anything, should come first and what should come later, as we must do when we design, build, and analyze ("reverse engineer") artefacts. In this respect, Kant says that organisms are – or at least must be grasped by us as – self-formative, bootstrapping operations, in which each part appears to be the joint product of all the other parts. This is what Kant means when he says that an organism is "a product of nature in which everything is both an end and also a means" and in which the parts are "reciprocally cause and effect of [one another's] form" (Kant 1793, ¶65–66. Ak. V, pp. 373, 376).

By contrast, the parts of a machine are *not* reciprocally related as cause and effect, or means and end, of one another. They are connected to one another in a looser, more "decomposable" way. True, the parts of a machine may all be necessary if the other parts are to perform their proper roles in the working of the whole object. The legs of a table, for example, must not be wobbly if the surface is to be stable. But none of these parts is a means that brings the others into existence. "In a watch," Kant writes, "one part is the instrument that makes the others move, but one gear is not the efficient cause that produces another gear" (Kant 1793, ¶65. Ak. V, p. 374). In this connection, Kant denies that organisms are even like the self-moving machines envisioned by Descartes. "A machine has only motive force," he writes, "whereas an organism has self-formative force" (*bildende Kraft*) (Kant 1793, ¶65. Ak. V, p. 374) – a statement from which we might well infer that epigeneticism is imprinted into Kant's very conception of an organism. Precisely because they are not self-formative, moreover, the parts of an artefact must be "the product of a rational cause distinct from the matter of the thing, that is, distinct from the thing's parts" (Kant 1793, ¶65. Ak V, p. 373). They compel an inference to a designer. The complex reciprocal causality of living things, on the other hand, requires us to disavow the very notion of an external agent. "We say far too little," Kant writes, "if we call [organisms] an analogue of art, for in that case we think of an artist apart from nature" (Kant 1793, ¶65. Ak. V, p. 375). Nor does it help if we transfer the idea of an external artificer into the nascent organism itself, as if there were some crypto-intentional agent lurking within it that steers it toward what it will require down the road.

Transferring the notion of an artificer from outside to inside in this way merely adds mystification to an already inappropriate analogy. "The organization [of living things]," Kant concludes, "infinitely surpasses our ability to exhibit anything similar through art" (Kant 1793, ¶68. Ak. V, p. 384).

The concept of a "self-formative force" refers explicitly to Blumenbach's concept of "formative force" ("*Bildungstrieb*"), as we shall see. In the "Critique of Teleological Reason," Kant attempts to clarify this idea by employing the related concept of "self-organization." An organism, he says, "must be both an organized and a *self-organizing* being" (Kant 1793, ¶65. Ak V, p. 374, our italics). By this, Kant means three things. First, he means that organisms make from inorganic matter, rather than merely find, the organic materials out of which they are formed as organisms, such as tissue, blood, and bone (Kant 1793, ¶64. Ak. V, pp. 371–2). Second, the process of ontogeny requires the pattern of reciprocal causation we have already noted. The self-organizing aspect of this process can be seen empirically, Kant argues, by considering the compensatory growth of one feature when another has been stunted in the developmental process. "If birth defects occur, or deformities come about during growth, certain parts ... form in an entirely new way, so as to preserve what is there and so produce an anomalous creature" (Kant 1793, ¶64. Ak. V, p. 372). Finally, organisms are said by Kant to be self-organizing in the sense in which a tree, for example, by reproducing itself, helps sustain the species-lineage that makes it possible (Kant 1793, ¶64. Ak. V, p. 371). Organisms are dissimilar to artefacts in this sense as well. They cannot be assembled from independent and exogenous parts, as artefacts can, but must come into existence by the reproductive activity of progenitors of the same lineage of which they are members, or even parts.

Nevertheless, Kant's denial of the stronger claim that organisms *are* artefacts, as well as of the weaker claim that organisms are in any important way *like* artefacts, did not prevent him from asserting that any hope we may have of alleviating the tension between finality and efficient causes in judging living beings does depend on our bringing to these puzzling beings a model of means-end causality that is drawn from our own experience as agents, particularly in contexts of making or production. Kant calls this model of causality "purposiveness." The goal-orientation and part-whole "fittingness" that allows us to identify and begin to investigate organisms means that they are "to be covered by a concept or idea that must determine a priori everything that the thing is to contain" (Kant 1793, ¶65. Ak. V, p. 373).[4] Unless this were so, we could have scarcely any awareness at all of these curious entities, which seem different from inanimate objects, let alone any clue to explaining how they work. But where, other than our own experience as designers, makers, and choosers, could we ever acquire the idea of a being in which "the concept of the whole determines the form of the parts" as fittingly, usefully, efficiently, even beautifully as we clearly observe in the case in organisms? (Kant 1936/1993. Ak. 22, p. 283). The very idea of an organism, accordingly, demands that, as in an artefact, an overall intention (*Absicht*) must be presumed to determine each constructive step, even though in organisms the parts come to be and come to be related to one another in ways that defy our experience with artefacts. "We cannot even think of them as organized things without also thinking that they were produced intentionally" (Kant 1793, ¶76. Ak. V, p. 398).

Kant's way of putting this crucial point is to say that even if we do not and cannot *know* that organisms are "purposes" – objects brought into existence in accord with an antecedent concept – we must admit that they cannot appear to us except as "purposive" (*zweckmässig*). "We call objects, states of mind, or acts purposive," he writes, "even if their possibility does not necessarily presuppose the presentation of a purpose … All that is required is that we be unable to understand their possibility except on the assumption that [they] were produced according to design" (Kant 1793, ¶10. Ak V, p. 220).[5] Thus Kant argues that organisms are "explicable" (*erklärbar*) and "cognizable" (*erkennbar*) only "by using the idea of purposes as a principle" (Kant 1793, ¶68. Ak. V, p. 383). We do this by using the teleological maxim, "Nature

---

[4] The term used here is "*Zweck*," which means "end" or "finality" as well as "purpose." Generally, we prefer to accept the conventional translation of the term as "purpose" to stress the semantic tie that Kant (unwisely in view of cybernetics) makes between the notion of an end-directed system and the structure of an intentional, means-end action.

[5] The "Critique of Teleological Judgment" is preceded in the *Critique of Judgment* by a "Critique of Aesthetic Judgment." In addition to being a contribution to the growing discipline of aesthetics in eighteenth-century thought, Kant's analysis of judgments of the beautiful seems meant minimally to establish, in the context of the overall aims of the *Critique of Judgment*, that a distinction can be drawn between purposiveness and purposes. In aesthetic judgments, and especially in judgments of the beautiful, purposiveness is ascribed without reference to purposes, and indeed in their complete absence. This prepares the way for Kant's ascription of purposiveness to living things, where purposes and purposiveness do not appear quite as separable.

does nothing in vain," to discover the suitability or fitness of each of the parts to bring about and sustain the whole, and vice versa. To the extent that organisms are identifiable and explicable only by means of this model – an epistemological necessity, even if not an ontological one – Kant is willing to say that "the [living] thing is a purpose [or end, *Zweck*], even if it is not an artefact" (Kant 1793, ¶65. Ak. V, p. 373).

We come now to a key point. All along, Kant has been asking in his "Critique of Teleological Judgment," not whether there are any living things, but whether there are any natural purposes (*Naturzwecke*). It is now clear that the answer is yes. Organisms are natural purposes. Kant says they are purposes in order to register our utter dependence in "grasping" them and "cognizing" them on an analogy between organisms and our own means-end reasoning activity, as we have just noted. The stress here is epistemic. Still, organisms are not merely purposes for Kant. They are, ontologically considered, *natural* purposes – the only case, in fact, in which Kant can finds natural purposes among objects in the world.

This characterization supports two important inferences. First, it allows Kant to invoke the nature-artefact contrast in order to stress that, as self-organizing and self-formative wholes, whose parts appear to be reciprocally means and ends of one another, organisms are not, and are not very much like, machines, and so cannot straightforwardly be assimilated to the design model or the physico-theological inference. Second, Kant's identification of organisms as natural purposes suggests that since organisms are more fully embedded in the natural order than, say, rational agents, there is every reason to try as hard as possible to bring natural laws to bear on explaining them. Characterizing organisms as "natural purposes," and not as artefacts, thus sounds a directive as well as a jarring, even potentially self-contradictory, note. "Nature," writes Kant in the *Prolegomena to Any Future Metaphysics*, "is the existence of things so far as it is determined according to universal laws" (Kant 1783, ¶14. Ak. IV, p. 294; see also Kant 1786b, "Preface." Ak IV, p. 467–8; and Kant 1788. Ak VIII, p. 159). The laws of nature by means of which we have scientific knowledge are mechanistic in the sense that "we regard a real whole of nature only as the joint effect of the motive forces of the parts" (Kant 1793, ¶77. Ak V, pp. 407–8). As natural beings, accordingly, organisms must clearly be under the sway of natural laws – presumably mechanistic laws – of various sorts. Yet Kant's point in characterizing organisms as natural purposes was to *deny* that as wholes they are under the control of parts in the way mechanisms are.

Oddly enough, it was probably this very tension that attracted Kant to his conception of organisms as "natural purposes." The reference to nature in this formula has the effect of tethering our inherent tendency to move illegitimately from seeing organisms as purposive beings to dogmatically and prematurely bringing God into the picture, a procedure that Kant believes will hinder rather than advance our pursuit of knowledge. On the one hand, the design inference puts too positive a spin on what the analogy with artefacts can achieve. On the other, it obscures the ways in which organisms are in fact deeply embedded within a system of natural laws. "The expression 'purpose of nature,'" Kant says,

> keeps us from mingling natural science, and the occasion it gives us to judge its objects teleologically, with...a theological derivation [of these objects]... We must carefully and modestly restrict ourselves to an expression that says no more than what we know, namely "natural purpose." For even before we inquire into the cause of nature itself, we find that nature contains such products and engages in their production. They are produced there in accord with known empirical laws; it is in terms of these laws that natural science must judge its objects...Hence the causality in terms of the rule of purposes...must be sought within natural science. Natural science must not leap over its boundary in order to absorb...something to whose concept no experience whatever can be adequate.
>
> Kant 1793, ¶68. Ak V, pp. 381–382

The Dialectic of Teleological Judgment

By following Kant's line of reasoning in the "Critique of Teleological Reason," we have come to the following realization: No matter how much of an enthusiast he was for assimilating systematic and secure knowledge to the methods and concepts of the natural sciences, it is hard to avoid the impression that Kant was delighted to find within the sphere of nature itself some objects – organisms, considered as "natural purposes" – that he was confident would forever resist full subsumption under mechanical laws. There are, it would seem, two main reasons for this apparent delight in limiting the claims of physical science on living beings.

First, the general project of the *Critique of Judgment* is to find a way of mediating the tension between theoretical reason, which places us fully within a deterministic causal order, and the freedom and responsibility that practical reason confers on us (Kant 1793, ¶9. Ak V, p. 196). The clash between these competing perspectives is at its most intense when we reflect that in a deterministic world, our virtuous actions cannot reliably be counted on to grant us happiness – a thought that is only partially alleviated by the faith that a congruence of virtue and happiness awaits us in a later life.[6] The mere possibility that nature itself contains, and indeed produces, beings that are as thoroughly purposive as organisms must appear to us as a sign, then, that the world as a whole might *conceivably* be arranged with a view to making it possible for us to realize our ends as rational beings. Kant writes:

> Organized beings are the only beings in nature that, even when considered by themselves and apart from any relation to other things, must still be thought of as possible only as purposes of nature. It is these beings, therefore, which first give objective reality to the concept of a purpose that is a purpose of nature rather than a practical purpose, and which therefore gives natural science the basis for a teleology.
>
> Kant 1793, ¶65. Ak V, pp. 375–6

Continually encountering, as we do, natural entities that cannot be grasped without construing them as purposes, Kant thinks we are entitled, even compelled, to reconsider even the idea of universal teleology, a notion that the advance of modern science had been discounting for some time:

> Once we have discovered that nature is able to make products that can be thought of only in terms of the concept of final causes, we...may thereupon judge its products as belonging to a *system of purposes* – even

---

[6] Since early in the nineteenth century, Kant has been interpreted as putting a premium on reconciling virtue and happiness in the afterlife. It was Kant's idealist successors, especially Schelling and Hegel, who revolted against this inference, which was insisted upon by their theological teachers, demanding instead that social and political changes that can either be anticipated or precipitated tend to create a world in which humans can be at home – that is, in which moral activity is not systematically at odds with happiness. In recent years, scholarship has suggested that Kant himself might not have as been much of a "Kantian" in this respect as he has been made out to be. This interpretation gives new salience to the *Critique of Judgment*'s project of reconciling theoretical and practical reason by finding a way to see mankind as the ultimate end of nature. See Guyer (2001).

if they...do not require us [in accounting for] their possibility to look
for a different purpose beyond the mechanism of blind efficient causes

Kant 1793, ¶7. Ak V, pp. 380–1, our italics

The claim is no longer that in order to support the moral destiny of
humans, the world must be purposive all the way up and all the way
down, as it was in the days of naive universal teleology. It is merely that
even as more and more aspects of the world system come to be explained
in terms of natural laws, the overall arrangement of that system can still
be thought of without contradiction, and indeed productively, as having
been hierarchically arranged to make the ends of beings such as ourselves
realizable. Viewed in this light, "the whole of nature appears as a system
in accordance with the rule of purposes, to which all mechanism of
nature...is now subordinated (at least for examining the appearances
of nature by means of it)" (Kant 1793, ¶67. Ak V, pp. 378–9).

The purposive systematicity of the world, Kant assures us, is not a
known, or a knowable, fact. Far from it. For the idea of universal teleology, or even the more basic notion of organisms as natural purposes,
cannot yield what Kant, in his technical vocabulary, calls "determinate"
judgments.[7] In other words, we cannot employ teleology to subsume
particular cases or facts deductively under general laws, thereby gener-
ating scientific knowledge, and we certainly cannot make a valid inference, on the basis of end-oriented cases, facts, and patterns, to a divine
maker. We cannot do these things because teleology of a determinate
sort simply does not fall under the sway of the a priori concepts that,
in unifying our sensory intuitions, also give us knowable objects and
natural laws. Instead, the teleological idea is merely a heuristic, or what
Kant calls a "regulative," idea: a construct that can help us think productively about particular objects and relations that we encounter, but
that we cannot subsume under law-covered explanation. Against this
background, Kant seems to have thought that the Aristotelian maxim,
"nature does nothing in vain," as well as maxims such as "nature takes
the shortest path (*lex parsimoniae*)," "makes no leaps (*lex continui in
natura*)," and is "parsimonious" in the sense of Ockham's razor (*prin-
cipia praeter necessitatem non sunt multiplicanda*),[8] might pay their way

---

[7] For Kant, the term "judgment" refers to the cognitive ability to relate the particular
to the universal. In determinate judgments, the direction is from securely established
generalizations to particulars subsumed under them. In reflective judgments, particulars
are given and a covering generalization is sought or entertained.

[8] On these maxims, see Kant 1793, "Introduction," ¶V. Ak V, p. 182.

by helping us discern empirical regularities that these guiding ideas happen to bring into view, as well as providing us with resources for drawing the various dimensions of our experience together into a whole, which is the overall aim of the *Critique of Judgment* (Kant 1793, Introduction ¶V. Ak V, p. 182).

In making this claim about the regulative use of teleology, Kant is presupposing the important distinction he makes in "The Critique of Teleological Judgment" between relative, or extrinsic, and intrinsic forms of teleological judging (Kant 1793, ¶63. A V, p 367; see ¶82, p. 425). Living as he did on the flat littoral of the Baltic Sea, Kant was aware of benign ecological arrangements that allow humans to make a living from the fertile soil and nutrient-laden waters that are washed down by rain-swollen rivers from mineral-rich mountains. Did these arrangements come into existence because they were, or would in the future be, useful to human beings dwelling, farming, and fishing on those alluvial floodplains and coastal waters? (1793, ¶63. Ak V, p. 367). Even more fundamentally, "Did Nature have a purpose in depositing ancient layers of sand, namely, to make spruce forests possible there?" (Kant 1793, ¶63. Ak V, p. 367).

There are dangers in giving a dogmatic answer either way. If we insist on an unqualified affirmation of the teleological perspective, we can easily end up canceling the scientific knowledge we actually have, or can plausibly acquire, about how the sand, the sea, the mountains, and everything else in such a sequence actually have come into existence, and how these elements are causally related to one another in a straightforward, when-then way (Kant 1793, ¶63. Ak V, pp. 367–8). For if we persist in pursuing a universal teleology in the full sense, we must be prepared to conclude that, as in an organism or natural purpose "each intermediate link" – the sand, the sea, the mountains, the trees, and anything else – "must be regarded as itself a purpose and its proximate cause as a means to it" (Kant 1793, ¶63. Ak V, p. 367). But in this case, we must drift either toward the absurd "Panglossian," if not Leibnizean, results that Voltaire satirized in *Candide*, or else fall into the error that Kant associates with Herder, who, in thinking of the whole world as a single purposive organic whole, ends up denying that matter itself is inherently inert, and so threatens the fundamental principles of physics that have been so carefully established.

On the other hand, an unqualified negative is equally suspect. In eschewing all cognition-oriented employment whatsoever of the teleological maxim, we fail to recognize the extent to which the world must appear to us to be an orderly, law-governed place, in which various sorts of beings are in fact systematically useful to one another, even to the point of being necessary for each others' existence. Having missed the evidence for these relationships, we might also forfeit the opportunity to track down the when-then patterns of cause and effect that undergird, say, the ecology of the Baltic littoral. Such explanations are devoutly to be wished for. Whenever they are possible, in fact, mechanistic explanations should be allowed to trump appeals to purposive causation. For they alone conform to the conditions necessary for scientific knowledge of objects, toward which our reason pushes us. (While Kant seems willing to envision at least two kinds of causality, purposive and mechanistic, he certainly appears to have limited the term "explanation" exclusively to the latter.[9])

We must, then, make use of the teleological idea in our pursuit of scientific explanations. To use the idea productively, however, it does not seem to be enough for Kant that we merely entertain it as a convenient fiction that might help us find mechanistic explanations. Kant's view is not that "pragmatic."[10] The appearance in our midst of beings that we must construe as thoroughly purposive – organisms – gives us just enough justification to at least think of the world as conceivably having been put together in such a way that mechanical laws serve as the means for constructing a cosmic hierarchy in which the well-being of its inhabitants, and especially of human beings, can best be realized. With the aid of the teleological idea, we might eventually unravel ecological systems that we discover into determinate chains of cause and effect. But if organisms – natural purposes – were ever broken down into

---

[9] Here we are skirting discussions in recent secondary literature about what precisely Kant meant by mechanism, whether he meant different things at different times, and whether at some stage he denied that mechanism is the only kind of causality that is consistent with the Second Analogy. Peter McLaughlin asserts that Kant, as he drifts under the influence of chemistry toward the *Opus postumum*, moves away from restricting empirical causality to mechanism, but that he never denies that, given our conceptual scheme, explanation is restricted to mechanism (McLaughlin 1990). Zumbach makes a case that Kant revised his view of explanation as well to accommodate what we now call teleological explanation (Zumbach 1984). Zumbach's intriguing suggestion seems to us to read into Kant a late nineteenth-century neo-Kantian distinction between explanation and understanding.

[10] Kant's regulative ideas, when applied to human affairs, are called pragmatic ideas. Kant's concept of a pragmatic idea may have inspired pragmatism. But he was no pragmatist. Ideas cannot be so arbitrary that they can be thrown away without threatening the knowledge acquired and stored by their means.

linear, deterministic chains of cause and effect among merely physical or chemical components (a contrary-to-fact conditional), Kant's justification for using the teleological conception for considering the bare possibility that the world is, in the final analysis, oriented toward purposes would, in his own view, crumble. Ironically, then, if we are to keep learning about nature, it is important that we rule out from the beginning a purely mechanistic view of organisms like that of Descartes, or even Buffon, and that we refrain from hoping or expecting that "perhaps some day another Newton might arise who would explain to us, in terms of natural laws unordered by any intention, how even a mere blade of grass is produced" (Kant 1793, ¶75. Ak V, p. 400).

Kant was aware that the history of thought is full of mistakes in both an overly teleological and an anti-teleological direction. The anti-teleological mistakes are of two sorts. In order to preserve the status of organisms as natural beings, the atomists, in one way, and Spinoza, in another, mismanage the concept of nature by holding that it contains no purposive items at all. Democritus and Epicurus take organisms to have come into existence by way of an improbable confluence of chance collisions. Kant regards this view as "manifestly absurd" (Kant 1793, ¶72 Ak V, p. 391). Far from offering us a path toward scientific explanation, these thinkers minimize, overlook, or even negate the high degree of lawful order that we find not only in organic development, but in experience as a whole. Spinoza, for his part, attempts to preserve the lawfulness of nature. But in his fatalism, he regards nature as a deterministically self-unfolding, self-positing system that moves inexorably in accordance with underlying principles that extend even to human actions. Where atomism exaggerates the role of chance in experience, fatalism underestimates it, making our experience as agents impossible. These are the two ways in which the teleological aspect of experience and its objects are unreasonably denied.

Other systems, meanwhile, attempt to preserve the purposive element in "natural purposes," but only by subverting the natural element. One of these we have already mentioned: Herder's attempt to resolve the conflict between mechanism and teleology by extending intrinsic teleology to nature as a whole, thereby obliterating the distinction Kant had drawn between them. Kant's opposition to this "hylozooic" interpretation (which he saw springing up in the proto-Romantic youth culture of his later decades) knew few bounds (Kant 1793, ¶¶72–3, pp. 392–4.)[11]

---

[11] On Herder and Kant, see Zammito 1992; 2002.

His horror at the idea was driven, in the first instance, by his conviction that such a view was no less atheistic than Democritean chance or Spinozistic fatalism. In the eighteenth century, in fact, atheistic materialism was more closely associated with the notion of living matter than with atomism. Hence, Kant was incensed at Herder's claim that "perhaps even in so-called dead things one and the same disposition (*Anlage*) for organization, only infinitely cruder and more muddled, might preside" (Kant 1785b. Ak VIII, p. 62). Kant's opposition was supported by his settled conviction that matter is by definition inert, capable only of inertial motion, whereas "hylozooists endow matter as mere matter with a property of life" (Kant 1793, ¶¶73, 65. Ak V, pp. 394, 374). To follow young, proto-Romantic intellectuals down this path would be to surrender much of the ground that the *Critique of Pure Reason* and the *Metaphysical Principles of Natural Science* had won.

The fourth system that Kant criticizes is "theism." Unlike the other three errors we have mentioned, Kant does not regard theism as incoherent or ideologically suspect. On the contrary, as we have mentioned, he respected this position for its utility in giving aid and comfort to the moral point of view. However, where Herder's hylozooism overextends the concept of internal teleology, theism, by way of its uncritical reliance on an analogy with artefacts, tends to introject external teleology into the very concept, and metaphysical status, of an organism. In failing to see the ways in which organisms are self-formative, self-organizing systems, in which every part is reciprocally cause and effect of every other, theism "does not make [their] manner of production a whit more comprehensible to us" (1793, ¶73. Ak V, p. 395). Indeed, as Hume pointed out in his posthumously published *Dialogues Concerning Natural Religion* (Hume 1779), theism, even if it were able to shed detailed light on how living things actually came into existence and on how they work, would still have to prove decisively that the apparent end-directedness of organisms "could not possibly result from a mere mechanism of nature" (1793, ¶73. Ak V, p. 395). As Hume argued, this is a tall order. It is difficult to see how theism could rule out the possibility of mechanism without begging the question not only against mechanism, but against a host of other possible, if unknown, accounts of the origin of living things – all of them probably beyond the range of knowledge.

It is nonetheless fairly easy to understand why, in spite of their various deficiencies, mistakes of all four sorts have been made, and doubtless will continue to be made. In being compelled by our cognitive weaknesses to construe organisms as natural purposes, we are inevitably pulled in two contrary directions, a mechanical and a teleological. How far in each direction can we go? If teleological ideas enable us to identity ecological patterns that we can trace to mechanical causes, who is to say that the same heuristic process cannot productively be projected into the *interior of the living being* itself? If so, might we not hope to find mechanisms that support embryological, physiological, and other organic functions in the same way that alluvial flows support plant, animal, and human life? At the limit of such inquiry, might we not stumble upon the point at which inert matter is worked up into living tissue in a mechanistic way? How, then, can we rule out the very possibility that a living thing might show itself to be "an effect of the motive forces of the parts"?

As we have already noted, the possibility that we might be able to penetrate the organism to this degree – the possibility, in other words, that there might be a "Newton of a blade of grass" – would be fatal to Kant's delicately poised regulative use of the teleological idea. Our justification for using that idea was said to be our experience of organisms as natural purposes. Finding that organisms are merely the result of the forces that govern their separate parts would undermine that support, plunging us headlong into materialism of one sort or another, and hence into various outrageous claims. It might seem preferable to tilt in the other, teleological, direction, as hylozooists and theists are said to do. But this, too, has its dangers. In this case, our anomalous experience of organisms as natural purposes will be interpreted as affording us a glimpse of how in organisms "the presentation of a whole also contains the basis that makes the connection of the parts possible" (Kant 1793, ¶77. Ak V, p. 407). Enjoying this kind of insight into the underlying nature of organisms depends, however, on affirming something that the *Critique of Pure Reason* categorically denies – that we have intuitive knowledge, however unclear, of a purposiveness that constitutively pervades organisms and the world order generally (Kant 1793, ¶77. Ak V, p. 407). For Kant, our intuitions are restricted to unorganized and disunified flows of sensory data, and to our pure, contentless intuitions of space and time. Our scientific knowledge is wholly constructed, with the help of the categories, from these forms of intuition. In moving from categories to laws, and from laws to explanations, we must presume that wholes are determined by forces operating at the level of parts. We have no intuitions, no matter how fleeting or feeble, of systems in which the moment we grasp the whole, we "intuit" how all the parts are brought into being and disposed. We have no intuitions, that is to say, of purposive systems.

In this way, or something like it, there arises what Kant calls "The Antinomy of Teleological Judgment." For Kant, an antinomy is an apparent contradiction: not a stupid mistake, to be sure, but instead a conflict that is subtle enough to arise over and over again by way of Reason's self-generated demand for answers to questions that go beyond "the bounds of sense" (Kant, 1787 [B], p. 490, in trans. Kant 1929, p. 422). The four cosmological antinomies that Kant discusses in the first *Critique* illustrate the general idea of an antinomy. Kant summarizes them as follows:

> Whether the world exists from eternity or has a beginning; whether cosmical space is filled with beings to infinitude, or is enclosed within certain limits; whether anything in the world is simple, or everything [is] infinitely divisible; whether there is generation and production through freedom, or whether everything depends on the chain of events in the natural order; and finally whether there exists any being completely unconditioned and necessary in itself, or whether everything is conditioned . . . and contingent.
>
> 1787 [B], p. 509, in trans. Kant 1929, pp. 433–4

Convincing arguments from seemingly secure first principles can be given for both alternatives in every case. Nonetheless, antinomies can be resolved either by bringing to light a hidden assumption under which both alternatives labor, or by shifting the ground from which the apparent contradiction arises. When this is done, the apparent contradictions turn out either to be contraries, in the logician's sense, which might both be false, or subcontraries, which might both be true. The apparent contradiction in the second cosmological antinomy is resolved, for example, in the first way; it could turn out that the world is neither finite nor infinite. In the case of the third antinomy, on the other hand, both alternatives might be true. The same events might be conditioned and free from different points of view: what, from a theoretical, scientific point of view, is determined by laws, including psychological laws, is revealed as free, from a practical point of view.

The systematic errors about organisms that Kant criticizes in the "Critique of Teleological Judgment" might be viewed as analogous to the transcendental illusions that arise when our speculations about cosmology push us beyond the sphere where our logic-chopping, concept-dependent minds can use secure principles to sort out sensory intuitions into true and false statements.[12] So viewed, errors about natural purposes can be represented as arising from the apparent contradiction between the following statements:

(1) All production of material things and their forms is possible in terms of merely mechanical laws.

and

(2) Some production of material things is impossible in terms of merely mechanical laws.

Kant 1793, ¶70. Ak V, p. 387

In view of the seeming intractability of this contradiction, it is easy to see why it might invite such desperate, and self-undermining, measures as it receives from the four contending parties that Kant has criticized. Democritus and Spinoza might be construed as attempting to secure (1) by denying (2), Herder and the theists as preserving (2) by denying (1). None of the advocates of the systems Kant criticizes seems to have realized, however, that organisms are not presented to us by way of seemingly secure lawful generalizations that conflict with one another when they are extended beyond the bounds of sense. Seen through the somewhat problematic lens of the notion of a "natural purpose," they are instead presented to us as puzzling particulars, whose governing laws we do not (yet) know, but which we hope to learn more about. That is to say, organisms are presented to our *reflective*, not to our *determinate*, judgment. Hence, although there may well be an antinomy of some sort between (1) and (2), it is not an antinomy of teleological judgment.[13] Given the way organisms are "grasped" by us and "cognizable" by us, the antinomy of teleological judgment holds instead between the

---

[12] Support for this way of explicating the antinomy of judgment as like a transcendental illusion is provided by Kant's claim that "going to the extreme of explaining everything merely mechanically must make reason fantasize and wander among chimeras of natural powers that are quite inconceivable, just as much as a merely teleological kind of explanation that takes no account whatever of the mechanism of nature makes reason rave" (1793, ¶78. Ak V, p. 411).

[13] Kant says an antinomy results from (1) and (2), yet "this antinomy would not be one of judgment, but a conflict in the legislation of reason" (1793, ¶70. Ak V, p. 387). It is not quite clear what he might have meant, but it is clear that (1) and (2) do not constitute the antinomy of teleological judgment.

following statements:

> (3) All products of material nature and their forms must be *judged* to be possible in terms of merely mechanical laws

and

> (4) Some products of material nature cannot be *judged* to be possible in terms of merely mechanical laws. (Judging them requires a quite different causal law, namely that of final causes.)
>
> Kant 1793, ¶70. Ak V, p. 387

From the reflective point of view, (3) and (4) are consistent after all. They are merely maxims that pull us in opposed directions when we think about organisms or ecological systems, and can even be made to do so productively if we manage them right. The first tells us that "we ought always to reflect in terms of the principle of mere mechanism of nature, and hence ought to investigate this principle as far as we can, because unless we presuppose it in our investigation of nature, we can have no cognition of nature at all in the proper sense of the term" (Kant 1793, ¶70. Ak V, p. 387) (We can have no cognition in the proper sense because explanation, it would seem, depends on mechanical causality.) "None of this," however, Kant goes on,

> goes against the second maxim – that...in dealing with certain natural forms (and, on their prompting, even with all of nature) we should probe these in terms of...the principle of final causes. For this leaves it undecided whether in the inner basis of nature itself, which we do not know, the physical-mechanical connection and the connection in terms of purposes may not, in the same things, be linked in one principle.
>
> Kant 1793, ¶70. Ak V, p. 388

Many commentators have claimed that the antinomy is resolved the moment we recognize the difference between reflective and determinate judgment.[14] However, this is unlikely. First, Kant himself says that

---

[14] This interpretation can be found in Adickes 1924–5, and, partly following Adickes, in Cassirer 1938, p. 341. McFarland 1970 finds the same inadequate interpretation in several other Kant scholars, and, in a qualified way, in Reinhard Löw 1980. Löw thinks that the chapters in which Kant treats the distinction between regulative and determinate judgment as a "preliminary" resolution of the antinomy, especially ¶¶70–71, do have the consequence of reducing both mechanism and teleology to regulative status. For Löw, however, these earlier chapters were actually written after the substantive resolution presented in ¶¶77–78, which should be regarded as definitive. For Löw, the considered doctrine of ¶¶ 77–78 has the consequence of making teleological judgment intimate the actual constitution of organisms as things in themselves, anticipating a

the move from (1)–(2) to (3)–(4) is only a "preliminary solution to the antinomy" – and he takes seven more chapters to reveal, in ¶78, "How the principle of universal mechanism of matter and the teleological principle can be reconciled."[15] Second, it has long been noted that merely shifting from a determinate to a reflective point of view for both sides of the antinomy has the effect of demoting (3) to a regulative principle – in effect retracting, or at the very least weakening, what Kant had said about the constitutive force of mechanistic causality and the determinate status of mechanistic explanation both in the *Critique of Pure Reason* and at other places in the *Critique of Judgment* itself. It seems clear, then, that the shift from a determinate to a reflective perspective is a necessary, but not yet a sufficient, condition for resolving the antinomy of teleological judgment. Something else needs to be done.

What needs to be done in addition is to demonstrate how the necessity under which we labor of "grasping" and "cognizing" organisms through reflective rather than determinate judgment implies, in and of itself, that "the same product and its possibility" appearing to us in two non-coincident causal frames – "the physical-mechanical connection and the connection in terms of purposes" – might conceivably be fused into a unity when organisms are considered as things in themselves. Kant's claim is not that from a "supersensible" perspective – the perspective of "things in themselves" – mechanical causality and explanation are merely regulative, or, conversely, that the teleological way of judging is uniquely true. It is that from this higher perspective, they are two approaches to what is substantively the same thing.

To make this argument, Kant – at least the Kant of the *Critique of Judgment* – begins by agreeing with the maxim, *individuum est ineffabile*. Even in determinate judgments, which subsume particulars under laws sufficiently well to produce scientific knowledge, the sensuously given particular can contain something "contingent," as Kant puts it, with respect to those laws, something that eludes their determination. "For the universal supplied by our (human) understanding does not determine the particular; even if different things agree in a common characteristic, the variety of ways in which they may come before our

---

turn toward a constitutively teleological view of the cosmos in the *Opus Postumum*. We are inclined to deny that the resolution of the antinomy has the consequence of making mechanistic judgments merely regulative, and hence to think that the doctrine of ¶¶ 77–78 is not as different from that of sections ¶¶ 70–71 as Löw takes it to be.

[15] These claims are embedded into the titles Kant provided for ¶¶71 and 78, respectively.

perception is contingent" (Kant 1793, ¶77. Ak V, p. 406).[16] This is more obviously the case when it comes to organisms. Their "contingency," or perhaps indeterminacy, with respect to general laws is apparent in our constitutional inability to bring the mechanistic and the purposive frames together. This might not be so if we had the intuitive knowledge that God presumably enjoys – a "power of spontaneity," Kant calls it, in which the cognizing of an object differs not a whit from its creation (Kant 1793, ¶77. Ak V, p. 406). If our minds were intuitive in this sense, our awareness of an organism would not only bring it into being, but would, for that very reason, contain and specify all its parts as expressions of a whole. As things are, however, the only kind of intuition we possess (other than that of the pure forms of space and time) is our capacity to receive sensory data, which the discursive, logic-chopping, non-intuitive apparatus of our mind works up into categorized objects. Indeed, our only clue to the way in which the parts of an organism are a priori integrated into, and expressive of, the whole is the way in which, as practical and productive agents, we subordinate means to ends. This may not get us very far. But the very fact that we must identify and learn about organisms by reflecting on them as "natural purposes," and must regard mechanistic laws as subserving those purposes in the way in which means serve the ends of an agent, presupposes that "in nature's supersensible substrate," mechanism and teleology might be fused. For in using one maxim to compensate for the deficiencies of the other, we must be presupposing the possibility of a deeper unity in which both frames are brought together to an intuitive intelligence.

It is easy to make either too little or too much of this argument. The appeal to the supersensible ground of the possibility of organisms is more than a mere logical possibility. It is a presupposition of our way of grasping and learning about living things. It falls short, however, of assertorically affirming such a transcendent unity. For it makes no claim that stands independently of our learning procedures. Nonetheless, the appeal to the possible unity of the teleological and the mechanistic in the supersensible constitution of organisms *is* strong enough to prevent Kant from defensively setting a limit beyond which the mechanistic notion cannot go, for fear that it would go too far. "It is quite undetermined," he writes, "and for our reason forever undeterminable, how much the mechanism of nature does as a means toward each final intention in

---

[16] We leave aside the issue of whether the claim that all human cognition contains traces of the "contingent" is consistent with the doctrine of the first *Critique*.

nature" (Kant 1793, ¶78. Ak V, p. 415). We can, indeed should, use mechanism, therefore, to penetrate as far as we can into the interior structure of organisms. We may do so, moreover, with little or no fear that we will end up as materialists. For the objects that we investigate in this way are given to us by means of an idea, "natural purpose," which by itself excludes the possibility of successful reduction on pain of losing our grasp on the entities we are talking about. From this perspective, proper method in biological inquiry is to manage our seemingly opposed maxims in such a way that teleological reasoning is used to identify biological phenomena and mechanism is used to explain the means by which they are realized. The explanatory work will never eliminate the referential and descriptive.

Kant and the Biologists

The influence of contemporary biological theorizing on Kant's "Critique of Teleological Reason" was not restricted to his views about the epigenesis-preformation issue. In fact, the interpretive framework set forth in this text was in part suggested by Kant's own interventions, over a period of several decades, into issues about the unity of the human species and the reality of different races that were being debated in the learned journals of the day, such as *Berliner Monatschrift, Teutsche Merkur*, and *Allgemeine Literarische Zeitung*. In fact, it is difficult to avoid the impression that the doctrines set forth in the "Critique of Teleological Judgment" were in part calculated to support Kant's positions in these controversies. In these occasional writings, Kant proposes to solve, or at least mediate, what look like empirical quarrels by offering conceptual distinctions and methodological preachments. This work of conceptual clarification had an impact on the formulation of methodological aspects of several research programs within German biology, especially at the University of Göttingen. In almost no case, however – including that of Göttingen – did the reception of Kant's interventions in biological and anthropological issues accord with his own intentions.

The origins of Kant's interactions with the emerging community of German biologists lie in the fact that, beginning in 1757, he regularly taught a course in "physical geography." Generally intended as a sort of "animal-vegetable-mineral" survey, Kant's first prospectus for this course proposed as its primary topic a survey of man, with special reference to the biogeographical distribution of human types in accordance with climatic considerations. Both Kant's prospectus and his subsequently published writings on these topics show direct familiarity with the successively issued volumes of Buffon's *Histoire naturelle* (1749–1767) – texts that had already stimulated Kant to develop his *General Natural History and Theory of the Heavens* (*Allgemeine Naturgeschichte and Theorie des Himmels*), in which he proposed in 1775 the "nebular" hypothesis about the formation of the solar system.

The influence of Buffon on Kant's lectures on physical geography shows itself in his efforts in his course, as well as in four articles published over a period ranging from 1775 to 1788, to refute the notion that humans belong to different subspecies, or even species (Kant 1775 [rev.1777], 1785a, 1786a, 1788). This notion, "polygenism," was from the start not devoid of racist implications, since Africans were invariably listed as departing furthest from the Caucasian norm, and as most likely to have had a different origin. It had begun to gain favor through use of the Linnaean classification scheme to pigeonhole different varieties of human beings as, in effect, subspecies. (Linnaeus himself recognized, at one point, an extinct or only residually extant subspecies of "troglodydes" in the genus *Homo*; see Sloan 1995, p. 12; and Chapter 11.) Kant, for his part, was deeply opposed to polygenism. In order to shore up the monogenetic alternative, he brought forward precisely what Linnaeus and others had denied – namely, Buffon's genealogical species concept and its implied criterion of species membership, according to which interbreeding organisms whose own offspring are fertile are, by definition, members of the same species (see Chapter 3). By this standard, all human beings are members of the same kind (*Art*); even potentially there are no distinct kinds (*Arten*) of our species.

This is not to say that Kant agreed with Buffon on every point connected with his species concept (and he certainly disagreed with his conception of ontogeny). Kant was less opposed than Buffon, for example, to efforts at systematic classification like those of Linnaeus (on Buffon's view, see Chapter 3). That is because, in one of his most important conceptual interventions, Kant distinguished, in a way that breaks with a tradition going back to Pliny, between natural history (*Naturgeschichte*) and natural description (*Naturbeschreibung*) (first articulated in Kant 1775. Ak II, p. 434). Natural description, Kant says, includes not only natural history in the traditional sense, but what Linneaus was up to – using the hierarchical categories that human language seems so spontaneously to generate to help us identify, compare, and otherwise orient ourselves in relation to the diversity of the living world. There is no harm in such an enterprise as long as it retains its fundamentally pragmatic status as a resource for what Kant calls "*Naturkunde*," or skill in dealing with nature, sometimes of a very humble, barnyard sort. The harm comes in conflating natural description with what Kant (not Pliny or Buffon) calls natural history. Natural history, practiced as he prescribes, extends to the origins and dispersion of biological kinds, and indeed to the history of the changing earth itself (Kant 1775. Ak II, p. 434). But Kant thinks very little of this curiosity can be satisfied (Kant 1775. Ak II, p. 434). True enough, illusory satisfactions abound. But they are often the result of confusing the notion of a lineage, on which natural history depends, and classificatory concepts that are appropriate solely to natural description.

The confusion arises, in the first instance, because species are both kinds (*Arten*) and lineages that are maintained by faithful chains of successful breeding (*Gattungen*).[17] The notion of lineages that breed true corresponds to the third sort of organic self-formation distinguished by Kant in the "Critique of Teleological Judgment," according to which species "form themselves" by renewing their membership through generation. Difficulties arise, however, when both concepts – the classificatory and the generative – are conflated and then projected together up the *scala naturae* to classificatory genera and higher taxa. In this case, we imagine a "consanguinity which originates from one single generative mother-womb," as Kant puts it in the "Critique of Teleological Judgment," alluding to Herder's *Ideen* (*Ideas for a Critical History of Mankind*) as well as to his own critical reviews of that work (Kant 1793, ¶80. Ak V, p. 419; see Kant 1785a. Ak VIII, p. 54). But in doing so, we imagine that our classificatory system (which is only a descriptive apparatus) is a mirror of the actual origin of taxonomic categories. That is to say, we imagine with Herder that kinds have origins in the same way that individuals do. In his reviews of Herder, Kant goes out of his way to say that he regards this as a "monstrous" idea (Kant 1785a. Ak VIII, p. 54). Given his distinction between natural history and description of nature, however, there really isn't much to worry about. The idea is a category mistake, pure and simple.

Kant argues that the confusion between natural description and history is just as mischievous when we project it downward to differences within a species – only this time the problem lies, not in treating classificatory distinctions as historical, but in treating historical

---

[17] The linguistic usage is a bit more complicated than this, but, on the whole, *Gattung* for Kant does not describe a classificatory genus, as in normal German, but has a historical ring that is intended to capture Buffon's species concept.

distinctions as if they were distinctions in kind. This does not mean that Kant, in his Buffon-inspired monogenism, did not recognize differences within the single human species. Indeed, Kant distinguishes rather sharply between races (*Rassen*), varieties or local variations (*besondere Schläge*), and evanescent sports (*Spielarten*) (Kant 1775).[18] He defines races as "deviations that are constantly preserved over many generations and that come about as a consequence of migration" (Kant 1775. Ak II, p. 430). They are departures (*Abartungen*) from the same historical stem (*Stamm*), caused by the expression long ago of certain inherited predispositions in a given environment, and the suppression of others (Kant 1775. Ak 2, p. 434). Following Linnaeus, Kant says that there are four such stable races, recognized by the skin colors white, yellow, red, and black (Kant 1775. Ak 2, p. 432; see also Kant 1785a, 1786a, 1788). There are, of course, plenty of mixes between the races, Kant acknowledges, referring to mulattos and other intermediates, and plenty of locally different populations with this or that quirk. But Kant argues that the very fact that these differences are less constant than races means that they are *not* races. "A specific diet can surely produce a stock of humans," he writes, "but the distinctions that identify such a stock as distinct quickly disappear when this stock is transplanted to another place" (Kant 1775. Ak 2, p. 431; on this topic, also see Chapter 11). Because he gave the concept of race a clearer definition than others had done, and because he treated races as more stable than local variations, Kant can be linked with the articulation of the ill-fated concept of race (Bernasconi 2001). However, to be fair, it should be acknowledged that Kant worked out his concept of race in order to support a strong, potentially "anti-racist" version of monogenism. His key idea is that races are not classificatory subspecies, as they were, at least potentially, for Linnaeus. For Kant, the concept of race becomes useful for reflection on the genuine questions posed by natural history, such as questions about the geographical distribution of populations of the same species, only when one abandons the tyranny of the classificatory impulse.

Herder ranks as high among Kant's opponents in issues about human diversity as he does in questions about the "monstrous" idea that new species come from a single "mother." In one of his critical reviews

---

[18] Kant makes the same distinctions in Kant 1785a, 1786a, and 1788. We cite the 1775 essay (revised in 1777; the 1777 text is available in an English translation by J. M. Mikkleson in Bernasconi and Lott 1999, pp. 8–22. Where the 1777 text is identical to the 1775 version, we use Mikkleson's translation.)

of Herder's *Ideen*, Kant writes, "The division of the human species into races does not find favor with our author" (Kant 1785a. Ak VIII, p. 62). For Herder, the distribution of human traits across different environments is simply a blur. It is more open to accidents than Kant would admit, and so, potentially, to changes in species boundaries. Not surprisingly, Herder entertains a very porous version of epigenesis; there are no species-specific germs (*Keime*). (Elsewhere, Kant remarks that Herder "rejects the system of evolution" (Kant 1785a. Ak VIII, p. 62); he means, of course, evolution in the original sense: the mechanical *rolling out* of the preformed organism.) For Kant, on the other hand, not only do germs keep ontogeny within species boundaries, but heritable predispositions (*Anlagen*) keep races adapted to specific environments. As he makes clear in his 1788 essay, the constancy of races also provides Kant with an opportunity to affirm his conviction of the indispensability of teleological reflection in biology and anthropology. In virtue of their rationality, bipedality, and related distinctions, human beings are far more physically mobile than other species. They need to become adapted to different climates. At the same time, human beings need to be protected from the potentially degenerative effects of novel climates (and, relatedly, of the fruits of intermarriage). The concept of race, and the notion of predispositions on which it depends, answers both needs. It keeps races inclined to remain in, or to find, congenial climates. At the same time, it limits the long-term, degenerative effects of whatever uncongenial conditions members of a specific race happen to find themselves in (Kant 1788. Ak VII, pp. 157–184).

Clearly, then, Kant's concept of inherited predispositions was developed in conjunction with his views about race. As early as his 1775 essay "Of the Different Races of Mankind," Kant was arguing that "numerous germs and predispositions must lie ready in human beings either to be developed or held back in such a way that we might be fitted to a particular place in the world" (Kant 1775. Ak II, p. 435). It was probably to sustain this analysis, too, that Kant entered into the thorny thickets of the preformation-epigenesis debate in the first place. Because his own notion of ontogeny was somewhat close to epigeneticism, Kant worried about the tendency of many epigeneticists, including Herder, to flirt with, or actually to embrace, materialism. To block this worry, he insisted that we must always begin our reflections (and end them as well) with the species as given, and with the individual organism as coming to be within a fixed context of species renewal, albeit with a wide variety of inborn variability that we must understand as purposively adapted to specific regions.

Against this background, it is not hard to understand why Kant thought he had found a kindred spirit in Johann Friedrich Blumenbach, a former student of Haller's and professor in the medical faculty at Göttingen. In his dissertation, *De Generis Humani Nativa Varietate* – which appeared later in the same year, 1775, as Kant's "Of the Different Human Races" – Blumenbach had come down roughly on the same side as Kant in the matter of races (although in later editions he added a fifth race, Malays; see Chapter 11). Moreover, Blumenbach, like Kant, wanted to find an interpretation of epigenesis that preserved something of the stress on heredity that had been one-sidedly embodied in the preformationist tradition. It was in this connection that Kant felt the influence of Blumenbach most fully. The second edition of Blumenbach's essay *Über den Bildungstrieb* ("On the Formative Impulse") appeared in 1789. (The first edition was published in 1784.) In the first edition of the *Critique of Judgment*, Kant wrote, "No one has done more by way of proving this theory of epigenesis than Privy Councilor Blumenbach" (Kant 1793, ¶81. Ak V, p. 424). Indeed, Blumenbach's proof of epigenesis makes use of the idea of a "formative impulse"; when Kant, in the "Critique of Teleological Judgment," sees organisms as self-forming and self-organizing, he is alluding to Blumenbach's *Bildungstrieb*. For both thinkers, too, this is a purely theoretical posit designed to account for facts such as these: Although the lopped-off parts of the polyp do indeed produce a whole organism, the organisms in question becomes smaller and smaller with each generation; and when, in the process of embryogenesis, one part is deformed, another tends to be built up in compensation. This is not the old concept of "soul," for which Kant had no use; soul would presumably not wax and wane as it does in generations of polyps. But neither is the *Bildungstrieb* anything as potentially capable of giving birth to kinds as well as individuals as Herder's "genetic power" (*genetische Kraft*). Instead, for both Blumenbach and for Kant, the formative force is very much like Newton's concept of gravity. It is known entirely by the effects it organizes and, to that extent, explains. Thus, with Herder probably in mind, Kant praises Blumenbach for

> avoiding too rash a use of [his *Bildungstrieb*] . . . He starts from organized matter. For he rightly declares it contrary to reason that crude matter on its own should have structured itself originally in terms of mechanical laws, that life could have sprung from the nature of what is lifeless, and that matter could have molded itself on its own into the form of a self-preserving purposiveness. Yet by appealing to this principle of original organization – a principle that is inscrutable to us – he leaves an indeterminable and yet unmistakeable share to natural mechanism. The ability

of matter in an organized body to take on this organization he calls a formative impulse (*Bildungstrieb*).

Kant 1793, ¶81. Ak V, p. 424

Soon, Kant and Blumenbach formed a mutual admiration society. Blumenbach was impressed enough by Kant's interpretation of his work to substitute in later editions of his *Über den Bildungstrieb* (Blumenbach 1789), and his influential *Handbuch der Naturgeschichte* (Blumenbach 1802, 1st ed. 1797), Kant's own formulations of what an organism is – something whose parts are reciprocally ends and means, and so forth – and of how we must investigate such things by combining mechanistic and teleological perspectives. In addition, Kant's way of distinguishing between, while also coupling together, the notions of *Keime* and *Anlagen* was taken up by Blumenbach, and through him became diffused not only among other Göttingen biologists, but among nineteenth-century German embryologists generally.

Yet, from the outset, Kant was also at cross-purposes with Blumenbach and his colleagues (see Richards 2000). Blumenbach and other members of the multi-generation research tradition, which, since Haller's day, had been centered at Göttingen – arguably, the premier research university of its time – treated Kant's notion of a regulative idea as little more than a heuristic method for arriving eventually at determinate explanations and actual causal mechanisms that might, in principle, displace the purposive conception of organisms from which we must admittedly begin. For Kant, on the other hand, we must remain forever wedded to the concept of the concrete "natural purposes" on which we reflect. This insistence is echoed in his resolution of the antinomy of teleological judgment. To attempt to reflect on matters outside the context of inquiry afforded by the concept "natural purpose" – on the origins of species, or on the point at which life meets its non-living substrate – is to lose touch with the very things we are talking about. In this way, Kant wanted to put some fundamental questions about natural history, in his sense of the phrase, off limits. A few years later, Kant's students heard him remark: "The system of epigenesis does not explain the origin of the human body. Rather it says much more: that about this we know (*wissen*) nothing" (Kant 1794, Ak XXVIII, p. 761).

The sources of these dissonances between Kant and the Göttingen biologists depend, in the first instance, on the fact that Göttingen ranks high among sites where the modern discipline of biology was being formed. The term "biologia" goes back at least as far as the Wolffian philosopher Michael Christoph Hanov's *Philosophia Naturalis*, the third volume of which, published in 1766, names it as the part of physics that studies living things (Hanov 1766).[19] Lamarck, too, used the term; in an unpublished manuscript, written in French in 1800, he speculated about a discipline he called *biologie*, which would find the exact point at which the "corps vivant" was unified with the "corps bruts" by "fluides invisibles" (Lamarck 1800, in Grassé 1944; see Caron 1988 for use of the term in eighteenth-century French medical schools, especially Montpellier). But it was Blumenbach's student, the Göttingen professor Georg Reinhold Treviranus, who, in a six-volume text that began to appear in 1802, gave the term *"Biologie"* both wider currency and the sense that it has borne ever since. In the title of his *Biologie, oder Philosophie der lebenden Natur für Naturforscher und Ärtze*, Treviranus named a comprehensive science that would gather together information about living things from a variety of special sciences, such as physiology, systematics, and comparative anatomy. Treviranus's *Biologie*, it would seem, is no longer part of physics, as it still was for Hanov; physics is now restricted to the study of *in*animate, or life-less, objects.

Treviranus's work implies an optimistic view about the prospects of biological knowledge. This is so, in part, because, although they were influenced deeply by Buffon, the Göttingen biologists, led by Blumenbach and Treviranus, rejected Buffon's skepticism about classification, and because they did not adopt Linneaus's "artificial" system, which was open to potential doubts, but instead worked toward a more realistic, or natural, classificatory scheme based on a multiplicity of traits – the so-called complete *"habitus"* of an organism.[20] Treviranus hypothesized

---

[19] On Hanov's early use of the term *biologia*, we are indebted to the researches of Gary Hatfield (perusal communication) It would appear that Hanov (and perhaps others that we do not know about) was led to posit *biologia* as a science of living beings because he rejected the traditional Aristotelian way of dividing ensouled (*psuchikos*) nature into the nutritive-reproductive soul of plants and the sensitive-locomotive soul of animals. For Hanov, only animals have souls. Hence a new term must be found to jointly name the two sorts of living, as opposed to inanimate, beings.

[20] Others who worked at or studied at Göttingen included Abraham Gotthelf Kaestner, who translated Buffon's *Histoire naturelle* I–III and wrote notes on it that formed the basis of many of the questions pursued by Göttingen biologists David Sigmund Büttner, Christian Wilhelm Büttner (no relation), and Christian Gottlof Heyne, who did some of the work on polyps on which Blumenbach depended. Blumenbach's students included some of the most illustrious biologists of their times. In addition to Treviranus, he trained the great Alexander von Humboldt, Carol Friedrich Kielmayer, Henrich Friedrich Link, and Johann Friedrich Menkel. See Lenoir 1981.

that questions falling within the scope of biology could be answered by using weighted statistical reports about the values of a number of variables that attend species, each defined by its total *habitus*: irritability (the capacity that, ever since Haller's idea of how the sperm awakened the potentials in the egg, had, at Göttingen, marked the line between living and non-living), sensibility, secretion, propulsion, and reproductive prowess. (Statistics was a specialty of Göttingen research into government.) The underlying hope was that by following how individual elements of the complete portrait of a species change with changing environments, definitive answers would eventually be found to questions that Kant would have classified as natural history, even those bearing on the origin of kinds and the material conditions for ontogeny. A fundamental basis of this program was Blumenbach's explicit rejection of the Great Chain of Being, and of classificatory practices dependent on it, such as those of Linnaeus. Close inspection of the *habitus* of species will reveal that there are in fact plenty of gaps in nature. "It is a great weakness," Blumenbach wrote, "to see in such pictures [as the Great Chain] the Plan of Creation... on the ground that 'nature makes no leap,' as natural theologians are wont to say" (Blumenbach 1802, p. 8; quoted in Lenoir 1981, p. 132).

Given the nature and ambitions of their project, the Göttingen biologists, Blumenbach among them, could not help failing to walk steadily along the tightrope that Kant had strung between the determinate and regulative uses of teleology. Their project explicitly eschewed Kant's distinction between natural history and description (as well as his retention of Linnaean classification in a descriptive context). Regulative ideas became methods for acquiring positive science, thus leaving open the possibility that somewhere down the road, the physical and chemical bases for the operation of the *Bildungstrieb* might be known.

These issues were brought out into the open when Georg Forster, who had studied at Göttingen, taught natural history at Kassel, and accompanied his naturalist father Johann on Captain Cook's voyages to the South Seas, attacked Kant in *Teutsche Merkur* (which at this time was still hostile to Kant's philosophy) (Forster 1786).[21] Natural history of the sort that Kant envisioned, Forster claimed, "would be a science for gods alone, not for men." By this standard, "we would be incapable of demonstrating the genealogical tree of so much as one single

---

[21] On this episode, see Sloan 1979, pp 131–137; Zammito 1992, pp. 207–213; Richards 2000.

variety up to its genus if it does not arise nearly in front of our eyes" (Forster 1786; quoted by Kant in Kant 1788, p. 161). Chafing at this limitation on biological knowledge, Forster went on not only to deny Kant's distinction between natural history and natural description, but to suggest (much to Kant's annoyance) that Kant had made the former so difficult precisely because he was an obscurantist, who was more interested in theology than in biology.

Kant's reply came in his 1788 essay *On the Use of Teleological Principles in Philosophy*, the text that forms the link between his writings on race and the "Critique of Teleological Judgment." Kant complained that Forster simply had not understood his principles (Kant 1788. AK VIII, pp. 160–1). In response, Kant reiterated his distinction between natural description and natural history, and reinforced the limits he had drawn on scientific knowledge of the latter by insisting that teleological ideas (of a rather external sort) are indispensable in natural history and never rise to the level of determinate, scientific knowledge. It seems to us more likely, however, that Forster understood all too well what Kant was saying. Indeed, Kant's relegation of classification conceived along Linnaean lines to mere natural description was reinforced, as he moved from this encounter to the *Critique of Judgment*, by a desire to damp down the expectation that much determinate knowledge would ever be found within the deeper recesses of natural history. Kant's reiterated insistence that there will never be a "Newton of a blade of grass" should be read in the light of the controversy we have just reviewed.

Eventually, some support for Kant was proffered by a former Göttingen student, Christoph Girtanner, who had become acquainted with Kant's philosophy through the work of Karl Reinhold, as well as from Girtanner's personal friend Johann Jachmann, who had served as Kant's secretary. It was Reinhold who first popularizied Kant's philosophy in the suddenly pro-Kantian *Teutsche Merkur*. In an essay in that journal, Girtanner proposed to mediate the quarrel between Kant and the Göttingen biologists (Girtanner 1796). He agreed with Kant and Buffon on the species concept, as well as with Kant's refinement of it, according to which a species consists of a number of distinctive germs with the potential for multiple realizations in different environments. But Girtanner concurred with his Göttingen teachers on classification by multiple characteristics. He proposed that the two ideas be together in such a way that comparative knowledge of traits could lead to knowledge of dispositions, which in turn could lead to knowledge of underlying species-specific germs.[22] But this interesting suggestion did little to bridge the gap. In some ways, it even combined the worst, not the best, of both sides. From the Göttingen perspective, it appeared to make positive biological knowledge harder to acquire than they would wish. For it would require them to find species-specific germs as grounds for their classifications. From the Königsberg perspective, Girtanner's proposal flirted with promising more than human beings could ever deliver. Kant made no response.

Kant's insistence on the limits of biological knowledge could only have become more pronounced if he had lived to see the full revolt against these limits by a group of young philosophers who had been brought to the University of Jena, in part through the influence of Göethe (who was looking for philosophical allies in his anti-Newtonian program for science). While still at the Tübingen seminary, Friedrich Schelling, the leader of the group, had made common cause with his classmates Georg W. F. Hegel and the poet Friedrich Hölderlin against Kant's restrictions of the knowable to mere appearance. They wished to turn what Kant had consigned to regulative ideas into constitutive knowledge, dreaming of a physics that "takes wings" by means of an aesthetic intuition, which, in true Romantic style, makes the poet the true philosopher and legislator, and that can be converted into knowledge by means of projecting one's imagination, feelings, and hypotheses onto a world presumed to be fitted to stimulating and satisfying all the human faculties.[23] With his careful delimitation of the spheres of aesthetic judgment from both moral duty and physical inquiry, this re-installation of full-scale intellectual intuitionism, developed ironically enough out of his own aesthetics, was precisely what Kant had wanted to prevent in his opposition to Herder. It was, in some sense, the return of the repressed.

Schelling's pitch was based on the proposed advantages of his "system" for unifying the unruly array of forces that, as we noted at the outset, had emerged in the final decades of the eighteenth century. Kant himself was preoccupied with these issues in his *Opus Postumum*. For his part, Schelling, having broken with Kant's theory of matter as inert stuff by redefining that concept in terms of a dynamic polarity between

---

[22] For an account of Girtanner's intervention, see Sloan 1979, pp. 137–143.

[23] This sentence summarizes the contents of the so-called "Oldest System-Program of German Idealism," which was jointly written by Schelling, Hegel, and Hölderlin in 1797. For an English translation, see Krell 1985.

attraction and repulsion, proposed to unify magnetism, electricity, and the vital force in terms of various proportions and disproportions between these dynamic poles (Schelling 1797/1988). This produced a marked shift toward Herder's hylozooism and toward his idea that the unity and diversity of nature's overall plan can be displayed as if all species sprang from a common, original matrix. This notion would have been disdained at Göttingen. It might even have constituted further evidence that Kant's concept of organisms as natural purposes and his insistence on retaining Linnaean classification merely served as a screen for recovering the universal teleology of the old hierarchical *scala naturae* under the guise of a regulative idea. Exactly the opposite reaction, however, took place at Jena. A number of comparative anatomists at that university, especially Lorenz Oken and Carl Gustav Carus, were swept up by Schelling's identity philosophy into enthusiasm for treating Herder's thought experiment as what Kant would have classified as determinate claims about the world – not, to be sure, as an evolutionary hypothesis, or even a devolutionist one along lines proposed by Buffon, but as a restatement of the classical *scala naturae*. For Schelling and his idealist followers, the restoration of the ancient and medieval world-view through aesthetic, rather than intellectual, intuition would provide an essential foundation for a chain of argument establishing that human beings are at home in the world (and shouldn't pine for another) because the world is perfectly fitted to the development and expression of their superior capacities. We may assume that Kant, who had set out to limit reason to make room for faith, would have disagreed.

✦

CHAPTER 5

Before Darwin I

A Continental Controversy

Introduction

In the first decades of the nineteenth century, the unquestioned center of work in the life sciences was the Muséum National d'Histoire Naturelle at the Jardin des Plantes in Paris. One of Darwin's early teachers, Robert Grant, spent time there, as did Darwin's friend, and later enemy, Richard Owen. It is an institution worth studying in itself, but the most notorious episode associated with it, on which we will focus here, was the debate in 1830 between two of its professors, Georges Cuvier, by then the unquestioned doyen of French science, and his long-time colleague Etienne Geoffroy St. Hilaire. Their disagreement had been building up for more than a decade, but it came to a head in the spring of 1830. We may put the nub of their quarrel very briefly, before providing a sketch of its historical background, and then returning to analyze their contrasting positions in a little more detail.[1]

Geoffroy was a man of one idea. As early as 1796, at the age of twenty-three, he had written, in an essay on a species of lemur (the "maki" or "macaco"): "It seems that nature has confined itself within certain limits, and has formed all living beings only on one unique plan" (Geoffroy 1796 in Le Guyader 1998, p. 35). Later, Geoffroy was to prefer the expression "unity of composition," but the thought of looking for widespread unities remained. Cuvier found this guiding principle offensive; it was attention to the harmonious adaptedness of each kind of living being, as distinct from every other, that governed his brilliant work in comparative anatomy as well as in paleontology. The unity

---

[1] The best account of the debate written in English is Appel 1987. The published text of the debate is reprinted in Le Guyader 1998/2004.

of composition, or of plan, was for Cuvier "only a principle subordinate to another much more exalted and much more fruitful, that of the conditions of existence, of the appropriateness of the parts, of their co-ordination for the role that the animal is to (*"doit"*) play in nature . . ." (Geoffroy 1830, in Le Guyader 1998, p. 165).

Only a few years later, Darwin was to underscore and comment on this passage; indeed, he commented on it in three different contexts.[2] He also marked it "Q" (for "Quote!"), and did in fact refer to it two decades later, first in *Natural Selection* (Darwin 1973), and then at the close of Chapter VI of the *Origin of Species*, the chapter entitled "Difficulties on Theory" (Darwin 1859, p. 206). How Darwin understood Cuvier's "conditions of existence" is another story, to which we shall return in Chapter 7. We may note, however, that one commentator renders Cuvier's "conditions d'existence" as "prerequisites of existence," a phrase that perhaps conveys Cuvier's meaning more plainly than does the literal version (Corsi 1988, p. 167).

But let us return to our texts, and try to place them in their historical context. Although the debate took place at the Royal Academy of Sciences, the long-term setting of the contest was the Muséum, in which both participants had worked and lived for nearly forty years. Geoffroy had been there at its founding, and Cuvier arrived only two years later. Remember that Buffon had been director ("intendant") of the King's Gardens and "Keeper of the King's Cabinet" for a good half of the eighteenth century. After he died in 1788, and a nonentity was appointed in his place, his staff – Buffon's former collaborator, the anatomist Louis-Jean-Marie Daubenton, the great plant taxonomist Antoine de Jussieu, Jean-Baptiste Lamarck (at this stage of his career also a specialist in botany), and others – objected to the presence of a royally appointed director. They proposed a more democratic structure, with twelve professors at a museum housing the collection and attached to the gardens. The professors and their assistants would give public lectures and guide people through the collections. By the time this proposal was set before the National Assembly, in September, 1790, the King's Garden had of course become the Jardin des Plantes, and the plan to democratize the affiliated institution must have seemed appropriate enough. Indeed, when in August the concerned parties had asked permission to make

---

[2] See Darwin 1987, B 112, n. 5, and DAR 74: 38, where Darwin summarizes Cuvier's claim of superiority of his principle over Geoffroy's, and writes in brackets: "I dispute this."

such a proposal, they had inquired, "Shall the tree of liberty be the only tree that cannot be naturalized at the Jardin des Plantes?" (Hamy 1893, pp. 97–100; quoted in Appel 1987, p. 17). Presumably, however, the National Assembly had other matters on its mind, and nothing was done with the proposal until the Convention was reminded of it after a visit to the Garden by a politician named Lakanal in 1793. He raised the matter the next day, June 10, 1793, and it passed unanimously with no particular attention – this in the midst of the Terror, and while most established intellectual or academic institutions, the University of Paris and all the academies among them, were being repressed. The new institution's budget included salaries for the professors as well as housing in the Garden's grounds. Both our protagonists would live there, with their families, from their appointments till their deaths. Think of the opportunities for close collaboration – and for intense rivalries and enmities! The professors were obligated to give forty public lectures a year, but they could assign substitutes to give their lectures, and could supplement their salaries with other positions – such as chairs at the Collège de France, for example – as they arose. Thus the new institution provided a unique opportunity for the pursuit of natural history independently of a medical career.

If it seems surprising that a successful and stable institution was founded at the height of the Terror, it should also be noticed that it was in part the events of the revolutionary years that made its eminence possible. The staff of the Garden had been in charge of the Cabinet of the King, which was housed on its grounds. Such cabinets, or natural history collections, had become increasingly popular throughout the eighteenth century. During the Revolution, collections belonging to various noblemen were confiscated and added to the formerly royal collection, and when the French armies moved out over Europe, they seized other such collections, in particular that of the Stadtholder of the Netherlands (Appel 1987, p. 35–6).[3] In this way, the collection of such objects at the Muséum became the leading center for the study of natural history and comparative anatomy.

Let us now move back to our leading actors. As a very young man, Etienne Geoffroy de Saint-Hilaire had come from the country, first as a seminarian, then as a law student, and finally as a student of medicine – a choice he used to facilitate his interest in the natural sciences, especially

---

[3] For the collection of curiosities in general, see Daston and Park 1998. The role of voyages in forming collections is discussed in Bourguet 1997.

in mineralogy. In particular, he was much taken with the work of René-Just Haüy, considered by historians to be the founder of crystallography. Haüy was a priest, who, with several colleagues, was imprisoned in the autumn of 1792. Geoffroy helped rescue him – and even assisted in a desperate escape for some of the others. In his gratitude, Haüy wrote from his refuge in the provinces, asking Daubenton to find a position for his savior. In this way, Geoffroy was appointed as an assistant sub-keeper of the Cabinet in March, 1793, and, lo and behold, three months later, this beginner in mineralogy found himself, at the age of twenty-one, professor of zoology at the new Muséum. He had the King's Cabinet to start from, and next year he also became keeper of the new menagerie established at the Jardin, positions he was to hold for half a century. From 1798 to 1802, however, Geoffroy participated in Napoleon's expedition to Egypt, where he was a member of the Institute of Egypt, founded in imitation of the Institute in Paris (the successor of the defunct Academy of Sciences). After the defeat of the French by the British in 1800, Geoffroy and his compatriots hoped to return home, but it took until January 1802 to effect this. At least (with the signal exception of the Rosetta Stone), Geoffroy managed to save the material he and other French scientists had collected, including an important collection of animal mummies.

As we have noted, before he left for Egypt, Geoffroy had already begun to develop an interest in patterns to be found across different kinds of organisms – analogies, as he called them, but homologies in the terminology that Richard Owen was to introduce some decades later. He continued to work in this direction, supporting his views by work in embryology and also in the study of monsters – or teratology, as his son Isidore would later call it. From research in both, he believed, it was possible to learn more about the unity of composition than from the study of normal adults alone. (At the same time, it should be noted, Geoffroy was also doing extensive work in the new science of comparative anatomy that Cuvier and others were developing. Not all his work was controversial; Cuvier himself would acknowledge his "friend and colleague's" contributions in his widely influential *Règne Animal*, published in 1817.) In 1807, after several abortive attempts, Geoffroy was finally elected to the Institute, and in 1808 he became Professor in the Faculty of Sciences in the newly founded University of France, a position he held until his death in 1844. The major work in which he announced his basic views was the *Philosophie anatomique*, published in two volumes (Geoffroy 1818; 1822).

In 1795, Geoffroy had invited to join the Muséum's staff a young man from the Duchy of Würtemberg, one Georges Cuvier, who was living in Normandy as tutor to the son of a French nobleman, and who had published some papers in natural history. Daubenton warned Geoffroy he would live to regret the invitation to Cuvier, and, indeed, Cuvier would come to dominate, not only the Muséum, but French natural history, and even French science as such. His skill in comparative anatomy plainly amounted to genius, and so did his administrative and political skill. He shrewdly refused to join the expedition to Egypt; by the time Geoffroy returned, Cuvier was already a member of the Institute; he became its permanent secretary in 1803. He came to hold many titles: ennobled as Chevalier in 1811, named Baron Cuvier in 1819, and made a Peer of France a year before his death in 1832. His major works – *Leçons d'anatomie comparée* (1800–1805), *Recherches sur les ossemens fossiles* (1812), and *Le règne animal* (1817) – were canonical for comparative anatomy, vertebrate paleontology, and systematics respectively. He was indeed "the Napoleon of science."

Two Approaches to Comparative Anatomy

The profound divergence between the two friends – who at first published some collaborative articles – is apparent at many levels. Take, for example, the very titles of their major works. With Cuvier, we have "Lessons in . . ." and "Researches on . . ." "The Animal Kingdom" is a less modest title, perhaps, but it lacks the sweep of Geoffroy's "Anatomical Philosophy," or that of the record of the debate itself, published by Geoffroy as *Principes de philosophie zoologique* (Geoffroy 1830). Sometimes, indeed, historians have called their difference – as Cuvier himself liked to do – a difference between "fact" and "speculation." This won't do, of course, since nobody sticks to "pure facts," and since Geoffroy did carry out observations and experiments, working, as we noted, in embryology and teratology in support of his theories – as well as contributing more routine anatomical reports. One way to approach the situation is to look at examples of their actual experimental work; the differences in their aims and interests show up clearly in their differing practice. Each one, it appears, was looking for unity, but in a very different locus.

Cuvier was in effect the founder, as well as the leader, of comparative anatomy. So much so, that when a hall of comparative anatomy was constructed in the Muséum, it was built with an entrance from the Cuvier family's residence. Both his supreme skill and his supreme self-confidence are illustrated in a report published in the Muséum's *Annals* in 1804 (Rudwick 1997, pp. 69–73).[4] As we have already noted, Cuvier was also a foundational worker in fossil as well as contemporary anatomy. In the 1804 report in question, he presented a description of "the nearly complete skeleton of a little quadruped of the genus of opossums, found in the plaster in the region of Paris." It had not been considered likely that fossils from a genus now found only in the New World would turn up in Europe. From examining the skeleton, Cuvier was confident that the organism had indeed been an opossum, and therefore a marsupial. But to prove this, he needed to find in his specimen the two bones that support the pouch. He sometimes boasted that he could identify an animal from a single bone; here he needed to find the single bones to confirm his diagnosis of the whole. To do this, he had to sacrifice some of the sacral vertebrae. As he reports: "I excavated carefully with a sharp steel point, and had the satisfaction of exposing to view the whole anterior portion of the pelvis, with the two supernumerary or marsupial bones that I was looking for, in their natural position and wholly similar to their analogues in the opossums" (Rudwick 1997, p. 71). The operation was carried out, he assures his readers, in the presence of reliable witnesses, to whom he had announced in advance his prediction that he would find the marsupial bones where they belonged – since, he proclaims, "the true hallmark of a theory is the power it provides to predict phenomena" (Rudwick 1997, p. 71). The question remained as to whether it was a marsupial closer to those of America or of Australasia. A careful examination of the foot led Cuvier to the conclusion that it was some kind of opossum, thus analogous to the marsupials of the New World, and a comparison with the species of living opossums being examined by his "learned colleague Geoffroy" (Rudwick 1997, p. 72) showed him that although it belonged in that genus, the species was not one of those known to be living then.

In conclusion, Cuvier celebrates the way in which his discovery helps to overturn the speculative "theories of the earth," according to which no American animals had ever extended their range to Europe. The first contradiction he had found for this view was the discovery of tapir teeth

---

[4] Quotations that follow are from Rudwick 1997.

in France. Here is the second, and, he confesses:

> . . . persuaded as I am of the futility of all these systems, I find myself pleased each time a well-established fact comes and destroys one of them. The greatest service one can render science is to make a clear space there, before constructing anything; to start with all those fantastic edifices that obstruct . . . the avenues [of progress], and that hinder from participation all those to whom the exact sciences have given the felicitous habit of acceding to evidence, or at least of ranking propositions according to their degree of probability. With this last precaution, there is no science that cannot become almost geometrical; the chemists have proved as much in recent times for theirs; and I hope the time is not far off when one will be able to say as much of the anatomists.
>
> Rudwick 1997, pp. 72–3

We might call this little report of Cuvier's "A Self-Portrait," for it illustrates not only his skill as a dissector and demonstrator, and his ingenuity in prediction, but his interest in what he calls "natural classi-fication," as well as his whole attitude to his calling. In the concluding claim, as his translator Martin Rudwick notes, he is clearly referring to Lavoisier's chemistry as well as to what the French considered Laplace's triumphant completion of Newtonian mechanics. He may also be al-luding (as Rudwick also suggests) to Kant's skepticism concerning the possibility that there could ever be a "Newton of a blade of grass." A "Newton of marsupials" would do as well! Indeed, in the introduction to his *Recherches sur les ossemens fossiles* of 1812, Cuvier straightfor-wardly asks, ". . . why should not natural history also have its Newton one day?" (Rudwick 1997, p. 185).

We shall return shortly to the question of Cuvier and classification, as well as to the implications (negative, he believed) of his work on fossils for the likelihood of any kind of transformism. What chiefly concerns us here, however, is the lesson this case can teach us about his observational and experimental interests. If he rejects the context of large speculative theories, and thus illustrates the sense in which, even in his theorizing, he wants to stay close to the "facts," we may also ask, what did he want from his 'facts'?

Obviously, Cuvier is not just describing one particular set of bones that he and his colleague Alexandre Brogniart discovered in the chalk basin of Paris. He wants to know just what *kind* of animal this is. William Coleman, in his classic work on Cuvier, insists on his subject's primary concern with "types" (Coleman 1964, pp. 74, 98–102). He admits that Cuvier seldom uses that term, and it seems better to avoid it if one can, since it suggests the later "archetype" of Richard Owen, which is surely derived as much from Geoffroy as from Cuvier. But it is indeed the unity of the organism, and therewith of a species of organism, that defines his chief concern. The skeleton, caught as it is, awkwardly, between two stones, looks to him like a marsupial, in fact like an opossum, and by careful dissection and observation he can prove that he is right. This is not just a collection of bones; it is the skeleton of an ancient mammal, a creature that gave birth to its young alive, but for some time carried and fed them in a pouch. And it is one of those animals more closely related to the species currently living in the New World than to those now living in the (to Europeans, still newer) world of Australia. It is the unity of this animal as a *specimen* of its species that Cuvier wants to understand. He is practicing comparative anatomy, as he insisted both early and late (in letters from Normandy to a friend back home in Würtemberg, and much later in the notorious Cuvier–Geoffroy debate of 1830) in the spirit of Aristotle. It was Aristotle, he insists, who founded comparative anatomy, and he is carrying it forward – presumably to the point of a Newtonian apotheosis – in the spirit of the ancient master.[5] Look at each kind of organism, whether you have to reconstruct it from bones or skins or have it living before you, in terms of its unified, harmonious nature, its life style (we might say). Clearly, in any organism, the parts cohere in a meaningful, well-adapted whole: Fishes live in the water – that's why they have gills, not lungs; birds live in the air – that's why they have light, air-filled bones. The skeleton found in the chalk basin belonged to a marsupial, so it had to have those bones to hold a protective pouch. It is the total adaptedness of just this creature for the way it lived, or lives, that governs, and should govern, the comparative anatomist's researches. Of course, one has to look at analogues of peculiarities of this specimen – of this species. Its near relatives show us where it belongs in a natural catalogue of organisms, and help us identify structures or behaviors we should expect to find in it. But what we are after is the unity of this kind, the way its parts function together to make just this kind of viable animal, adapted to its due place in a marvelously organized nature.

Not that each species was wholly unrelated to every other; as we will note, Cuvier was in search of a natural classification. In his

---

[5] See Cuvier's statement in Geoffroy 1830, in Le Guyader 1998, p. 164. Cuvier refers to "a special science called comparative anatomy, but which is far from being a modern science, since its author is Aristotle."

Règne animal (1817), he would describe the members of four large *"embranchements"*: the vertebrates, mollusks, articulates, and radiates, which together embraced, as he insisted, the whole animal kingdom. But, he also insisted, there could be absolutely no overlap between the members of any one of these overarching groups. So there is unity only within separation.

We shall look more carefully in the next section at the philosophical implications of Cuvier's methods. But in the meantime, we should warn our readers that when Cuvier looks at his specimens in relation to related species, he is examining them from neither of the perspectives we might now recognize. Relation, before Darwin, does not entail genealogy. There is an order of nature that can be examined simply as the order it is, without any suggestion that it came about through any kind of descent with modification. At the same time, that does not make Cuvier a "special creationist" in the sense of American fundamentalism. Although he was apparently a theist, indeed the leader of Protestantism in France, Cuvier did not, for the most part, mix science with religion. At least once, when he was angry, he made a pronouncement concerning the divine ground of nature (Cuvier 1825). But that was unusual, and did not, in any case, evince a close connection between his natural-theological speculations and his scientific practice. As we shall see in the next chapter, it was the British, with their love (or the love of some of them) for natural theology, who served up Cuvier's "Revolutions of the Globe" (the "Preface" to his *Recherches sur les ossemens fossiles*) in that pious context. We should try to understand Cuvier's enterprise outside of those alternate Whiggish contexts.

Exhibiting Cuvier's methods through a single example is perhaps like trying to identify, or even to construct, an organism from a single bone. It does provide, or so we hope, some sense of the astonishingly detailed work that Cuvier, and to some extent his colleagues and followers, were carrying on in this remarkable period. It was a time of growth and expansion in a discipline or group of related disciplines: comparative anatomy, taxonomy, and paleontology, analogous to the explosion of knowledge in genetics and molecular biology in the twentieth century. And Cuvier's aim in all this work was also authoritative for many of those who succeeded him. However, what we want to do here is present a contrast between Cuvier and his "learned colleague, Geoffroy."

It would be nicely symmetrical if Geoffroy's study of opossums exhibited his talents and his interests as plainly as the case of the Paris fossil does Cuvier's. But that does not appear to be the case.[6] Instead, we shall consider a series of three memoirs presented by Geoffroy in 1807, followed by a fourth in 1817, and then by the first volume of the *Philosophie anatomique* in 1818.[7] Geoffroy, like Cuvier and their colleagues, did a great deal of detailed work in comparative anatomy, but, as we noted at the start of this chapter, always in search, not so much of the unity of this organism, as of the unity of plan, or of composition, across a wide range of organisms, in this case, fishes compared with other vertebrates. In the first memoir of 1807, he displayed an analogue (homologue, we would say) of the pectoral fin of fishes and the arm of other vertebrates. Aristotle had made a gesture in this direction, but had not carried out the comparison in detail. Geoffroy also announced a homology between the furcula – wishbone – of birds and corresponding bones (or their absence!) in fishes, a comparison he continued in more detail in the second memoir. Cuvier and his disciples, of whom there were many, would never look in fishes for a bone found in birds and so clearly intended to assist in flight. But Geoffroy declared: "Without a direct object in swimming animals, without a utility determined in advance, and thrown, so to speak, by chance into the field of organization, the furcula enters into connection with the organs near it; and according to the manner in which this association is formed, it takes

---

[6] The article "Marsupiaux," which he published in 1823 in the *Dictionnaire des sciences Naturelles*, may serve as a summary of Geoffroy's interest in this subject. Although he mentions his leading long-term themes, he does so briefly and non-aggressively. The bulk of the article reports the work of others, as well as some of his own, especially on the vexed question of marsupial reproduction. The question was whether the young of pouched mammals were "born" from the uterus or the teats of the mother. Geoffroy's conclusion was that they are cast at a medusa-like level of organization into the pouch, where they develop attached to the mammary glands. Thus the pouch serves as a second uterus. Geoffroy states his conclusions as a challenge to other naturalists to investigate further this still controversial subject.

[7] The three memoirs of 1807 are: "Premier mémoire sur les poissons, où l'on compare les pièces osseuses de leurs nageoires pectorales avec les os de l'extrémité antérieur des autres animaux à vertèbres," *Annales du Muséum d'Histoire Naturelle* 9 (1807), pp. 357–372; "Second mémoire sur les poissons: considérations sur l'os furculaire, une des pièces de la nageoire pectorale," *Annales* 9, pp. 413–427; "Troisième mémoire sur les poissons: où l'on traite de leur sternum sous le point de vue de sa détermination et de ses formes générales," *Annales* 10 (1807), pp. 249–264. The memoir of 1817 is: "Du squelette des poissons ramené dans toutes ses parties à la charpente osseuse des autres animaux vertébrés, et premièrement de l'opercule des poissons," *Bulletin de la Société Philomatique* (1817), pp. 125–127. This text is repeated in the *Philosophie anatomique*, pp. 29–30. We are here following the detailed account in Appel 1987.

on uses which are in some sense prescribed by them" – that is, by the neighboring organs (Geoffroy 1807a, p. 370, in Appel 1987, p. 87). What the comparative anatomist should look for, in Geoffroy's view, is the same bones differently arranged in different organisms. Function will be served of course – "fish gotta swim, birds gotta fly!" But nature provides the same components in different shapes, sizes, and arrangements in different organisms. So even if in some fishes, the furcula is absent – well, it was there, or is there, in spirit, so to speak. We all have some idea how this works in the most familiar example of homology: the vertebrate limb. You can make a list of components of *a* limb, or, as Owen was famously to do, of *the* limb, and find them differently disposed in the bat's wing, the whale's flipper, the bird's wing, the primate's arm. These arrangements can be very odd: The horse walks about on its middle finger nail, for instance; four of its digits have become vestigial. Remember, though, for a comparative anatomist in the first few decades of the nineteenth century – and for most Frenchmen for longer than that – "become" here is not historical. There is no implication of transformism. There is an order of nature that we are studying. Let us do so "philosophically," looking for unifying principles across kinds rather than a clutter of this, that, and the other.

In his third memoir, Geoffroy identified a piscine sternum – following on the vain attempts of other anatomists (including Cuvier) to make this identification. He showed his peculiar investigative bent, moreover, in his use of embryological data to support his conclusions. However, one peculiar piscine structure he still had failed to homologize: the operculum or gill cover. Only in 1817, after examining an attempt by Henri Blainville, a protegé of Cuvier, did the solution come to him. In his enumeration of the bones of the vertebrate skull, Cuvier had omitted the bones of the inner ear – and those were indeed, Geoffroy concluded, the homologues of the operculum. Triumphantly, he proceeded (safely isolated in his country home) to compose the first volume of the *Philosophie anatomique*, which would formulate in style his methodological and theoretical conclusions.

If it seems unfair to balance one particular investigation of Cuvier's against a series by Geoffroy, it is difficult to grasp any one of the latter's researches without watching his guiding principle – the unity of composition – steer him through a number of steps. We could look simply at the culminating discovery of 1817; but then we would be, chronologically, far in advance of Cuvier 1804. And we wanted to give some sense of their working alongside one another throughout the period preceding, and leading up to, the debate of 1830. It seems not unfair to say that we see Cuvier working to find unity within separation, and Geoffroy to conquer separation by discovering unity.

Two Philosophies of Science

With samples of their work before us, let us look more carefully at Cuvier's and Geoffroy's guiding principles as they themselves pronounced them. What matters most for Cuvier is what he called "conditions of existence." What does that mean? When, at the close of Chapter Six of the *Origin*, Darwin cites this expression "so often insisted on by the illustrious Cuvier," we think of the conditions in the environment of an organism that favor the survival in the next generation of one slightly differing variant over another – say, very slightly better aerated bones in an avian ancestor, or rather, in a population of avian ancestors (Darwin 1859, p. 206). However, as we can see from the case of the Parisian marsupial, that is not at all what Cuvier intended. He did include some reference to the organism's habitat – birds in the air, fish in the sea – but what concerned him first and foremost was the integrated, harmonious coordination of all the parts, each functioning to produce a functioning whole. Kant's formula – each part as both end and means – seems too mechanical for as inspired a naturalist as Cuvier, whose ruling passion from his youth had been things living or once living (though he did at least once refer to Kant's insistence that we refer to the whole organism [Cuvier 1800–1805, I, pp. 5–6]). In the end, as in the beginning, that's what matters: the *organization* of this bird, this fish, this opossum. Granted, each individual is characteristic of its kind, and one kind can even stand in for others. Thus, in his massive work on fishes, he takes the common European perch as typical of a whole range of species (Cuvier and Valenciennes 1828–1833). But this is, so to speak, one life-style exemplified with differences, in a (limited) number of cases. The notion of slightly differing small variations that would ultimately add up to a new species would be simply inconceivable on Cuvier's premises. That's why the phrase "prerequisites of existence" seems to render his chief principle more adequately than its literal translation. It's a question of what is needed to make this kind of animal live its life in its allotted place in nature.

Indeed, his phrase, "conditions of existence," Cuvier tells us, is what people commonly refer to as "final cause" (Cuvier 1817, vol. I, p. 6). At first hearing, that sounds very strange. A final cause is "that for the sake of which," the end served by the steps leading up to it, as an adult male is for Aristotle the final cause of the seed and its development. But how can conditions be ends? They are what is needed if the end is to be achieved, not the end itself. Perhaps we can say that understanding what is meant by conditions of existence means reasoning in terms of final causes rather than in terms of when-then causes and effects. The end is the very existence, in its proper habitat, of this closely integrated entity, beautifully suited to that habitat and no other. The conditions are the harmonious, well-integrated means to that end.

In contrast, Geoffroy's guiding principle, as we noted, is "unity of plan" or "unity of composition." And that means unity of composition *from* one species, one genus, one family, *to* another. Although Geoffroy had not been spontaneously devoted to natural history, but had been thrust into it by a set of curious chances, he did do many excellent dissections, for example, of Egyptian fish and crocodiles, or, as we saw Cuvier acknowledging, of marsupials. So he took kindly to his new profession. But always – though he sometimes tried to control this inclination – he wanted to find generalizations that would hold at least for a large group – all vertebrates, and finally all animals. Indeed, both early, in Egypt, and later, in his final years, Geoffroy liked to indulge in speculations about reality itself, enunciating a principle of like-for-like that would hold absolutely everywhere. Even when more restrained, it was generalization he was always looking for. "Unity of composition" meant not the unity of composition – as in a painting – of this entity, but unity of the materials forming a range of entities. It was a "theory of analogues" (again, of what we would call homologues) that he hoped to apply wherever he could. Why call it a dog's paw and a man's foot? A man's arm and a bird's wing? It's the same thing, used differently – and even built differently, but of the same materials, in an analogous arrangement.

It is often said that Geoffroy stressed structure, whereas Cuvier insists on the primacy of function.[8] But that is too simple a contrast. For one thing, even Cuvier bases his general divisions on the structures characteristic of different groups.[9] On the other hand, if Geoffroy finds in our arm and the bird's wing the "same" organ with a different function, the two are also differently shaped. *Both* form and function, in his view, are subordinate to the unity of composition, which shapes differently in

[8] See the now classic (but confusing) Russell 1916.
[9] See *Le règne animal*, passim.

different circumstances components that are basically the same. So when, in the passage we have already mentioned, Darwin contrasts Cuvier's "Conditions of Existence" with the opposing principle (presumably Geoffroy's) of "Unity of Type," he is being a little unfair. It is Cuvier who insists on the unity of *this* type, Geoffroy who wants to find unity *across* types – a unity that Darwin, of course, ascribes to descent.[10] Geoffroy, as we suggested earlier, does not need to make the connection between unity across types and descent, although, as we shall see, his approach is not in principle incompatible with some kind of transformism, as Cuvier's is; and he did in fact later present, in a couple of articles, a more general transformist view (see Le Guyader 1998, pp. 123–128, 240–243.).

Perhaps it may help us to understand Geoffroy's ruling passion a little better if we go back for a moment to his starting point. It seems to us odd that a student fascinated by mineralogy would turn to the comparative anatomy of animals – but it was the job he was offered, and he was to apply himself to it with enthusiasm for half a century. We should notice, however, that "natural history" had so far included three kingdoms, animal, vegetable, and mineral – as in the children's game. And Geoffroy's friend and teacher, Haüy, was in effect the founder of crystallography. He once happened to drop an item in a cabinet of crystals, found it shattered along regular faces, and proceeded to elaborate a doctrine of crystalline structure that has persisted ever since. What one does there is to enumerate certain basic constituent shapes that are arranged in different minerals in differing characteristic patterns. Something like that is what Geoffroy was trying to do for the constituents of animal groups, and ultimately for all of them (Le Guyader 1988, pp. 79ff).

To summarize: We have Cuvier's conditions (or prerequisites) of existence opposed to Geoffroy's unity of composition as divergent guiding principles. Further, each of these is in turn supported by two corollaries. These are, for Cuvier, the correlation of parts and the subordination of characters, and for Geoffroy, the principle of connection and the balance of parts.

Cuvier's two subordinate principles both follow plainly from the first. If our chief aim is to demonstrate the delicate adaptedness of the entire animal for its peculiar life style, we will have to show how its constituent tissues, organs, and systems are mutually related for this end. Thus, for

---

[10] For the connection between unity and descent, see Darwin 1987, B112, n. 5, and Darwin 1859, p. 206.

example, fishes in their aqueous medium will have a harder time breathing than animals on land or in the air. To get them enough air, therefore, their circulation will have to be especially rapid. Subordination of characters is a bit trickier. Which characters are most important? The great botanical systematist Antoine-Laurent de Jussieu had used constancy of characters as his guide. Cuvier sometimes followed this method, but sometimes insisted on a functional criterion of importance (see Coleman 1964, pp. 78–82).

With Geoffroy, the first corollary – the principle of connections – again follows plainly from his guiding conception, the unity of composition. As we have seen, it is not the unity of each organism Geoffroy is primarily concerned with, but the unity of materials used, sometimes for different functions, across a number of species. What is constant is neither the whole structure nor the particular shape and size of each constituent part, but their *arrangement*: how they are situated in relation to one another. A horse's residual four digits are there, on each side of the surviving middle digit, on which he puts his weight. Their order is the same as in other animals, like ourselves, which use all five. And the balance (*balancement*) of parts seems equally obvious: Our middle finger is a bit longer than the others – and of course, in our case, the opposing thumb appears especially important. The exaggeration, or repositioning, of one part is balanced by a relative subordination of others. In their public controversy, for example, Geoffroy put forward the case of the hyoid bone in the howling monkey, which was outsized, whereas other components of the complex had been reduced to mere ligaments. Cuvier ridiculed this notion: The howling monkey needs a huge hyoid for its howling – that's why it's there. Why try to find "analogues" for bones that other animals need but this one doesn't? But in view of the unity of composition – and the principle of connections – it seemed to Geoffroy only reasonable to look for the same basic number of components, larger or smaller or even turned this way or that in one or another context. No final causes here, except in subordination to available structural components.

We must add one more contrast: their approaches to development and reproduction. Cuvier evaded this question in the main, asserting his adherence to some kind of preformation, but declaring generation to be a great mystery, insoluble by the human intellect (Cuvier 1817, pp. 17–20). Geoffroy, with his interest in embryology, was decidedly an epigeneticist. Indeed, in a footnote to a memoir on the organization of insects, delivered ten years before the debate with Cuvier (Geoffroy

1820, in Le Guyader 1998, pp. 71–79), Geoffroy takes their disagreement on this matter as fundamental to the difference between them (Le Guyader 1998, p. 306, n.18). Given how peripheral this issue was for Cuvier, that seems to be an unlikely thesis. What Geoffroy had said in the text to which this note was appended is perhaps a more reasonable summary (this ten years before the debate!). After regretting that "the greatest naturalist of our age" had pronounced "an absolute condemnation" of his views, he continued:

> What M. Cuvier condemns in this moment is the entirety of my views, that is my whole philosophy, which he has already made the subject of his criticisms in the analyses of the academic sessions of 1817 and 1818. But what does this divergence of opinions really prove? Only that M. Cuvier and I think differently about theories: in this we are doing nothing but reproducing the one and the other of the two forms in which the human spirit has always proceeded. The nuances of these two ways of seeing and feeling the facts appear everywhere where the judgment of men is involved.
>
> Le Guyader 1998, p. 79[11]

Cuvier and Classification

In our post-Darwinian day, we find homology important in relation to taxonomy – since, as Darwin noted both in the *Origin* and in his copy of the published debate, "unity of type" derives from descent, and so gives us useful clues for identifying relations among various forms (Darwin 1987, B112, n. 5; 1859, p. 206). For Geoffroy, however, despite his obsession with homology, taxonomy, or classification, was not an overriding interest. He did start on a taxonomy of mammals, but never finished or published it (Appel 1987, p. 83). For Cuvier, on the contrary, the development of a "natural classification" was always a predominant aim. In a way, as the *Règne animal* shows, that was the ultimate goal of his immense and intensive labors in comparative anatomy, or at least a major step on the way to the completion of that task.

The search for a natural classification of the plant and animal kingdoms is a much more complicated story than we are trying to tell. It is clear, however, that by Cuvier's time, on the one hand, despite the

[11] The note we cited follows the term "nuances."

artificiality of Linnaean nomenclature, its usefulness had made its acceptance universal. On the other hand, as we noted earlier, Linnaeus himself had been caught between the convenience of a relatively artificial system and his desire to reflect the natural affinities of plants or animals in such a way as to produce some kind of "rational" systematics. The longing for a natural system continued to worry taxonomists throughout the eighteenth and early nineteenth centuries. However, as Cuvier himself avers, by 1816, when he wrote the preface to the *Règne animal*, the time was ripe for the fulfillment of that long-felt desire. For plants, the work of the Jussieu family, culminating in Antoine-Laurent de Jussieu's *Genera Plantarum*, had provided a model for a similar effort for the animal kingdom. Further, as Cuvier acknowledges, the work of many naturalists, both at home and abroad – for example, Geoffroy's work on quadrupeds, or Lamarck's magnificent subdivisions of the many kinds of organisms that Linnaeus had listed simply as "insects or worms" – had added immeasurably to the knowledge available to would-be taxonomists. Of great importance also was the enrichment of the Muséum's collections, which provided invaluable new materials to aid in such an enterprise. There were also reliable reports of voyages, but Cuvier emphasizes the fact that wherever possible he has verified any reports by others in terms of his own observations and dissections. Even in the section on insects (a major branch of the Articulates) he has added some information of his own to the text supplied by his colleague Pierre-André Latreille. The range of his knowledge and the sheer quantity of work he has done is astonishing – especially considering the time he had to spend on administrative duties outside the Muséum. In fact, he remarks that the kind of methodical attention demanded by natural history is the best preparation a man can have to handle with the appropriate efficiency any organizational task demanded of him!

Conceptually most important, perhaps, for Cuvier's taxonomic labors was the discovery and application of the principle of subordination of characters. He had long distinguished the major organ systems of various animals: their respiratory, circulatory, muscular, nervous systems, and so on. Which was the most significant of these for purposes of classification? From the writings of J. J. Virey he had gotten the idea of taking the nervous system as fundamental, and on that ground he had come to the notion of his four *embranchements*: vertebrates, mollusks, articulates, and radiates (or zoophytes). As he wrote in his announcement of this discovery in 1812:

In considering the animal kingdom from this point of view and being concerned only with the animals themselves, and not with their size, usefulness, our knowledge of them, great and small, or any other accessory circumstances, I have found that there exist four principal forms, four general plans, upon which all of the animals seem to have been modeled and whose lesser divisions, no matter what names naturalists have dignified them with, are only modifications superficially founded on development of or on the addition of certain parts, but which in no way change the essence of the plan.

Cuvier 1812a; quoted in Coleman 1964a, p. 92

To make his system easier to follow, moreover, he proceeded from these overarching groups to each smaller unit – class, order, family, genus, sub-genus, species – without repeating for any of the narrower units the "important" characters he had already listed, with the level of each group distinguished orthographically as well as verbally. And in each case, of course, we arrive at last at the species, the ultimate unit, with its essential prerequisites of existence, the indissoluble constant with which, in the last analysis, every naturalist is concerned.

To begin with, vertebrates, for example, are divided into four classes: mammals, birds, reptiles, and fishes. Let's look briefly at birds: "oviparous vertebrates with double circulation and respiration, organized for flight" (Cuvier 1817b, vol. I, p. 290). There is first a fairly lengthy introduction, describing what birds have in common, both anatomically and in their way of life. Cuvier then proceeds to describe several orders of birds, distinguished chiefly by their organs of mastication (beaks) and of prehension (beaks and feet). Each order is subdivided into families and genera, and finally, particular species are described under each genus. The first of six orders, for example, are the birds of prey, among which we find the genus of falcons, in turn divided into two groups (sub-genera?) of "noble" and "ignoble." The ignoble falcons, in turn, include the eagles, characterized by "a very strong beak, straight at its base and curved only toward its point. It is among them that the largest species of the genus and the most powerful of all the birds of prey are found" (Cuvier 1817b, vol. I, p. 313). Then we come to eagles strictly speaking, with the subscript "Cuv," which presumably means that Cuvier himself examined them. They have "the shank feathered up to the root of the digits; they live in the mountains, and hunt birds and quadrupeds; their wings are as long as the tail, their flight as high as it is swift, and their courage surpasses that of all other birds" (Cuvier 1817b, vol. I, p. 313). So we have a kind of mental geography of the class of birds, showing how each group is distinguished within a common pattern. And in each case, the division reflects what one might call a point of functional anatomy: Beaks and feet are anatomical units, but what really distinguishes them is their use, for chewing or snatching. What the natural historian is after is a grasp of the manifold ways that nature works. What best distinguishes him is his sense of the complex unity of each of these as distinct from every other.

The Problem of Transformism

The Cuvier–Geoffroy debate took place in 1830. In 1837, Darwin opened his first transmutation notebook (the B notebook). Almost a century and a half into the Darwinian tradition, we are bound to ask – or at any rate many commentators *have* asked – how does the Paris controversy relate to evolution? Directly, it shows no such relation at all; at least transformism, as it was then called, played no obvious role. But, as we noted earlier, Darwin commented on the debate, marked his copy of Geoffroy's published account, and referred to it again in the *Origin*. Besides, there was Lamarck, who died in 1829 and who, since 1802, had been announcing and elaborating a full-fledged theory of evolution (in the later, transformist sense of that term). He had disciples, even in Britain (see Chapter 6). And there were others, too, who, while not adopting the Lamarckian view, showed some sympathy for the notion that species might develop, one from another (see Burkhardt 1997). So let us look, at least sketchily, at the status of what we call evolutionary ideas among naturalists, or in particular comparative anatomists, in the period we have been considering.

The standard view – the received view, we might call it – certainly entailed the permanence of species. Cuvier, of course, was instrumental in discovering many extinct species; he blamed their demise on global catastrophes, as we will have occasion to see in the next chapter. But each one, while it lasted, was what it was and nothing else. His position, and that, we believe, of the majority of his contemporaries, is well exemplified in one of Richard Owen's *Hunterian Lectures*, delivered in May 1837, several months before Darwin opened his B notebook. Owen has been reviling the notions of recapitulation that were then circulating: various forms of the thesis that embryos of "higher" forms pass through stages corresponding to adult form of "lower" types. "The doctrine of Transmutation of forms during the Embryonal phases," he tells us:

... is closely allied to that still more objectionable one, the transmutation of Species. Both propositions are crushed in an instant when disrobed of the figurative expressions in which they are often enveloped, and examined in the light of a severe logic.

Owen 1992, p. 192

The "severe logic" in question has been expressed two paragraphs earlier, with respect to the notion of recapitulation:

> Such propositions you will at once, Sir, perceive, imply that there exists in the Animal Sphere a Scale of Structure differing in degree alone: – nay, they imply the possibility of an individual, at certain periods of its development, laying down its individuality, and assuming that of another Animal; – which would, in fact, abolish its existence as a determinate concrete reality.

Owen 1837, p. 192[12]

This is – as Cuvier would have been happy to admit – the old Aristotelian insight that every interesting *this* is a *this-such*: a specimen of a kind, not to be mixed with other kinds. To think otherwise would be to resuscitate Empedocles' fig-bearing vines. Imagine, if you can, a bird hatching out of a reptile's egg: that would put an end to species altogether – for we know like produces like; there would be no order left, no natural history, only chaos. Think of Cuvier dissecting his fossil opossum: He's not looking at these bones, he's inspecting a new kind of animal – an old, long-perished kind, but new to him and to his peers, a kind with its own characteristic anatomy, physiology, and life-style. It's that sort of process, finding one unique form of life as distinct from others, that marks the daily practice as well as the long-term aims of the comparative anatomist. The Cambridge zoologist C. F. A. Pantin spoke of something he called "aesthetic intuition." "A taxonomist may say about a new species," he wrote, "'Whatever it is, I feel sure it is, I feel it is a member of the *Metrididae*,' even if he has great difficulty in saying why it is so" (Pantin 1968, p. 83). But if species don't stay put, if snakes hatch out birds or vines start bearing figs, surely the naturalist has lost his calling. It was not only his wife Emma's piety that made

---

[12] Although Owen is chiefly notorious as the enemy of Darwin, and especially of Huxley, he did ultimately admit some change in species, though not on the Darwinian model. We are here referring only to the situation in 1837, two decades before Darwin's great work. See Rupke 1994 for a careful examination of Owen's complex, and changing, views.

Darwin feel he was committing a murder. He would be asserting, as Cuvier said of Lamarck in the notorious eulogy delivered after his colleague's death, that "species do not exist in nature." Cuvier concluded: "A system established on such foundations may amuse the imagination of a poet; a metaphysician may derive from it an entirely new species of systems; but it cannot for a moment bear the examination of any one who has dissected a hand, a viscus, or even a feather" (Burkhardt 1984, pp. 446–7).

Both Owen's and Cuvier's condemnations are indeed sweeping. There were, however, others among their contemporaries who entertained notions of organic mutability. Among these was Geoffroy himself. Geoffroy, unlike Cuvier, did sometimes refer to Lamarck – his senior colleague – with admiration and respect. He may even have entitled his report of the debate "Principles of Zoological Philosophy" in memory of the recently deceased naturalist. Admittedly, in contrast to Lamarck, at least in his major works Geoffroy did not develop a global theory of species mutability. At the same time, as we noted earlier, Geoffroy's principle of unity of composition did indeed leave room for transformist ideas where they seemed appropriate.

This had become evident, in particular, in a controversy with Cuvier about the crocodiles of Caen a few years before the great debate. Cuvier had identified specimens of a fossil species that he considered a member of the still existing group of gavials. Geoffroy showed, by careful examination of the evidence, that the animal in question was not a member of the modern group. He named it *Teleosaurus*, because (mistakenly) he thought it a probable ancestor of mammals – so "perfect saurian," presumably in anticipation of the yet more perfect mammals. Although he was wrong in his broader genealogical speculation, he was correct in his comparative anatomy. And he certainly did suggest that the fossils of Caen were ancestors of modern species – a claim Cuvier would not have dreamt of making about the many fossil forms he and Brongniart had discovered in the Paris chalk. In fact, as we noted earlier, Geoffroy also published several papers on this issue in the year after the 1830 debate. He also expressed views about the probable causes of the origin of new species. They were, he believed, effects of the environment, often on the developing foetus. Here, as elsewhere, his work in embryology and in particular in teratology were important influences.[13] Moreover,

---

[13] For a summary of the crocodilian controversy, see Appel 1987, pp. 130–136; Geoffroy's later publications on transformism are discussed in Le Guyader 1998, pp. 240–244.

in 1833, he published a more general transformist memoir in which he contrasted "two zoologies, that of the antediluvian epochs and that of the present world."[14] Yet, despite these interests or the study of cases such as the crocodilian, he did not develop publicly a general theory like that of Lamarck.

So we may, after all, consider Lamarck's work as the most conspicuous example in this period, and this milieu, of a fully developed evolutionary theory. Cuvier, through his savage obituary memoir, would have us believe that no one paid serious attention to it. Although it would take us too far afield to go in any detail into the problem of Lamarckism, its meaning and its destiny, we had better look at the situation a little more closely before we leave the Muséum and its controversies. Very briefly, we should note three points in this connection.

To begin with, it should be remembered that Lamarck's theory of organic change was only one (relatively limited!) theory among others that he espoused. He developed a chemistry to refute Lavoisier – a move that did not earn him respect from contemporary chemists! He also set out what he called a "terrestrial physics" – a discipline intended to include geology and meteorology as well as the science that, in an unpublished sketch, he had named biology (see Chapter 4).[15] All this was already anathema to Cuvier and his adherents, with their continuing dislike of systems and system-builders. And if you object that classifiers such as Cuvier are building systems, they are doing it on the ground of their detailed knowledge of a multitude of particular organisms, which they are endeavoring to place in their carefully discriminated, "natural," places alongside other organisms that are significantly similar to them. They are not looking for strange, imagined consequences of wildly inclusive invented theories. Lamarck was doing that, not only in his late work on organic change, but anyhow and anywhere. This in contrast to his superb work on the classification of invertebrates, which every one acknowledged.

---

[14] The full title of Geoffroy 1833 is worth noting: "Considérations sur des ossements fossiles la plupart inconnjus trouvés et observés dans les bassins de l'auvergne, accompagnées de notes ou sont exposées les rapports et les différences des deux zoologies, celle des époques antédiluviennes et celle du monde actuel." ("Considerations on the fossil bones, chiefly unknown, found and observed in the basins of the Auvergne, accompanied by notes in which are explained the relations and the differences of the two zoologies, that of antediluivan epochs and that of the present world".) Quoted in Le Guyader 1998, p. 336, n. 10.

[15] On Lamarck's "Biologie," see Burkhardt 1997, pp. 11–39; also p. xvi and accompanying bibliography. See also in Chapter 4, the section on "Kant and the Biologists."

Second, Lamarck's evolutionary theory itself was easy to parody. A bird steps into the water, gets its feet stuck in the mud, pulls hard so that its feet begin to stretch, passes on this elongation to its offspring, and by and by we have cranes and herons and such, with their long legs suitable for wading. Cuvier's parody has persisted; "the inheritance of acquired characters" is still what most people associate with Lamarck. True, he did believe in the retention of habit from one generation to the next; but that was the second of two principles he set at the base of his theory. The other was a tendency of life itself to develop progressively more complex forms. That was, for him, the primary explanation of life's history. The retention of habits sallies out sideways, so to speak, from the main direction. To start things off, Lamarck had also to defend spontaneous generation, another unpopular move. Heat itself, he thought, could initiate life – it had happened more than once. Oddly enough, moreover, he disliked the very thought of extinction: Life's creativity shouldn't really allow its products just to disappear. So we have the paradox of a great student of extinct forms – Cuvier – who firmly denies any organic mutability beyond the species, and an evolutionist (as we would call him) who doesn't want to face the fact of extinction: for us, a primary piece of evidence that things have changed (Burkhardt 1997, pp. 128–136).

Third, we should ask whether Lamarck was as completely ignored as Cuvier declared. We have seen that there were others who, in a more limited fashion, did accept some kind of transformism as possible, or even probable. And, as we noted, Geoffroy, although he did not adopt his system (and thought the environment a more powerful influence than habit), did show some respect for the aging Lamarck. Later, as Cuvier could not have known, Lamarck's evolutionary views would prove influential among radical thinkers, especially in Great Britain, often in close connection with a radical reading of Geoffroy (Desmond 1989; and Chapter 6). Still, Lamarck certainly was not in his time an authoritative figure like Cuvier, or even Geoffroy, whose principle of the unity of composition, with its stress on homologies, did have its influence both at home and abroad.

After that hasty glance at Lamarck, let us look again in conclusion at our two chief actors, and in particular at the question of the relationship of their debate to what we now call "evolution." Directly, we said, there was really none. However, if we look at some of the basic concepts involved in the Cuvier–Geoffroy conflict in the light of Darwinian theory to come, some illuminating paradoxes appear. Darwin was certainly an adaptationist; so was Cuvier. But Cuvier's adaptation-ism by its very nature excluded the possibility of an origin of species, while Darwin's demanded it. Geoffroy's use of analogies, which Owen would distinguish as homologies, permitted the naturalist to envision the descent of one species from another. Yet in his reference to their differing principles in the *Origin*, Darwin elevates the principle of the anti-evolutionist Cuvier – the conditions of existence – to the status of a "higher law" than the unity of type – the expression by which he appears to be referring to Geoffroy's principle of the unity of composition, a principle that does at least leave open the possibility of descent with modification.

If we include Lamarck in our comparison, the situation is even more obscure. Lamarck is certainly an adaptationist, of an extreme variety, holding that organisms can adapt themselves, and therewith their descendants, to the circumstances in which they find themselves. Darwin refers to Lamarck fairly often in the notebooks, ridiculing his use of "willing," yet quoting with approval both the French theorist's stress on the continuity of development in nature, and the importance of his enunciation of fundamental laws of growth, and even, at one stage, stressing the priority of habit to structure.[16] Moreover, the Lamarck-ian concepts of use and disuse persist all the way to the *Origin* and beyond.[17] At the same time, under the guidance of natural selection, adaptation seems to return in a thoroughly non-Lamarckian fashion. It seems even more emphatically anti-Cuvierian, since for Cuvier it is the whole integrated organism that is, now and forever, adapted to its proper niche in nature. You couldn't alter little bits of it – make the wings a little more feathered or the bones a little airier – and come up with a new kind of animal. That would be to destroy not only the naturalist's profession, but nature itself. And although Geoffroy's principle of connections, entailed by his unity of type, will seem so useful to the evolutionist, it takes a subordinate place.

How does this come about? We will try to answer that question in more detail when we consider Darwin's (ambiguous) attitude to teleology. For now, we may try a two-part answer. Darwin's marginalia in his copy of the Geoffroy's *Principes de philosophie zoologique* are

---

[16] References are listed in Darwin 1987, p. 975. For his approval of Lamarck's achievement, see especially C 119. See Kohn 1980.

[17] See the subject index of the facsimile edition of the first edition of *Origin of Species*, and Ernst Mayr's introductory comments to the facsimile.

difficult to interpret; but in extracts he made from the *Annales d'histoire naturelle* for 1830, when the Geoffroy–Cuvier exchange was also published, it seems clear that at this point he did not find Cuvier's principle superior to Geoffroy's.[18]

That was in 1837, or earlier, when Darwin was already thinking in terms of the mutability of species, but before he had come upon the idea of natural selection. Twenty-two years later, in following his own advice to quote the passage, he again accepted Cuvier's subordination of "unity of form" to "conditions of existence," still taking the former to be due to inheritance (or descent) – as he had done earlier – but deriving the latter now from natural selection. It seems at first sight that this transforms altogether Cuvier's nice cozy principle of the adaptedness of

---

[18] Here is the passage in which Cuvier contests Geoffroy: " . . . ce n'est qu'un principe subordonné à un autre bien plus elevé et bien plus fécond . . . à celui des conditons d'existence, de la convenance des parties, de leur coordination pour le rôle que l'animal doit jouer dans la nature, voilà le vrai principe philosophique d'où découlent les possibilités de certaines ressemblances et l'impossibilité de certaines autres; voilà le principe rationel d'où celui des analogies de plan et de composition se déduit, et dans lequel, en même temps, il trouve ces limites qu'on veut méconnaître." Geoffroy 1830, in Le Guyader 1988, p. 165. As we mentioned earlier, Darwin wrote in the margin "Q" for "Quote!" He also wrote in the other, right-hand margin, "Cuvier's words," and placed an opening quotation mark before "c'est ne . . ." Then he seems to have underlined "bien plus élevé et plus fécond." And he placed lines to the margin of the passage down to the end of the text on p. 66. After "nature" and before the first "voilà," he noted: "[A]ll this will follow from relation," and at the right top corner of the page he wrote: "The unity of course due to inheritance." In the bottom margin of p. 65, he wrote: "I demur to this alone." What was he demurring to? David Kohn suggests that what Darwin objected to was taking Cuvier's principle alone, outside the transmutation context into which he was setting it. That must be right. It is clearer if we read the marking of the passage a little differently, taking the seeming underlining of "bien plus . . ." as marking the beginning of the passage "demurred to" and the mark after "nature" as a closing mark rather than just a caret – it seems to look like both – and then a separate caret beneath the line. Whichever way we read Darwin's notations, however, it is clear, that in the 1830s, Darwin already wanted to set Cuvier's principle, as well as Geoffroy's, in the context of transmutation. However, the comment "I dispute this" in the extracts from the *Annales* sheds yet another light on this matter. It does seem that at this period Darwin was skeptical about the superiority of Cuvier's principle. We are grateful to David Kohn and to Duncan Porter for assistance with interpreting Darwin's markings on his copy. Following Kohn's careful analysis of Darwin's early theorizing (Kohn 1980), we gather that in Darwin's insertion "relation" means what Kohn calls Darwin's sexual theory – that is, the belief that sexual reproduction tends, as such, to produce variability and hence adaptation, whereas "inheritance" refers to the principle of "reproductive inertia."

the whole organism, making it refer to an indefinite series of just slightly differing states of adaptedness of very partial and particular characters now or some time long ago to just slightly different circumstances. At the same time, it should be noted that, in his notebooks as well as in the correspondence, Darwin speaks as much *for* as *against* final causality, and we cannot take it that he did not interpret Cuvier's principle in a sense at least allied to that of its author. True, he transforms it from static to historical, and from holistic to piecemeal. Yet a decade earlier, he had referred to it as "the idea of design."[19] We will reserve this puzzle for Chapter 7.

In conclusion, we have been looking at the Muséum in the early decades of the nineteenth century, particularly at the debate of its two most celebrated members, first, in its historical setting, then in terms of the respective approaches of Cuvier and Geoffroy to the practice of comparative anatomy, and of their differing philosophies of science. Next, we noted Cuvier's construction of a "natural" classification. And finally, with a sideways glance at Lamarck, we asked about the status of transformist thinking at this time, particularly in relation to Darwin's comments on the debate itself. As far as their influence goes, we would, admittedly, have told a very different story if we had asked about the reception of Geoffroy in literary circles in France, where he became a symbol of progress against the reactionary, and too powerful, Cuvier, or about the reception of Geoffroy, along with Lamarck, among medical reformers in Britain (see Desmond 1989). As the titles of this and the next chapter indicate, we are looking at the nineteenth century "before Darwin." That is, if you like, a conventional and limited point of view; but all points of view are limited somehow, and this is the one we have chosen.

---

[19] Commenting on Owen (1992, lectures delivered in 1849), Darwin wrote: "Some think, falsely. I argue that conformity of plan is opposed to idea of design." Darwin 1990, p. 655. For further evidence of Darwin's attitude to teleology, see Chapter 7.

✦

CHAPTER 6

Before Darwin II

British Controversies about Geology and Natural Theology

British Geology in the 1830s

British science came of age in the first half of the nineteenth century. It was then that associations devoted to this or that particular scientific field began to spring up – the Geological Society, the Zoological Society, and the Linnean Society, to speak only of those closely connected with our subject. These organizations – more like elite clubs than modern academic societies – afforded a meeting place in which learned clerics from Cambridge and Oxford could consult with urbane gentleman-scientists such as the lawyer-turned-geologist Charles Lyell. The societies afforded no financial support to their members. They merely received scientific reports from the field – from people such as the field geologist William Smith, for example, whose lack of social standing and wealth precluded him from membership in the Geological Society, even though he was later lionized by it – and then engaged in formal debates about these reports. Institutions like these, and the networks of friendship and animosity they fostered, figure prominently as forums in which the issues raised in this chapter were discussed. They also formed the matrix in which the young Charles Darwin made his mark, first as the protégé of the Cambridge professionals John Henslow and Adam Sedgwick, and then of Lyell.

The decade of the 1830s, during the first half of which Darwin was circumnavigating the globe aboard the *Beagle*, was an especially important moment in the development of modern British scientific institutions. The British Association for the Advancement of Science (BAAS) was founded in 1831. At once more professional and more open to the general public than the discipline-based societies we have mentioned, the BAAS (which was modeled on a similar institution in Germany, the *Gesellschaft Deutscher Naturforscher und Ärzte*, founded in 1822) met in a different provincial city each summer. Its purpose was to assess the state of inquiry in the various scientific disciplines, and thereby to diffuse a modern, self-consciously scientific point of view to a middle class that was just then intent on every sort of reform. (The first English Reform bill was passed in 1830.) Women, excluded from associations such as the Geological Society, sometimes attended the sessions of the BAAS.[1]

It was in connection with the founding of the BAAS that the polymath William Whewell, Master of Trinity College, Cambridge, coined the term "scientist." He did so to mark a distinction between those who worked directly in this or that scientific field, or "section," as BAAS lingo had it, and those who, like himself, were dominantly methodologists and philosophers of science.[2] Certainly there remained a good deal of overlap between scientists so defined and men who thought about the nature and methods of science. John Herschel, for example, was both an eminent astrophysicist (as was his father before him) and author of an important treatise on scientific method, the *Preliminary Discourse on the Study of Natural Philosophy* (Herschel 1830). As the very title of Herschel's essay suggests, however, both roles – scientist and methodologist – were covered prior to Whewell's suggestion by the old name "natural philosopher." (Aboard the *Beagle*, Captain Robert FitzRoy and his crew referred to Darwin, albeit in friendly jest, as "ship's philosopher.") It is worth mentioning in this connection that Whewell seems also to have been the first person to use the term "philosophy of biology" (Whewell 1840, I, IX, 1).[3] He meant the phrase to refer to study of the underlying concepts and questions (such as "what is life?") and to methodological norms (such as appeal to final causes, of which

---

[1] Socially somewhat below the BAAS were societies promoting popular science to the self-improving working class in the hope of deterring them from radical politics, notably the Society for the Diffusion of Useful Knowledge (SDUK), founded in 1826. See Secord 2000, pp. 48–51.

[2] Whewell wrote in 1840, "We need very much to describe a cultivator of science in general. I should incline to call him a scientist" (Whewell 1840, I, Introduction, p. 113). The need for such a term had already been raised in the nascent BAAS as an alternative to "natural philosopher."

[3] References to Whewell 1840 refer, first to volumes in Roman numerals, then to books within volumes, also in Roman numbers, then to chapters within books in Arabic, and finally to sections in Arabic, preceded by ¶. Page numbers to Whewell 1840, when given, refer throughout to Whewell 1967.

Whewell enthusiastically approved) that were presupposed by scientists who worked in the life sciences, even if these professionals did not think much, or particularly well, about these norms on their own.

During the period in question, no science was developing as fast or provoking more intense debates than geology. During the first half of the nineteenth century, geologists were consolidating a picture of earth history, which, in its most general outlines, we still share. As early as 1723, John Woodward, an English geologist, had reported that

> terrestrial matter in France, Flanders, Holland, Spain, Italy, Germany, Denmark, and Sweden is distinguished into strata or layers, as in England; that these strata are divided by parallel fissures; and that there are enclosed in the stone and all the other denser kinds of terrestrial matter great numbers of shells, and other productions of the sea, in the same manner as in that of this island.
>
> Woodward 1695; quoted in Whewell 1840, II, XVIII, 2

By the 1840s, a variety of these strata of world-wide uniformity and succession had been distinguished by means of protracted and intense efforts at mapping, comparison, and correlation. (That is why one will find Jurassic formations today in North America, and not just in the Jura mountains; Cambrian remains in all sorts of places other than Wales; and the Devonian in many more places than Devonshire.) What undergirded this achievement was a steady move away from using mineralogical characteristics as criteria for further differentiating the late eighteenth-century division between primary, secondary, and tertiary strata – Greywacke [from the German *Grauwacke*] for the "transition" strata between the primary and secondary, for example, or Old Red Sandstone for the boundary between the transition and the secondary – toward using instead distinctive fossils, or patterns and proportions of fossils, as touchstones for identifying and differentiating strata.

As recently as the middle of the seventeenth century, the notion that fossils – the word means "dug up things" – are the remains of living beings was still a novelty. The first person to seriously argue for it was Nicholas Steno, whose views found support among the members of the newly founded Royal Society in England, especially the naturalist and natural theologian John Ray.[4] By the 1830s, however, no one doubted

---

[4] For an account of how fossils were construed at various periods, see Rudwick 1972, rev. 1976.

that, as the Oxford geologist William Buckland put it:

> the study of organic remains forms the peculiar feature and basis of modern geology, and is the main claim of the progress that this science has made since the commencement of the present century. [For] the surest test of the identity of time is afforded by correspondence of the organic remains.
>
> Buckland 1836, I, p. 92[5]

By this criterion, the primary, granitic rocks were almost by stipulative definition lifeless. Presumably that was because the early earth was too hot to sustain life. (Buffon and Cuvier shared the view that the earth had been slowly and progressively cooling off; most people did, at least until Lyell suggested otherwise in 1830.) The so-called Transition zones, dubbed "Paleozoic" in 1841 by John Phillips of the British Geological Survey, were characterized by the remains of marine invertebrates, as well as the occasional primitive fish. It was the decoding of these Transition zones that conferred on the fossil-based method of stratigraphy its high repute.[6] After Sedgwick had identified the Cambrian as the oldest stratum of the Transition, and Roderick Murchison, a London-based gentleman-scientist like Lyell, had postulated the Silurian as lying just above it, these two geologists buried their differences about the sharpness of the breaks between eras in order to win international recognition for the Devonian (which corresponded with the Old Red Sandstone) as the most recent Transition stratum.[7]

The Secondary formations lying above the Transition zones contained, among other things, the "coal measures," which in Germany and England had long provided much impetus for geological exploration.[8]

---

[5] In a similar vein, Whewell remarked, "All who have the slightest acquaintance with the recent additions to our knowledge of the earth, either in this or other countries, know well that the study of organic remains, more than any other single class of facts, has instructed and can instruct us on questions of the contemporaneous or successive origin of mineral deposits" (Whewell 1832, p. 104.)

[6] Buckland spoke uncontroversially when he defined geology as "comprehending the history of unorganized mineral material ... and the past history of the animal and vegetable kingdoms, and the successive modification which these two great departments of nature have undergone" (Buckland 1836, p. 159). For an account of early methods of establishing and sorting out the Transition, and of the rise of paleontology from mineralogy, see Laudan 1987.

[7] For a definitive and lively account of the debates leading to the recognition of the Devonian, see Rudwick 1985.

[8] On coal measures, see Laudan 1987.

These were the remains of the abundant plant life that had once flourished in a warm, tropical atmosphere. During the Secondary, land alternatively rose from and disappeared back into shallow seas. The Secondary seas were filled with fish, as well as with crustaceans, testaceans, and molluscs. The land was populated by reptiles, saurians, and, at the shore, amphibia.[9] There were a few mammals in the secondary, but these were small and uniformly marsupial. It was the Tertiary, by contrast, that was the age of mammals, placental as well as marsupial. Lyell's proposal to separate the three major strata of the Tertiary into the Eocene, Miocene, and Pliocene reflected his confidence in using fossil patterns as a "geological chronometer." For these concepts depended on whether a given Tertiary stratum contained fossils of only a few currently extant mammalian species, a middling quantity of them, or many such remains (Pliocene, from the Greek, *pleion*, many).

Lyell's efforts reflect the influence of Cuvier's and Brongniart's paleontological work on the Tertiary (see Chapter 5). Cuvier's conclusion was that in fairly recent times, by geological standards at least, a revolution or catastrophe – Cuvier freely uses both terms – did away with the huge mastodons, giant sloths (*megatheria*), and other quadrupeds that had once roamed the Pliocene. "None of the large species of quadrupeds whose remains are now found imbedded in regular rocky strata," Cuvier argued, "are at all similar to the known living species." (Cuvier 1817a, p. 87). What is more,

> the human race did not exist...at the epoch when these bones were covered up...I do not presume to conclude that man did not exist at all before these epochs. He may have then inhabited some narrow regions...However this may have been, the establishment of mankind in the countries in which the fossil bones of land-animals have been found, that is to say, in the greatest part of Europe, Asia, and America, must necessarily have been posterior not only to the revolutions which covered up these bones, but also posterior to the other revolutions by which the strata containing the bones have been laid bare.
>
> Cuvier 1817a, p. 131

Cuvier's inference greatly discomfited those in Great Britain who had hitherto been sure that whatever geology turned up was likely to help interpret, and so corroborate, the historicity of the Biblical story of Noah's flood. The discomfited were not restricted, moreover, to the merely pious. In England and Scotland, the Biblical text had long been

---

[9] Saurians were distinguished from reptiles in the literature of the time.

used to guide what eventually became professional geological inquiry. Thomas Burnet's *The Sacred Theory of the Earth* (Burnet 1691) attempted to show that Noah's flood, which he took to result from a collapse of the earth's crust into "the waters below," was responsible for the present, chaotic state of the earth. While denying Burnet's postulation of an originally homogeneous and featureless earth, Ray, in his *Three Physico-theological Discourses*, appealed to the same "waters below" to explain how marine fossils (which he acknowledged to be organic) had been transported through hidden springs to the tops of mountains (Ray 1693). Later, once the massive extent of these annihilations had been acknowledged, Ray used the Biblical approach to explain extinctions. In the light of Cuvier's work, however, the illustrious Buckland was now forced to acknowledge that

> the large preponderance of extinct species among the animals we find in the [Pliocene] caves and in superficial deposits of diluvium, and the non-discovery of human bones among them, afford . . . strong reasons for referring these species to a period anterior to the creation of man.
>
> Buckland 1836, p. 81

The clear implication was that the Biblical flood, which took place *after* the creation of humankind – it was supposed to be punishment for human wickedness – was not the cause of the sudden break that marked the end of the Tertiary, and that, contrary to the cosmic importance ascribed to it by Scripture, Noah's deluge must have been, as Buckland himself conceded it was, "gradual and of short duration" (Buckland 1836, p. 81), since it did not result in any increase of extinct species *after* the end of the Tertiary.

Growing scientific consensus about these matters soon fused with anxiety about their implications for the rather cozy view of the world that the British had entertained throughout the eighteenth century. The British establishment, Tory and Whig, repudiated both the would-be political absolutism of the Stuarts and the religious "enthusiasm" of the Puritans. They did so by combining respect for science (on terms laid down by the Royal Society) with a religious view of the world that was to be kept self-consciously moderate by the demand that revealed religion, with its potentially fanatical, even regicidal, appeal to faith, must be built upon and constrained by natural religion. Natural religion – the religion that all decent, reasonable human beings were presumed to be capable of arriving at and cultivating, even in the absence of revelation – was backed in turn by the argument for the existence of a creator

God from the design of the natural world.[10] Except among a narrow band of heterodox deists, accordingly, the design argument was generally taken in Britain not as a philosopher's replacement for irrational faith, but as an inducement to accept a moderate version of *revealed*, providential religion – and the moderate political order that it backed (Brooke 1979, pp. 39–64). For even if things physical and chemical are generally governed by mechanical laws – the "secondary causes" by which God governs his world from a distance – the coadaptedness of organs to organisms and of organisms to their environmental niches assures us, in the words of William Paley, the argument's most influential exponent at the end of the eighteenth century, that "the degree of beneficence . . . exercised in the construction of sensitive beings favors a ruling providence . . . our life is passed in [God's] constant presence" (Paley 1802, Ch. 26). No mechanistic law, Paley purported to prove in his *Natural Theology* (1802), could conceivably have produced the exquisite fit between means and ends that we find in the structure of every organic kind, or their webs of mutual dependence and support. Accordingly, each species must have been crafted by a Being whose benevolence had arranged a hierarchy of inanimate and animate things to provide a suitable habitation for each species, and especially for the species He valued most, human beings.

Under these conditions, any threat to the mutual support that science, natural theology, and revealed religion were supposed to afford one another might well reopen wounds that were assumed to have been healed. Unfortunately, geology posed just such a threat. For the discovery of "deep time," in which species loss was more than occasional, episodic, or marginal, threatened to push religion and science apart again. "The wreck of former life is awe-inspiring," Buckland noted with Cuvier, and human beings, when at last, not very long ago they arrived, arrived on an earth already scarred by geological upheavals (Buckland 1836, pp. 94,

---

[10] For natural religion and its relation to the argument from design in England, see Byrne 1989. The tradition did not run very deep in France. Montaigne had devastated an early version of it in its Ciceronian form in "Apologie de Raimond Sebond," the longest of his *Essays* (Montaigne 1965, pp. 138–351; 1580/1582/1588, 415–589). What there is of it is found in the eighteenth century, and is directly inspired by English works. The best instance is the nine-volume work of Abbé Noël Pluche, *Le Spectacle de la Nature* (1732–1750; see Pluche 1740, a bestseller that seems, according to Roger, to have helped provoke Buffon to produce a more naturalistic version of natural history (Roger 1989/1997, p. 75).

115).[11] This didn't sound much like *Genesis* I. To make matters worse, this growing split was occurring just when populist, evangelical religion, born in part of the miseries of urbanization and early industrialization, was formally severing its ties with the established churches – Anglican in England, Presbyterian in Scotland – in the name of an emotional, faith-based pietism, and when, in part to meet this threat, some parts of the clerical establishment were tempted to reject efforts to make science serve the cause of religion, including the tradition of natural theology.[12]

The geologically centered controversies that form the subject of this chapter took place within this context. The most important of these controversies divided uniformitarians from catastrophists. The same challenges also divided fixists from transmutationists. And they also divided those who sought to rework the argument from design to accommodate the new geology, such as the authors of the so-called *Bridgewater Treatises*, and, in a different key, Richard Owen, from writers such as Lyell, who sponsored a geological theory that was at once less theoretically threatening to standing versions of the argument and at the same time markedly indifferent to it in practice.

Given the high stakes, advocates of all these positions tended to invoke conceptual and methodological dicta as often as they appealed to empirical data. Indeed, one marked characteristic of these debates – and a large part of what makes them philosophically interesting – was the willingness of those who participated in them to use conceptions of what did or did not count as good science as a stick with which to beat opponents over the head. Lyell, for example, so vigorously protected his substantive opinions by appealing to a particular view of scientific methodology – the view promoted by his friend Herschel in his

---

[11] Recognition of the extent of extinctions also impressed the Göttingen biologists whom we discussed in Chapter 4. In his *Handbook of Natural History*, Blumenbach wrote: "Every paving stone in Göttingen is proof that species, or rather whole genera, of creatures must have disappeared. Our limestone likewise swarms with numerous kinds of lapidified marine creatures, among which, as far as I know, there is only one single species that so much resembles any one of the present kinds that it may be considered as the original of it." (Blumenbach 1806, in Blumenbach 1865, p. 283.) Blumenbach's interest in these facts was not at all in their consequences for natural theology, but in the support they afforded for the Göttingen school's rejection of the *scala naturae*. "Nature ... will not go to pieces," Blumenbach wrote, "if one species of creature dies out and another is newly created."

[12] On resistance at Oxford to natural theology, see Secord 2000, pp. 253–258; for similar tendencies at Cambridge, see Garland 1980.

Preliminary Discourse (Herschel 1830) – that any theory opposed to his own was branded not as bad science, but as not science at all. For his part, Whewell tried to help the cause of catastrophism and the cause of religion, both natural and revealed, by setting out in his *Philosophy of the Inductive Sciences* criteria for theory acceptance quite different from those advanced by Lyell and Herschel.

Uniformitarians and Catastrophists

The terms "catastrophism" and "uniformitarianism," like many names for things scientific in this period, were coined by Whewell. Catastrophism, in his usage, referred to the opinion that "great changes of a kind and *intensity* quite different from the common course of events, and which may therefore properly be called catastrophes, have taken place upon the earth's surface." (Whewell 1837, II, II, 8, ¶1).[13] Uniformitarianism, by contrast, referred to the claim that all geological changes must be, and presumably could be, explained in terms of forces whose nature and intensity could be observed in the present geological era. To all intents and purposes, uniformitarianism meant the doctrine advocated in Lyell's three-volume *Principles of Geology*, which appeared between 1830 and 1833. Catastrophism, by contrast, was a more heterogeneous category.

It might seem that Cuvier could pass as a paradigmatic catastrophist. His authority in using the methods of comparative anatomy to reconstruct whole animals from a few fossil bones had precipitated agreement about the extent of extinctions, their characteristic suddenness, and the youthfulness of the present geological era.[14] Prior to Cuvier's work with

---

[13] References to Whewell 1837 are, first, to volumes in Roman numerals, then to books within volumes, also in Roman numbers, then to chapters within books in Arabic, and finally to sections in Arabic numerals preceded by ¶.

[14] Cuvier wrote in his *Essay on the Theory of the Earth*: "Comparative anatomy possesses a principle which, when properly developed, enables us to surmount all obstacles. The principle consists in the mutual relation of forms in organized beings by means of which each species may be determined with perfect certainty by any fragment of its parts." Cuvier uses this method to determine that the animals whose bones he found in the Paris chalk "left their remains in the strata of our earth after the last retreat of the sea but one . . . All of the genera which are now unknown . . . belong to the oldest of the formations of which we are now speaking . . . I have been led to conclude that there has been at least one, and very probably two, successions in the class of quadrupeds previous to that which at the present day peoples the surface of the earth." (Cuvier 1817a, pp. 83, 98, 102.) See Chapter 5 for Cuvier's model.

Brongniart, it was not only possible, but customary, to interpret bone deposits in caves and digging sites as chance assemblages or the haphazard remains of the prey of omnivorous animals. Now that Cuvier had reconstructed these assemblages as the bones of single animals from extinct species, this position was no longer likely. Cuvier's view, however, carried commitments with it. As we saw in the last chapter, the validity of Cuvier's method of comparative anatomy depended on the harmonious "conditions of existence" exemplified by each type of organism. The fact that organisms belonging to kinds reconstructed in this way were no longer to be found on the earth clearly implied, then, that different circumstances must have once obtained in a given place, such as the Paris chalk, since, conversely, different circumstances had supported a different, but equally well-ordered array of kinds. In addition, the suddenness with which geological breaks occurred, together with the equally sudden changes in fossil patterns by which such breaks were identified, implied, in Cuvier's words, that "most of the catastrophes which have occasioned them have been sudden" (Cuvier 1817a, p. 16).

By these standards, Cuvier surely *was* a catastrophist. However, Cuvier was not Lyell's target. In part, this was because Cuvier usually refrained from Biblical and theological remarks, and especially from appealing to miracles to make good the loss of species at the boundaries between geological eras. (The habit of making such appeals was decidedly an English and Scottish failing, not a French one.) In addition, Lyell wished to retain Cuvier's assumptions about the perfect adaptedness of animal kinds (see Chapter 5). So he spared Cuvier by appealing to an analogy with the paradigm science of astrophysics. He distinguished between descriptive, or what Whewell sometimes called "phenomenological geology" [Whewell 1840, II, XVIII, Introduction], and geological dynamics. That is to say, Lyell viewed Cuvier's data as affording *descriptions* of the phenomena that Lyell's own geological dynamics would *explain* by currently operating forces rather than by appealing to catastrophes, or, worse yet, divine intervention.

It is just this distinction that Lyell did not think was respected by Cuvier's most prominent British follower, Robert Jameson, whose 1817 translation of Cuvier's *Essays on the Theory of the Earth* (from which we have been quoting) had put a Scriptural spin on the master's work, thereby preempting the search for a genuinely scientific dynamics. Nor did Lyell think that the distinction between description and causal explanation was sufficiently respected by Adam Sedgwick or by Buckland, respectively Cambridge's and Oxford's most eminent geologists. Sedgwick and Buckland conceded that the record of the rocks was playing havoc with existing interpretations of *Genesis*. Unlike Cuvier, however, they saw in catastrophes (at least until evolutionary speculations in the 1840s scared them off) the steady rise of ever more complex and perfect life-forms. "I think," Sedgwick told the Geological Society in 1831, "that the approach to the present system of things has been gradual, and that there has been a progressive development of organic structures subservient to the purposes of life" (Sedgwick 1831, p. 167). Geological and biological history was ordered enough, directional enough, even progressive enough, accordingly, to inspire both new readings of scripture and new versions of the argument from design. These were combined to show – in the Bridgewater Treatises, a series of volumes that had been commissioned by a repentant rake, the Earl of Bridgewater, to defend the argument from design in the face of the new geology – that a providential power had presided over and periodically intervened in the whole unfolding process in order to prepare the earth for human habitation.[15] Lyell was utterly contemptuous of this line of thought. However, it was not until Sedgwick and Buckland joined Whewell in acknowledging Lyell's distinction between description and explanation, and went on explicitly to *deny* Lyell's dynamical contention that present forces working at present intensities are competent to account for all the facts, that they became catastrophists *sensu stricto*. For it is *this* claim that Whewell called catastrophism, and the claim about which argument was soon raging (Whewell 1840, I, X, 3, ¶3).

Contemporary scholars generally follow Martin Rudwick in distinguishing three components in Lyell's "uniformitarianism," which neither Lyell nor Whewell bothered to separate. These are: actualism, gradualism, and steady-state recurrence or cyclicalism (Rudwick 1972 [rev. 1976]). Actualism is a methodological notion that forbids appeals in good science, including geology, to anything but forces currently in play, currently observable, and operating at roughly the same intensity as they do now. The idea was that invoking anything other than these forces must involve "mere hypothesizing," which would violate the empirical and inductive canons of modern science. In contrast to actualism, gradualism means that igneous forces such as volcanoes, and the earthquakes associated with them, as well as aqueous forces such as erosion, are at work more or less constantly, although certainly not as uniformly or as smoothly and gradually as a planetary system (in spite

---

[15] Buckland argues for this view in his Bridgewater treatise (Buckland 1836).

of the fact that the latter was in some respects Lyell's model). Finally, and most problematically, there is a notion of steady-state recurrence in Lyell's work. This means that earth history works on a repeatable cycle that extends as far back into the past as one can see, and so can reliably be projected as far into the future as one can imagine. In promoting this cyclical view, Lyell was explicitly updating James Hutton's model of geological dynamics, which had been put forward in a natural theological context before the challenges of contemporary geology had arisen and had been promoted by Hutton's commentator, William Playfair (Hutton 1795; Playfair 1802).[16] Lyell's sponsorship of Hutton's "steady-state" view was well motivated within his system. It gave him a lot of time in which known forces (actualism) operating at current intensities (gradualism) could achieve the effects Lyell attributed to them. (It is this heavy reliance on time, time, time, together with other points of agreement, that made Darwin Lyell's faithful disciple in spite of their differences.)

Lyell's argument, roughly summarized, is this. Elevation and subsidence combine to produce a recurrent seesaw between the surfacing and submerging of land on a worldwide scale. The land rises under the influence of volcanic forces, especially earthquakes; once risen, it sinks under the forces of erosion. The land also moves around the globe atop a molten core "like a hostess circulating at a party," to use Jonathan Hodge's memorable analogy (Hodge 1982, p. 7). Since land absorbs more heat than water, global climate warms when the greater mass of it happens to cluster around the equator. Conversely, when the preponderance of the land mass shifts toward the poles, as it did during the most recent ice age, global temperature drops. Sooner or later, however, these climatological eras will reverse, as they have in the past. The world, and everything in it, is, in the long run, constant.

Today, when we have what seems to us sufficient reason to affirm actualism, to question gradualism, and to deny that the earth runs in eternally recurrent cycles, it can seem that Lyell was blind to the highly contingent connection that exists among the three elements of his uniformitarianism. But this is not entirely fair. Lyell was aware of the distinctions. His aim was to *argue* that the forces countenanced by actualism, which can be observed to work gradually, are empirically competent to account for all known geological facts, and then to show that any other approach – catastrophism, for example – must involve arbitrary

---

[16] For Lyell's relationship to Hutton, see Hodge 1982, pp. 6–7.

hypothesizing that is neither called for by the facts nor in accordance with the actualist norms that have thus far characterized successful inductive science.

To see the power of this argument, as well as the lasting influence it exercised over the emerging professional field of geology (and methodologically over Darwin), we must recognize that Lyell's actualism and gradualism were not meant to suggest that the earth is now, always has been, and always will be a peaceable kingdom. On the contrary, Lyell's visits to Southern Italy, where he did much of his field research, made him aware of the devastating effects earthquakes and volcanoes could have on the terrain and on local populations. Lyell's claim was simply that over a long enough period of time, earthquakes and volcanic eruptions at intensities he and many others had observed would add up to the alternations of elevation and subsidence, the shifting of the relative position of landmasses on the globe, and the consequent climatological changes that would explain the data of geology. If Northern European geologists had not seen this, opting instead for unobserved instantaneous catastrophes, it was in no small measure because "they inhabit about a fourth part of the [earth's] surface, and that portion is almost exclusively the theater of decay" (Lyell 1830–1833, vol. I, p. 81). Italian geologists, with whose work Lyell was familiar, seldom found the sharp breaks between geological eras that their Northern colleagues had rashly postulated.

Here, as everywhere in Lyell, asking the reader to shift the spatial and temporal perspective from which a phenomenon is viewed carries the weight of the argument. Geologists scratch at the earth only at temporally and spatially discontinuous points. They then wrongly infer that the fossil record itself is discontinuous. "To a mind unconscious of the intermediate links in the chain of events, the passage from one state of things to another must appear so violent that the idea of revolution in the system inevitably suggests itself" (Lyell 1830–33, vol. I, p. 160). Geologists hastily conclude, for example, that a single proto-mammalian fossil – a small piece of a small jaw found in the Secondary at a place called Stonesfield, England, apparently belonging to a sort of opossum – marks the exact point of origin of a class of animals that would rule the earth in the Tertiary.[17] For Lyell, the correct inference, if there is one, should be that one fossil does not a special creation make. There must be many mammals still in the Secondary, and even if there are not,

[17] For an account of the Stonesfield mammal, see Desmond 1989, pp. 308 ff.

their disappearance should be attributed to unfavorable conditions for preservation, not their actual absence. Lyell's confidence on this point depended explicitly on Cuvier's assumption that, as Lyell put it, "there has never been a departure from the conditions necessary for the existence of certain unaltered types of organization" (Lyell 1830–33, vol. I, p. 161). Cuvier meant the conditions that would always sustain *all four* of his *embranchements*.

To be sure, Lyell did not deny extinctions at the species level. What he did was to shift the focus from *sudden mass* extinctions of whole biota to what is now called biogeography, especially to migrations induced by geological and climatological change (see Hodge 1982). It is an actual fact, caused by a currently observable force – a "true cause" (*vera causa*), to use a phrase that went back to the revered Newton's "rules of reasoning" about natural philosophy – that under the competitive stresses induced by geological and climatological pressures a species will be induced to migrate from its customary range. As a population moves, it must either displace or be displaced by a competing species. (Lyell is echoing Augustin De Candolle's insistence on the ongoing war of species against species – a quite different notion from Darwin's notion of competition within species.) The great plasticity and adaptivity, which, according to Lyell, God built into the fixed nature of most species, often enables them to adjust to new circumstances, and induces their competitors to adjust as well (Lyell, 1830–33, vol. II, pp. 24–28).[18]

It is true that under these pressures, some species are likely to go extinct. "In the vast interval of time," Lyell wrote, "the physical condition of the earth may by slow and insensible modifications have become entirely altered... One or more races of organic beings may have passed away, and yet left no trace of their existence" (Lyell 1830–33, vol. II, p. 179). Lyell conceded, accordingly, that now and then God might have to make good the losses – within a given genus, of course, species of all of which Lyell (cheered by the Stonesfield fossil) thought could be found at least *somewhere* on earth in every geological era.

It might seem to violate Lyell's actualism, of course, that no one has actually seen a new species come into existence. Lyell's response is yet another display of his characteristic way of arguing by asking his readers to change their angle of vision. It is not *likely* that anyone ever has or will observe this process. A new species will start out as a small, isolated population, perhaps a single pair. Its biogeographical

[18] On "preadaptation" in Lyell, see Hodge 1982, pp. 72–75.

spread – what today is called adaptive radiation – will be subject to all the contingencies that any population must meet (Lyell 1830–33, vol. II, pp. 175–179). To be sure, the likelihood of its increase and dispersal will be enhanced by the fact that after any given episode of extinction, God would have preadapted a new species to new "conditions of existence." Still, Lyell thinks it is unlikely that we will ever see the process of creation at work, hidden as it is in the nooks and crannies of the world. Like Newton's, Lyell's God is discreet. Except where He must, He works by secondary causes, and when He intervenes, He makes sure no one is looking.

Methodological Issues: Herschel versus Whewell

Lyell's uniformitarianism gained immediate support from Herschel, who saw in the *Principles of Geology* his own stress in sound inductive science on actually observable causes – as well as a laudable displacement of God to an honored place on the sidelines, so that in the space cleared for human agency, a purely professional science of "secondary causes" might flourish. In a wildly enthusiastic letter, some of which was published, Herschel went beyond Lyell's reserve by claiming that the latter had placed geology at last on a par with astrophysics, Herschel's forte. For Herschel, Lyell's geological principles are grounded in physical dynamics, which keep the earth's motions tuned to the "state of undisturbed equilibrium" that necessarily attends any system that rides smoothly atop a "wonderfully inert" liquid interior (Herschel to Lyell, February 20, 1836, in Cannon 1961, p. 306). Herschel's stress on a unified science also led him to suggest that eventually the occasional birth of a new species "would be found to be a natural in contradistinction to a miraculous process," as the precessions of the equinoxes, which in Newton's view, required occasional divine intervention, were later found to be purely natural (Herschel to Lyell, February 20, 1836, in Cannon 1961, p. 306).[19]

---

[19] Lyell permitted Babbage to reprint this portion of Herschel's letter to him, but, as he explained in a letter to Herschel, only because Babbage "introduced it as a counterpart of a passage from Bishop Butler [a respected natural theologian] and ... in such company no one could be otherwise than correct and orthodox" (Lyell to Herschel, May 24, 1837, reprinted in Cannon 1961, p. 312). This sentence is characteristic of Lyell's anxieties about both political and theological orthodoxy.

Herschel's enthusiasm for Lyell's work reflected his conviction that the very point of inductive science is to find the inviolable, uniform laws that structure the natural world, and, having done so, to identify the "true causes" (*verae causae*) that guide the workings of these laws. For Herschel, a causal realist rather than a Humean phenomenalist, laws are not mere constant conjunctions. Laws of nature are *laws* of nature because they are able to "sustain counterfactuals," as we would put it today. In stating an invariant relation, one must be able to say what can and cannot happen, not merely what does or does not happen.[20] When a genuine law looks as if it is being violated, one is not forced to take it back, as in the case of merely empirical regularities. Rather, one must look for what is masking its action. Laws have "nomic" necessity. Given their importance, Herschel argued that laws of nature cannot be established by "gratuitous theorizing." Although hypothesizing may sometimes seem to give plausible explanations by inference from an "arbitrary principle" – Herschel was sensitive about contemporary pseudo-sciences such as phrenology – hypothetical reasoning cannot find the *verae causae* that make the laws of nature into *laws*, and hence make nature into the kind of system that it is. Only when such a *vera causa* has been independently established by induction and analogy from observed regularities is one entitled to use deduction to elicit a law's implications.

In a display of enthusiasm for the *Principles of Geology*, Herschel cited Lyell's geological work, which had been published only a year before, and might hardly be expected already to count as secure knowledge, as a paradigmatic example of the eliciting of a *vera causa*. Herschel begins by throwing cold water on the "mere hypothesis" – Buffon's and Cuvier's, presumably – of a cooling earth:

> Some consider the whole globe as having gradually cooled from absolute fusion; some regard the immensely superior activity of former volcanoes, and consequent more copious communication of internal heat to the surface in former ages as the cause. Neither of these can be regarded as real causes in the sense here intended; for we do not know that the globe has so cooled from fusion, nor are we sure that such supposed greater activity of the former than of present volcanoes really did exist.

---

[20] Herschel writes, "Every law is a provision for cases that may occur, and has relation to an infinite number of cases which may not occur, and never will" (Herschel 1830, p. 36).

Herschel continues:

> A cause possessing the essential requirements of a *vera causa* has been
> brought forward in the varying influence of the distribution of land and
> sea over the surface of the globe: a change of such distribution in the
> lapse of ages by the degradation of the old continents and the elevation
> of new, being a demonstrated fact; and the influence of such a change on
> the climates of particular regions, if not the whole globe, being a perfectly
> fair conclusion from what we know... Here we have, at least, a cause on
> which a philosopher may consent to reason.
>
> Herschel 1830, pp. 146–7

Since the implication was precisely the one advanced by Lyell
himself – that to be a uniformitarian was to be scientific, and to be
scientific was to be a uniformitarian – it may readily be imagined how
poorly this argument sat with the Oxbridge establishment, and especially with Whewell. If accepted, Herschel's philosophy of science would
have discredited catastrophism as a scientific theory, thereby widening
the growing gap between religion and science by aligning catastrophism
with Jameson's and Buckland's Biblicalism. Moreover, Herschel's view
would make the ideal of inductive science so dependent on direct observation that it would call into question the very possibility of what
Whewell called "palaetiological" sciences – sciences of events whose
cause-effect relations could not be directly observed because they are
not cyclical, but are either sequential or unique. Accordingly, in reviews of Lyell's *Principles of Geology* and Herschel's *Preliminary Discourse* which appeared in the conservative *Quarterly Review* in 1831–2,
Whewell began to sound themes that would eventually blossom into an
alternative account of inductive science designed to accommodate the
historical sciences, including geology and its sub-science, paleontology.

In his review of the *Principles of Geology*, Whewell admitted that
Lyell was right to insist that one should try observable causes before
being forced to invoke others. Nor did he dispute that "Lyell has made
it probable... that some species do become extinct by the operation of
causes now in action." (Whewell 1832, p. 121). Nonetheless, Whewell
observed, "the manner and rate of extinctions is very far from being
settled by such reasoning" (Whewell 1832, p. 121). He went on omi-
nously to suggest that Lyell's view of extinction would seem to call for
an equally naturalistic view of speciation, as Herschel's speculation to
Lyell on this very point showed. Yet for such a theory of speciation,
Whewell notes, "there is not a bit of evidence" – hardly the thing for an

observation-obsessed inductivist to embrace (Whewell 1832, p. 129). Whewell then deployed this sort of table-turning strategy on Lyell's argument about the absence of evidence for the creation of species. By Lyell's own criteria, absence of evidence can hardly count as solid, anti-hypothetical inductive science. If anything, a genuine "philosophy of the inductive sciences" will testify mostly in favor of catastrophism. For given the paltry evidence currently available, it "would be rash to suppose, as the uniformitarian does, that the information which we at present possess concerning the course of physical occurrences...is sufficient to enable us to construct classifications which shall include all the past under the categories of the present" (Whewell 1832, p. 126). In the light of mass extinctions spread across quite different geological eras, Whewell thought it decidedly unscientific for Lyell to shoehorn the history of life into categories cobbled up to fit contemporary life forms. What was needed in the palaetiological sciences was a "natural" classification scheme that would follow the history of life rather than impose itself externally on it, as Linnaeus's taxonomic categories did.[21] In sum, uniformitarianism was for Whewell a constraint on the development of concepts that would eventually extend the reach of inductive science, and so could be criticized on methodological grounds as actually standing in the path of scientific progress!

In saying these things, Whewell was pointing to the weakness of Lyell's steady state cyclicalism, which he took to be based on a misplaced analogy between earth science and celestial mechanics. As for gradualism, Whewell was willing to grant Lyell plenty of time. But force – and therewith, cause – was another matter. Whewell was unwilling to say that various conjunctions of past circumstances *could not* on occasion have produced such extraordinary combinations of forces that even if these forces were confined in kind to those currently observable, they could not by their *joint intensity* effectively have sealed one geological epoch off from another, caused mass extinctions, and rendered the present, seemingly placid state of the earth but a poor guide to its chaotic past. Here, Whewell was invoking against Lyell (and Herschel) a strong sense of the necessary character of scientific knowledge, which in his opinion their uniformitarianism did not live up to.

---

[21] Whewell on Linneaus's artificial system, Whewell 1837, II, XVI, 4; on a natural method in botanical systematics, 1837, II, XVI, 6; on Whewell's preference for a "natural system" in zoology, based on physiology, tracing the issue from Aristotle to Cuvier, 1837, II, XVI, 6.

Yet, as Michael Ruse has observed, Whewell's main target was neither Lyell's cyclicalism nor his gradualism. He was after what might seem the most plausible part of Lyell's philosophy of science, his actualism – and after Herschel's methodological benediction of it as well (Ruse 1979, p. 47). It was not in the polite pages of a review of the *Principles of Geology* that Whewell would press this argument, however, but in two massive tomes on the *Philosophy of the Inductive Sciences* (1840) and two more on the *History of the Inductive Sciences* (1837).[22] Inductive science, Whewell conceded, is indeed about "the common process of collecting general truths from particular observed facts" (Whewell 1840, I, I, Introduction). But that does not mean that science is about eliciting laws from raw perceptual data. "Observation," he claims, "already involves reasoning." For, he chimes in (more or less) with Kant, "in order to be connected by our ideas, sensations must be things or objects, and things or objects already include ideas" (Whewell 1840. I. I, 2, ¶10).[23] It follows that "facts are bound together" ["colligated"] by the aid of suitable conceptions" (Whewell 1840, II, XI, 4, ¶1). It also follows that facts are theories writ small and theories are facts writ large. For "the apprehension of a fact implies assumptions which may with equal justice be called theory"; and "if proven true, such theories become facts" (Whewell 1840, I, I, 2, ¶11). Again, it follows that inductive science is about more than finding invariant laws, even *verae causae*. Because human beings are placed in the world "as interpreting the phenomena" they see (Whewell 1840, I, I, 2, ¶10), induction is about the slow development and articulation of concepts that are so apposite to their precise subject matters – gravity to mechanics, for example, or life to biology – that they result in knowledge of what necessarily makes the entities under discussion what they are. This is a stronger, more essentialistic – if

---

[22] It has long seemed odd to scholars that Whewell would be generally so laudatory of Herschel in his review of the *Preliminary Discourse* in the *Quarterly Review* when his own philosophy of science disagreed with Herschel's so fundamentally (Whewell 1831). It should be recognized that Whewell and Herschel had long been friends, having worked together to reform the British teaching of calculus and having shared a desire to expunge Paley from the required texts at Cambridge. So personal delicacy may have had something to do with it. It is more likely, though, that Whewell's philosophy of the inductive sciences was not fully worked out in the early 1830s, when he wrote his review of Herschel. See Fisch 1991, pp. 79–81.

[23] This is not a misprint. Whewell actually says that sensations must be, and not merely be about, things or objects. He says this in order to press the case against Locke's twofold mistake of thinking of sensations as ideas, and then reducing ideas to sensation. Whewell thinks of sensations as directly giving the solid objects to which our ideas refer. They are matter; ideas are their form.

you will, more *antique* – conception of necessity than Herschel had postulated. The concepts of objects governed by laws of nature underwrite the very existence of the objects to which these laws refer. For Whewell, nothing less than necessity of this sort can yield secure scientific *knowledge*. This conception certainly assumes that the world is more constant than we moderns assume. What is distinctive about Whewell in this tradition, however, and grounds his claim that he really is talking about *inductive* science, is his insistence that a new science will always come to maturity in and through the discussions and controversies that over time clarify its fundamental defining concepts. In spite of his reliance on Kant, Whewell's method is historical as well as transcendental.

To exhibit this process at work, Whewell, unlike Herschel, wrote a history of the inductive sciences. In this history, what Whewell called the "consilience of inductions" – the process by which a hypothesis is found to "explain and determine cases of a kind different from those which were contemplated in the formation of the hypothesis" (Whewell 1840, II, XI, 5, ¶11) – is treated less as a way of confirming or corroborating a law by applying it to new, unexpected cases than as a way of further determining the very *meaning* of the concept in question. Whewell says that "every step of an induction must be confirmed by rigorous deductive reasoning." Indeed, he asserts that "the hypothesis of the deductive reasoning *is* the inference of the inductive process" (Whewell 1840, II, XI, 6, ¶18)." Still, any piece of hypothetical deductive reasoning *might* be false. Only if and when reasoning from hypotheses makes ideas "leap together" – the literal meaning of "consilience" – into a necessary truth that has been made evident by a concept precisely apposite to its subject matter does it constitute a true induction.[24] At the risk of turning contingent claims into definitional truths, Whewell professes himself

---

[24] It is in this respect that Whewell took himself to differ from Aristotle. Both thought that science consists of necessary truths, both privileged pure theory over practical application, and both recognized that different sciences require different concepts and methods. Rather than thinking of the mathematical and logical form of a finished science as what makes it count as demonstrative knowledge, however, Whewell thought of (hypothetical) deduction as a means of finding and grounding, by intensively pursued discussions, the first principles of an inductive science. It was Aristotle's tendency to hug the shores of common sense – his tendency to ground what is on what "we" say – that, according to Whewell, had prevented him and others from establishing inductive science in antiquity; it was not syllogistic form, as myth would have it, that stood in the way of inductivism, but the foreshortening of inquiry caused by this defect. See Whewell 1850, pp. 63–72.

certain that "there is no example in the whole history of science in which the consilience of inductions has given testimony in favor of an hypothesis afterwards discovered to be false" (Whewell 1840, II, XI, 5, ¶11).

For Whewell, these abstract pronouncements had immediate import. The concepts that confer necessity on our knowledge of earth history and the history of life were, he thought, still in the process of emerging. They were doing so by way of a protracted discursive process (concentrated on quarrels about proper nomenclature[25]) which, if successful, would transcend the concepts of the most recently triumphant science. "Every attempt to build up a new science by application of principles which belong to an old one," he wrote, "will lead to frivolous and barren speculation." (Whewell 1840, II, XI, 2, ¶13). Accordingly, in Whewell's view, Lyell and Herschel were in no position to assimilate geology, palaeontology, or biology to concepts, models, and methods that had admittedly conferred certainty on physics, and perhaps chemistry, but were inappropriate to the historical or "palaeetiological" sciences. Whewell then used this principle to defend the inductive bona fides of catastrophism. It is true that he looked forward to a time when, as the relevant concepts emerged, the quarrel between catastrophists and uniformitarians would disappear into consensus, as had the quarrel between those stressing igneous and aqueous sources of geological change that had attended the birth of geology itself (Whewell 1837, II, XVIII, 8, ¶2).[26] Nonetheless, Whewell was confident that in any

---

[25] Whewell touches on this point in his 1831 review of Herschel's *Preliminary Discourse*, when he briefly laments Herschel's insufficient attention to the subject of nomenclature and concept formation. "Mr. Herschel," he writes, "has made many judicious observations upon ... the subject of nomenclature: perhaps, however, he has too much restricted his views to its uses as shown in the process of classification ... But the effects of a well-chosen nomenclature, or at least of a distinct terminology, extends as fully to the expression of laws as to the recognition of objects" (Whewell 1831, p. 391). The only other negative note in the review is about Herschel's Bacon-like paean to the practical benefits of science; Whewell thought of this utilitarian attitude toward science as a constraint on the progress of pure, theoretical knowledge and concept formation. In both *Philosophy* and *History*, these two issues are expanded in ways that suggest that Whewell was only beginning to formulate his own philosophy of science when he reviewed Herschel.

[26] Given his stress on the discursive causes of scientific progress, it is not surprising that, according to Whewell, the long quarrel between the "vulcanists" and "neptunists" was responsible for the development of a sophisticated geology that included both igneous and aqueous forces. It is taken by Whewell as a model for what might, and should,

final theory, the insistence of catastrophists on the unique, unrepeatable historicity of the processes with which earth historians deal would be strengthened.

Whewell's doctrine of concept formation as conceptual clarification also led him to become an early advocate of the methodological and conceptual autonomy of biology, and a determined anti-reductionist. The very concept of life, he argued, and of the distinctive "vital force" that governs it, is still very much in the process of formation. "What a vast advance would that philosopher make," he wrote, "who should establish a precise, tenable, consistent conception of life" (Whewell 1840, II, XI, 2). To be sure, life depends on chemical affinities, and even on electrical bonding and loosening. But that does not mean that life is essentially a matter of chemistry or electricity. "The vital powers . . . involve a reference to ends and purposes, in short, to manifest final causes" (Whewell 1840, II, XIII, 10, ¶6). Far from jettisoning teleology, the science of biology, when it matured, would finally succeed in fully clarifying a notion that was coeval with the historical birth of scientific reflection on life (Whewell 1840, I, IX, 6). (The concept of life had been crucial to Cuvier's exemplary work, for example, but, Whewell argued, had not yet been fully analyzed either by Cuvier or by anyone else.) When at last the concept of life has been articulated, Whewell intimates that the seemingly miraculous appearance of a new species will not look like the gross violation of the natural order that it does when conceptions dragged into "the philosophy of biology" from physics are taken as touchstones (Whewell 1845).

Reviewing the *History* and *Philosophy of the Inductive Sciences* in the *Quarterly Review* in 1841, Herschel accused Whewell of pressing "the doctrine and assumption [of "truths antecedent to experience"] to a far greater extent than [can be found except] perhaps in the writings of some of the later German metaphysicians." Politely, Herschel confessed

---

happen to catastrophism and uniformitarianism; both approaches will be recognized in a higher synthesis that clarifies fundamental, essential concepts. (Neptunism was the brainchild of Abraham Gottlief Werner, Professor of Geology at Freiburg, who thought of minerals as precipitates formed out of the action of water on more or less soluble chemical elements. In arguing for this position, Werner was explicitly contesting Linnaeus's notion that minerals arrange themselves into a hierarchy of invariant genera and species in much the same way that plants and animals do. The influence of Werner's "Neptunism" was quite large. It was diffused widely in other European countries by way of an energetic international movement of "Wernerians." See Laudan 1987 on this history.)

to "a leaning, though we trust not a bigoted one, to the other side" (Herschel 1841, p. 151). This is a little tendentious. Whewell's necessary ideas developed over time and after much discussion, rather like the analytic a posteriori truths about the essence of this or that natural kind favored by contemporary realists such as Saul Kripke or, on some occasions, Hilary Putnam – or, better yet, like Charles Sanders Peirce before them[27]). Nonetheless, the subsequent history of English and American philosophy of science has clearly backed Herschel's empiricism. John Stuart Mill went even further in this direction. In *A System of Logic* (1851), Mill supported empiricism by putting a markedly phenomenalistic, positivistic, anti-realistic spin on it. Whatever his failures, Mill succeeded in forging such a strong link between inductivism and empiricism that we can scarcely see how Whewell, who was decidedly not an empiricist, could plausibly think of himself as defending the *inductive* sciences at all. From an empiricist point of view, Whewell's "consilience of inductions" looks like a mere fecundity of hypothetical deductions, and so a violation of Newton's strictures against hypothesizing. That is, in fact, how most philosophers of scientific method today interpret Whewell, and the grounds on which they criticize him for rashly concluding that a theory can be turned into a scientific truth just because it is consilient with a large number of facts and laws. From this empiricist perspective, too, Whewell's insistence on an antique concept of scientific knowledge as restricted to necessary truths, the paradigm of which is an axiomized mathematical system, looks like a constraint on the growth of the very sciences that Whewell was the first to chronicle. In all such views, however, there lurks a certain empiricist dogmatism that fails to do justice to questions about what a *historical* natural science would look like.

Fixism versus Transmutation

Thus far we have been considering the challenges to theology, as well as the opportunities for philosophical adjudication, that geology and its sub-science paleontology afforded British scientists and methodologists

---

[27] Peirce defended Whewell's view of the inductive sciences by conceding the term "induction" to the empiricist opposition and defending what Whewell called induction under a new term of his own devising, "abductive inference." Peirce gave Whewell credit for stimulating his thought in his Harvard lectures, but hardly anywhere else. See Fisch 1991, pp. 109–110.

of science in the first half of the nineteenth century. Yet throughout this entire period, another way of dealing with the problems posed by "deep time" lay open. According to Lamarck's hypothesis of *transformisme*, expressed in his 1809 *Philosophie Zoologique* (which we discussed briefly in Chapter 5), lineages under pressure from a shift in their conditions of existence only *appear* to go extinct. Rather than dying out under the pressure of geological change, they can be presumed to have transmuted themselves into new forms by way of an inherent tendency in matter to complexify combined with the efforts of organisms to adjust to new environmental circumstances. As we remarked, it is a striking view for an evolutionist – not to have believed in extinction! Still, that was Lamarck's view.

On the face of it, it might seem that the geological controversies of the 1830s and 1840s would have afforded fertile ground for the diffusion of transmutationism in Britain. Under the pressures of a sometimes radically changing environment, why could not species that found themselves under stress have resisted extinction (and so have obviated the need for compensatory spurts of special creation) by transmuting themselves into something new and better adapted? Why could not scientists such as Lyell, who objected to God's scurrying about in what the transmutationist Robert Chambers called "an inconceivably paltry exercise of creative power" (Chambers 1844/1994; quoted by Sedgwick 1845, p. 1), have allowed the creator to work on the history of life by respectable "secondary causes," as he did in physics? For a variety of reasons, however, some empirical, others ideological, and still others – those most important to us – conceptual, the transmutationist alternative remained anathema to most professional British scientists, lay as well as clerical, throughout the 1830s and 1840s.

Lamarck's transmutationist hypothesis had been combined in France with Geoffroy's unity of type by some of Geoffroy's disciples. By the 1820s, this approach had crossed the channel by means of several comparative anatomists whose anti-Cuverian convictions had provoked them to look for evidence of transmutation at nodal points between invertebrate and vertebrate body plans. Among these advocates of transmutation – and supposedly of the French politics that seemed to travel with it (Desmond 1989) – were the disreputable Edinburgh anatomy lecturer Robert Knox (disreputable because he unwittingly received corpses from the notorious grave robbers Burke and Hare) and the only slightly more genteel Robert Grant, who before moving to the new University of London in 1826 as the country's greatest expert on sponges, corals, and other such invertebrates had served as Charles Darwin's mentor in Edinburgh.

For his part, Lyell had become anxiously aware of the staying power of the Lamarck–Geoffroy idea while he was in Italy and France in the late twenties; he did not pick it up from the likes of Knox or Grant (see Corsi 1978, especially p. 224). Driven by the new geology, Lyell read Lamarckism precisely as a hypothesis about how to deflect the problem of extinction and special creation. The temptations of transmutationism were strong enough to provoke Lyell to devote part of the second volume of his *Principles of Geology* to refuting it. Lyell laid on empirical objections aplenty. Although the extent of adaptability to change in animal kinds is presumably greater than we might earlier have realized, Lyell asserted that "there are limits beyond which the descendants of common parents can never deviate from a common type" (Lyell 1830–33, II, p. 23). The most ancient species that have been preserved, such as grains of wheat found in Egyptian tombs, he argued (in blithe indifference to his own insistence on time, time, time) are to all intents and purposes identical with those grown today. Apart from practical difficulties, moreover, even species that humans have been manipulating for millennia – dogs, cattle, cats, pigeons – are still capable of producing fertile offspring and of carrying on their kind. If there is any tendency in nature it is toward degeneration rather than the opposite. Domesticated animals revert to type very quickly after being returned to the wild. Hybrids, meanwhile – a presumed source of new species – are seldom stable. Given Lyell's geology, finally, there is very little *need* for new species. Migration can do most of the work in a world far less catastrophic than Cuvier's British followers imagined, and full of species with considerable preadapted plasticity. Nor was there any reason to think that life history is any more progressive than geology, as Lamarck assumes. The Stonesfield fossil shows that mammals existed in the Secondary. If this opossum-like creature had not been adapted to its conditions of existence, it would not have been found; if it was so adapted, there is no reason to think there had not been others of its class.

Objections like these were put forward in support of ideological worries later displayed in Lyell's discomfort at Darwin's *Origin of Species*, a work Lyell encouraged its author to write and that in many ways treads in his footsteps (see Hodge 1982; Bartholemew 1973). These concerns can already be found in private notebooks that Lyell began keeping on the species question in the early 1830s (Wilson 1970). Lamarck was identified by the British as a materialist because his presumption that matter has an internal tendency toward self-motion, complexification, and life contravened a notion that was basic to British creationism: that matter is inherently inert. Self-sustaining matter is what materialism had meant in Great Britain ever since the seventeenth century, and what it still meant in the early nineteenth, when French claims about upwardly mobile matter intersected with the rhetoric of revolution (Desmond 1989). Like Whewell, Lyell could not separate the general idea of transmutation from Lamarck's version of it. Nor could Lyell separate Lamarck's version of transmutation from his own personal horror at its materialist metaphysics and presumably its politics, or from his fear that any reduction of the distance between man and animal would undermine the moral norms on which civilized society was founded.

Apart from ideology, however, two difficulties made transmutation seem as internally incoherent to Lyell as it did to his catastrophist opponent Whewell, who pronounced the idea "extravagant" and "arbitrary."[28] First, both Lyell and Whewell saw Lamarck as radicalizing Buffon's doubts about the objectivity of genera and higher taxonomic categories by spreading these doubts to species. The transmutationist, Lyell wrote, "regards every part of the animate creation as devoid of stability and in a state of continual flux" (Lyell 1830–33, vol. II, p. 20). For his part, Whewell declared, "The hypothesis of transmutation denies the reality of species" (Whewell 1845, p. 50). Without an anchor in the reality of species, however, the very notion of a systematic zoology and botany (on which Lamarck, author of a well-regarded work in invertebrate classification, depended as much as anyone else) would be arbitrary and senseless. In *Geology and Mineralogy with Reference to Natural Theology*, a Bridgewater Treatise, Buckland took Lyell's demonstration that "species have a real existence in nature" as sufficient to refute "the argument advanced in support of transmutation of species." (Buckland 1836, p. 51).

The second supposed incoherence was about adaptedness. To sustain the doctrine of transmutation, Lyell argued, one would have to abandon Cuvier's principle of perfect teleology in the organization of living things, with its consequence of ideal adaptedness to their environments. For the presumed cause of transmutation, he claimed, is *lack* of adaptation. But it was Cuvier's principle that brought to light the paleontological

---

[28] Whewell called transmutationism "extravagant" in Whewell 1832, p. 126; he called it "arbitrary" in Whewell 1837, II, XVIII, 6, ¶4.

facts that the transmutationist sought to explain in the first place. So the supposed explanation undermines the very thing that is to be explained. In reviewing Lyell, Whewell mentioned this objection to transmutation with approval. "Because we now see nothing but adaptation in the organic world," he wrote in an ironic vein, "therefore the sole and universal agent has been the *lack* of adaptation" (Whewell 1832, p. 115, our stress).

We may infer from the positive response to Lyell's second volume that agreement on transmutation was well-nigh universal among parties who otherwise fought tooth and nail. In spite of this consensus, however, the smoldering issue of transmutation suddenly burst into flame in 1844 with the publication of a work called *Vestiges of Creation* (Chambers 1844/1994). The anonymous author, who was eventually identified as the Scots publisher and phrenological enthusiast Robert Chambers, went beyond Lamarck's and Geoffroy's British disciples by arguing that modern geology and embryology were now corroborating transmutationist conclusions that had hitherto rested solely on comparative anatomy. Geological history, Chambers wrote, correlates *exactly* with the successive appearance of zoophytes, radiata, mollusca, and articulata; then of fish; then of land animals, of which the first are reptiles; then of birds; and finally of mammals, of which the first are marsupials. This steady ascent through the geological eras, Chambers argued, is caused by a species-generating prolongation of the period of gestation under conditions of environmental stress. His evidence was the notion of recapitulation – the claim that in their development, organisms run through the adult stages of all earlier kinds. To make matters even more alarming for those who saw in transmutation a self-moving and self-perfecting matter, Chambers framed his account of the ascent of life within an account of cosmic evolution beginning with a universe that had formed itself out of a gaseous nebula and ending with a phrenological psychology that tended to reduce mind to brain. Cheekily, however, all these materialist-sounding commitments were set forth as benevolent secondary causes, and as evidence of God's design!

Vestiges was at first disdained by the scientific establishment. The accomplished comparative anatomist Richard Owen, who was sometimes called "the British Cuvier" – in part because he had bested Geoffroy himself in an argument about the platypus[29] – declined to

---

[29] A detailed account of this controversy can be found in Desmond 1989, pp. 279–288; for the controversy in France, see Chapter 5.

write a review. Owen had come to believe that biological order consists of systematic transformations of "archetypes" that are presumed to dwell as Platonic ideas in the mind of God, but that appear successively in time as each Cuvierian *embranchement* is structurally adjusted both to their environments and to the demands of phylogenetic progress. Perhaps just because he held this rather heterodox view, rather than a version of the traditional argument from design, Owen was vehemently opposed to evolution on naturalistic terms. He would not dignify Chambers with a response. Whewell, too, approached the matter gingerly, contenting himself with a dismissive talk at the BAAS meeting in 1845 and with repackaging some passages from his *History*, *Philosophy*, and his Bridgewater Treatise (Whewell 1833) in *Indications of the Creator* (Whewell 1845). Eventually, however, *Vestiges* created such a "sensation" that the scientific establishment was forced to make a reply (Secord 2000). It was Sedgwick who at last became sufficiently incensed by the growing influence of *Vestiges* to write a long, ill-tempered attack in *The Edinburgh Review* (Sedgwick 1845).

Vestiges, Sedgwick wrote, is "intensely hypothetical...The author has a mind incapable of inductive reasoning" (Sedgwick 1845, p. 2). So weak was the reasoning (and "charming" the style) that Sedgwick confesses to having thought when he began to read the book that a woman must have written it! (Sedgwick 1845, p. 3).[30] There were also factual errors. The history of life does not map onto successive geological eras at all, as Lyell and others had shown. "It is not true that only the lowest forms of animal life are found in the lowest fossil beds...The highly developed crustaceans are among the very oldest fossils" (Sedgwick 1845, pp. 30–31). Out once again came the Stonesfield fossil to show that mammals existed in the Secondary, and by extension that all four Cuvierian *embranchements* were represented in every era since the Cambrian (Sedgwick 1845, p. 43). Moreover, species do not grade into each other as we move through strata. They are always found fully formed and fully adapted or not at all (Sedgwick 1845, p. 32). As for embryology, higher kinds do not in fact recapitulate the adult stages of less developed organisms; embryos of all species are indeterminate until

---

[30] As Secord points out, it was for a long time rumored that Ada Lovelace, Lord Byron's daughter, was the "Vestiginarian." Secord also notes that Darwin's privately hostile reaction to Sedgwick's review was partly occasioned by the pain that his disparagement of women caused Darwin's wife Emma. See Secord 2000, p. 431.

they reach their own, well-adapted terminus (Sedgwick 1845, p. 75).[31] The alleged conceptual errors we have mentioned also made an appearance in Sedgwick's review. If each species "is perfect of its kind, and its parts are so related and fitted to each other that the existence of one part. . . . implies the existence of the rest," – Cuvier's principle again – it makes no sense to talk of transmutation as driven by lack of adaptation (Sedgwick 1845, p. 9). All this pious talk about secondary causes and design is mere cant anyway. For the very notion of a secondary cause excludes the notion of a self-moving, self-organizing, self-ascending matter.

If Lyell's refutation of Lamarck had afforded an occasion for the geological establishment to hide their disagreements, the effort to erect a common front against *Vestiges* seems to have initiated some substantive shifts of opinion among them. Sedgwick, who had earlier used catastrophism and even progressivism to support his creationism, now began to sound like a uniformitarian. In insisting against the author of *Vestiges*, for example, that species do not start and stop at the boundaries between strata – "Silurian species," Sedgwick wrote, "rise into the Devonian system" (Sedgwick 1845, p. 34) – he was tacking back toward a claim that his collaborator and rival Roderick Murchison had espoused and that he himself had earlier resisted. More important for us, however, were the effects of the consensus about *Vestiges* on methodological issues. Herschel's assumption that factual issues can be cleanly separated from conceptual ones, and that theory-independent facts can be appealed to to refute concepts such as transmutation, tended to displace Whewell's more complicated approach. Whewell seems to have sensed it coming. In a letter, he remarked of Sedgwick's review, "To me the material appears excellent, but the workmanship bad" (Whewell to R. Jones, July 18, 1845, in Todhunter 1876, vol. II, pp. 326–7).

---

[31] In Whewell 1837, II, XVIII, 6, ¶4, Whewell mentions that the Estonian embryologist Karl Ernst von Baer had demonstrated that strict recapitulation is false. Organisms in the course of their development do not go through adult stages of other, presumably less perfect kinds; at most there is an analogy between their immature stages, which disappears as the organism becomes the precise kind of thing it is. The passage is reprinted in *Indications of the Creator*, in the context of responding to Chambers. Sedgwick is alluding to this important discovery in his review. The point would be a key to Darwin's own rejection of strong recapitulationism. Chambers, for his part, cheerfully incorporated this correction into new editions without any sense that he was contradicting himself or undermining his case. See Secord 2000, p. 250.

We have earlier noted that in the same year in which *Vestiges* appeared, 1844, Darwin was writing out his own transmutationist essay (on the basis of a sketch he had made two years earlier). Not surprisingly, then, he read *Vestiges* in a state of great anxiety about having been scooped on the basic question of transmutation. He was relieved to find that Chambers had employed a different mechanism – and a bad one (Darwin to Hooker, January 7, 1845, in Darwin 1985–, vol. III, p. 108; on Darwin's anxiety, see Browne 1996, p. 461; Secord 2000, pp. 429–31). But Darwin was even more disturbed by the brutally negative reaction to *Vestiges* among those whose support for transmutation he hoped to acquire (Secord 2000, p. 431; Browne 1995, pp. 457–65). In a letter to Lyell, he confesses that he read Sedgwick's review "with fear and trembling" (C. Darwin to C. Lyell, Oct 8, 1845, in Darwin 1985–, p. 258). Surely this negative reaction was one of the chief reasons for Darwin's silence between 1844 and 1859; for his obsessive collection of facts, facts, facts (mostly, at first, about barnacles); for his scrupulous attention in drafting the *Origin* to methodological norms as framed by Herschel;[32] for his presentation of natural selection as a strictly uniformitarian mechanism for transmutation; and not least for repeating in a marginal notation to a later edition of *Vestiges* a warning he had penned to himself in his early notebooks: "Never say higher or lower" (Darwin 1990, vol. 1, pp. 163–4).[33]

In the end, however, all this caution proved insufficient. When the *Origin* was published its message was quickly assimilated to the image of transmutation that *Vestiges* had engraved on the public mind. All the same objections were trotted out again – mummified Egyptian pets and kernels of wheat, the Stonesfield mammal, and so on. When the tide turned in favor of "Darwinism," moreover, as we will see in Chapter 8,

---

[32] It has been argued that Darwin was influenced methodologically as much by Whewell as by Herschel, mostly on the ground that Darwin used hypothetical deduction in framing his main argument and used what looks like "consilience" to organize the last part of the *Origin* (Ruse 1975; 1979, pp. 179–180). Hodge has contested this view by construing Darwin as a Herschelian who was trying to complete Lyell's geology (Hodge 1982). In this chapter, we have argued that hypothetical deduction is not a sufficient condition for Whewell's notion of consilience. From this perspective, the difference between Herschel and Whewell bulks larger than it does on a more empiricist view of inductive science. If Darwin was influenced by Whewell, it was by a very watered down Whewell. See Chapter 7.

[33] Darwin wrote to himself in 1838: "It is absurd to talk of one animal being higher than another" (Darwin 1987, B 214).

it was a Darwinism muddied by the progressivism that "evolution" still suggests to non-expert English-speaking audiences, from which, if truth be told, Darwin never quite freed himself. This way of interpreting Darwin is not as odd as it might seem. Almost until the end of the nineteenth century, *Vestiges* was still outselling the *Origin* (Secord 2000, p. 526).

Geology and Natural Theology

In view of its importance in eighteenth- and nineteenth-century Britain, we cannot end this chapter without briefly mentioning the effects of the developments in geology and methodology (as well as the ever-looming specter of transmutation) on the argument from design, the lynchpin of the comforting British conviction that science is compatible with, and indeed smoothly leads to, a tempered faith. To all intents and purposes, this means judging the effects of these developments on the version of the design argument presented in Archdeacon William Paley's *Natural Theology* (Paley 1802). For by then, his popular and influential book had displaced earlier instances of its well-plowed genre, not only because it was well-written, but because of its innovative way of getting around objections to earlier versions of the argument, including the version that Hume attacked in his posthumously published *Dialogues Concerning Natural Religion*.

The oldest versions of the argument for the existence of a creator God from apparent design assume a *kosmos* thoroughly permeated with end-directedness. Although the argument can be traced as far back as Plato's *Laws* and Xenophon's *Memorabilia*, the version transmitted to Europe through Cicero's *De Natura Deorum* was the most influential. In this dialogue, which Hume used as his literary model for his *Dialogues Concerning Natural Religion*, the Stoic representative of the argument portrays the world as a single living substance integrated and organized by a "world soul." This being assumed, there was no pressing need to distinguish between external and internal teleology, or between the present use of a trait and its future utility; in the *kosmos* as a whole, no less than in the individual body, nature does nothing in vain. This organicist view was rendered problematic, however, by pluralism about substances and even more by Christian creationism, which made God transcendent to the world. Thomas Aquinas, in his "fifth way," still portrays elements and planets as moving in an orderly fashion because they are seeking an end-point, as an arrow seeks the target. But Aquinas infers that this could not possibly occur without the divine analogue of an archer.[34] By this route, teleology came to mean intentional design by an agent who is separate from his product. Final causes were thereby removed from the organism, their primary locus for Aristotle, and ascribed to a maker – with the result that living beings were assimilated to craft-objects, and ultimately to machines.

The mechanization of the world picture, with its attendant conviction that matter is inherently inert, put further stress on this argument. Descartes, Hume, Kant, and many others agreed that it is improper to think of merely physical objects as being directed toward ends at all; the laws governing their behavior are mechanistic. The most that could be said is that God had set up these laws as vicarious, secondary instruments. Hume's refutation showed, however, that there was no compelling reason to think even this; a whole raft of alternative hypotheses cannot be excluded. "For aught we can know *a priori*," Hume's character Philo says, "matter may contain the source or spring of order originally within itself . . . there is no more difficulty in conceiving that the several elements, from an internal unknown cause, may fall into the most exquisite arrangement than to conceive that their ideas, in the great, universal mind, fall into that arrangement" (Hume 1779, Part II). Why, moreover, should conscious, intentional thought, which is only a part of nature, be considered the cause of the whole? "Why select so minute, so weak, so bounded a principle as the reason and design of animals as cause of the whole universe?" (Hume 1779, Part II).

What was novel about Paley's *Natural Theology* was that it cleverly shifted the ground. Paley acknowledges, probably for reasons just alluded to, that "astronomy [and by extension physics generally] is not the best medium with which to prove the agency of an intelligent creator" (Paley 1802, ch. 22).[35] Instead, the entire work is devoted to cataloguing ingenious adaptations of organic parts and behaviors to organisms and ways in which organisms of one species are co-adapted to one another – cabbages and moths, for example, or flowers and insects. Mechanical laws, we are given to understand, could not conceivably account for such phenomena; their patent functional organization precludes that. Accordingly, Paley's argument is not the one Hume had rendered so vulnerable. It does not, for one thing, infer the designed nature of

[34] *Summa Theologica* pars. I, quest. II, art 3.
[35] Paley goes on in the same context to say that once the hypothesis of a creator has been accepted on other grounds, "astronomy shows God's brilliance."

the whole universe from its parts. Instead, it concentrates entirely on living things, remaining judiciously silent on whether the laws of physics must be construed as secondary causes. Since its *explanandum* is not the whole *kosmos*, moreover, its persuasive force does not rest on the sort of induction that Hume calls into question. Paley insists that a single case of organic adaptedness is sufficient to guide the mind to infer the existence of a creator-god. Piling on case after case, as he does, merely induces subjective assent to a conclusion that logically rests on each particular instance (Paley 1802, ch. 6).[36]

The problem was that Paley, who was still an eighteenth-century thinker, assumed a world very unlike the one being revealed by geology. Paley's species have all presumably been there since the dawn of creation. Even the rocks, he says in his famous opening gambit about stumbling over a watch in a stipulatively uninhabited and unvisited country, "may for anything I know to the contrary have lain there for ever" (Paley 1802, ch. I). Geology tells us, however, not only that species, and probably many higher taxa, have come and gone, but that the very rocks beneath our feet have not in fact lain there for all eternity. One who would still like to assent to Paley's conclusion must, therefore, interpret creation as an ongoing, episodic, interventionist sort of affair. Yet the perfect adaptedness that Paley catalogues is threatened by this interpretation. Clearly something must have gone wrong if these objects of divine craftsmanship didn't make it through geological changes. Moreover, the God who made them seems less than sublime as he scurries about the universe patching up weaknesses in his original design. The transmutationist hypothesis, with its rejection of perfect adaptedness, insinuates itself into these worries. "The *Vestiges* warns us, if any proof were required," wrote a reviewer of that work in *The Guardian*, an

---

[36] Commentators have often wondered why Paley does not address himself to Hume directly. It is clear that he read the *Dialogues*; he refers to them in *Natural Theology* itself (Chapter XXVI). Some have argued that the fact that "Paley utterly ignored Hume's objections and the tremendous popularity of his work offers further testimony to the commonplace but forgotten assertion that Hume was an isolated figure in his own time" (LeMatieu 1976, p. 30). Another, related, idea is that it would perhaps have elevated Hume's text to an importance it did not have if Paley had mentioned him. Still another notion is that Paley was too philosophically superficial to notice how devastating Hume's argument was (Hurlbutt, 1985). Another possibility is that Paley noticed Hume, but tried to make a popular argument that would draw attention away from him. A more likely reason is that Paley's argument is not Hume's, as we have argued, and so he does not need to mention it.

organ of the "half-popish" Oxford movement, "of the vanity of those boasts which great men used to make that science naturally leads to religion" (Church 1847; quoted in Secord 2000, p. 256; and Brooke 1979, p. 50).

The Bridgewater Treatises, eight volumes in all, were written by the leading lights of the day in order to diffuse these worries by rescuing the argument from design. They do so by deploying a variety of strategies. First, the relationship between organisms and their parts is sunk more deeply into the soil of up-to-date chemistry (William Prout), physiology (Peter Mark Roget, of *Thesaurus* fame), and anatomy (Charles Bell), thereby presenting nature not as an efficient cause in its own right, as Lamarck would have it, but as full of law-governed materials that are ingeniously shaped into adaptations by a designing God who works *with* natural laws, not in miraculous opposition to them. Second, the functional organization of organs to organisms and the adaptedness of organisms to one another – Paley's strongest suit – is shown not to be threatened by the record of the rocks. Here, Cuvier's principle, the importance of which we have stressed throughout this chapter, is used by Buckland to show, for example, that crocodiles are ugly "because they are fitted to endure the turbulence and continual convulsion of the unquiet surface of our ancient world"; and that "in all the characters of the gigantic, herbivorous, aquatic, quadrupeds we recognize adaptations to the lacustrine conditions of the earth during that portion of the tertiary to which the existence of these seemingly anomalous creatures appears to have been limited" (Buckland 1836, pp. 164, 112).

Whewell, for his part, developed a third line of argument. He had few tears to shed for Paley. Together with other influential figures, including Sedgwick, he had long campaigned, in fact, to have Paley's *Principles of Moral and Political Philosophy* removed from the list of required undergraduate readings at Cambridge; it was, they all argued, replete with the grossly simple sorts of utilitarianism that secularists such as Bentham were beginning to turn against a theocentric world-view.[37] Paley's *Natural Theology* was almost as well-read as his *Principles*, although it was not itself required for examinations. (In his *Autobiography*, Darwin, for example, portrays himself as having all but memorized it during his Cambridge days [Darwin 1958].) For Whewell, however, *Natural Theology* was doubly suspect because it projected onto the

---

[37] Sedgwick preached a famous sermon against Paley's moral philosophy in the Trinity College Chapel in 1932. It was the opening shot in a long war. See Garland 1980.

natural world the same simple-minded utilitarianism that the earlier book had projected onto the moral and political plane.

Whewell's alternative approach to the design inference is aimed at reducing the gap between mechanistic physical laws and the "purpose and provision" (Whewell 1833, p. 13) that we can hardly help ourselves from seeing in living things by suggesting that the laws of physics themselves are not as devoid of "purpose and prevision" as Herschel and Lyell, and a fortiori the transmutationists, might think:

> It will be our business to show that the laws which really prevail in nature are, by their form, that is, by the nature of the connection which they establish among the quantities and properties which they regulate, remarkably adapted to the office assigned to them; and thus offer evidence of selection, design, and goodness in the power by which they were established . . . The law may be the same, while the quantities to which it applies are different . . . Not only is there agreement between the nature of the laws which govern the organic and inorganic world, but also a coincidence between the arbitrary magnitudes which such laws involve . . . The length of the year [for example] is the most important and obvious of the periods which occur in the organic world . . . In the existing state of things the duration of the earth's revolution around the sun, and the duration of the revolution of the vegetable functions of most plants are equal [and] adjusted to each other . . . Such an adjustment must surely be accepted as a proof of design.
>
> Whewell 1833, pp. 9, 18[38]

Whewell's interest in this line of reasoning springs from his fascination with what today is called "the strong anthropic principle." Improbably enough, the myriad cosmological constants are just those that permit the eventual flourishing of beings like ourselves. "Can this be chance?" he asks (Whewell 1833, p. 29). Whewell supposes that chance cannot be the cause because he cannot imagine any possibilities other than either pure chaos or explicit design. That is a little short-sighted. But Whewell was enthusiastic about developing a new argument from design along these lines in large measure because he wanted to close the rift between mechanistic physics and finalistic biology without regressing to the ancient view that the world is a single living thing, and, at the same time, without flirting with materialist reductionism, to which he was deeply opposed.

---

[38] All quotations and citations from Whewell 1833 are from the 1839, 4th edition.

How conclusive did Whewell take his argument about cosmological constants to be? Whewell acknowledges with Kant that we cannot dispense with thinking about organisms in terms of final causes. His innovation is to extend the same pattern of thinking to physics itself. Whewell was as well aware as Kant or Hume, however, that his argument could not *prove* the existence of God, at least by way of an explicit, step-by-step argument. Instead, he deploys notions of "colligation" and "consilience" (which would appear in a more developed form in his *Philosophy of the Inductive Sciences*) to urge that

> when we collect design and purpose from the arrangements of the universe, we do not arrive at our conclusion by a train of deductive reasoning, but by the conviction which such combinations as we perceive immediately and directly impress upon the mind: "Design must have had a designer."
>
> Whewell 1833, p. 344

This inference, and the providential theism it embodies, are for Whewell conditions of experiencing the world, not logical inferences from it. "Such a principle," he writes, "... is not, therefore, at the end, but at the beginning of our syllogisms" (Whewell 1833, p. 344). It is a priori, although, as we have seen, for Whewell this means that it must have been discursively articulated and refined by scientific inquiry (Whewell 1833, p. 349). Still, he admits that his principle "can be of no avail to one whom the contemplation or description of the world does not impress with the perception of design" (Whewell 1833, p. 345).[39] This is not much of a proof.

Almost all *The Bridgewater Treatises* are pervaded by a fourth, markedly dubious line of argument that also figures in Whewell's contribution to the cause. In the era of modern geology, any design argument must account for changes over time. The *Bridgewater* authors tried to satisfy this demand by appealing to *future* utility to justify earlier states of earth history and the history of life. All the apparent waste and wreckage of former ages, we are now given to understand, is after all "full of purpose and prevision." This Panglossian way of thinking is

---

[39] Whewell could not help becoming entangled in a conflict between the voluntarism of his argument for God's intentions in setting up the constants by which his laws of nature do their "secondary" work and the strong necessitarianism of his own criteria for what counts as good science. Ruse 1977 argues for a contradiction (Ruse 1977); Fisch argues against it (Fisch 1991, pp. 156 and 200, n. 18).

at work in Buckland's claim about the coal deposits of the Secondary. "However remote may have been the periods at which these materials of future beneficial dispensations were laid up in store," Buckland writes, "we may fairly assume, beside the immediate purposes effected at or before the time of this deposition... an ulterior, prospective view of the future uses of man" (Buckland 1836, pp. 497–8). Buckland admits that "the production of a soil fitted for agriculture and the general dispersion of metals" were not "conducted solely and exclusively with a view to the benefit of man." Nonetheless, he is sure that they "were all foreseen and comprehended in the plans of the great architect of the globe, which, in his appointed time, was destined to become the scene of human habitation" (Buckland 1836, p. 4). Even in the philosophically unsophisticated Buckland, however, there is as much reticence about what the design inference proves as there is in Whewell. Like other post-Paleyeans, Buckland sometimes confined his "proofs" to the claim that the orientation of organic life toward human uses is "entirely in accord with the best feelings of our nature" and "the greatness and goodness of the creator" (Buckland 1836, p. 10). Again, not much of a proof.

Surprisingly, we find a rather large quotient of this sort of "prevision and provision" thinking even in Lyell. "It seems fair to presume," Lyell writes, "that the capability in the instinct of the horse to be modified was given to enable the species to render greater services to man" (Lyell 1830–33, vol. II, p. 42). The subsequent rise of a secular scientific philosophy that has lionized Lyell as a professional has made it difficult to see that he stood well within the orthodox matrix of natural law, natural religion, and moderate providentialism that had long characterized British thought (see Lyell 1830, vol. I, ch. 4–5). He was concerned to protect both the professional status of the emerging science of geology and the old religious consensus by forbidding arguments, such as those he saw in Jameson, Buckland, and Sedgwick, that made scripture a guide to science, rather than the other way around. To this end, Lyell resisted interpretations of the argument from design, which emphasize the blessings that one species confers on another. He insisted that "it is surely more consistent with sound philosophy to consider each animal as having been created first for its own sake" (Lyell 1830–3, vol. II, p. 99). Perhaps he meant to suggest by remarks like these that his own claims about the horse being created for subsequent human use, and other claims in this vein, should not be taken at face value. However that may be, Lyell's chief contribution to the effort to keep the

traditional theodicy in balance with the new geology is his effort to en-
sure, in uniformitarianism, that the earth stays as constant and abiding
as Paley presumed that it does, even if it changes a good deal more than
Paley thought, and his decision stipulatively to build a vast amount of
adaptive variability into each species in order to make sure that, come
what may, it, or at least its genus, stays adjusted to changing environmental conditions. Lyell's attempts to push the argument from design
to the margins of professional biological thought are genuine enough.
But they should not obscure the fact that the argument that he thus
marginalized is pretty much the same argument that Paley and the
Bridgewater authors had rendered suspect precisely by their attempts to
defend it.

CHAPTER 7

✦

Darwin

Interpreting Darwin

Prior to the twentieth century, no prominent scientist has left so detailed a record of his life and thought as has Charles Darwin. Between notebooks, marginalia, unpublished as well as published works, and a vast correspondence, he has provided an almost endless set of data for scholars to assimilate and interpret. Traditionally, most students of Darwin followed what Hodge has called the "Franciscan" view – that is, a view supported by his life and letters as Darwin's son Francis assembled and published them (Hodge 1985, p. 207), and also by the brief autobiography Darwin himself wrote for the benefit of his family (Darwin 1958). Since then, a great deal of further material has entered the public domain: Darwin's sketch of his theory in 1842 and his essay of 1844, the unfinished draft of his "big book," *Natural Selection*, the transmutation notebooks, and a more complete correspondence, carefully annotated, now available through the year 1863 (as of the present writing).[1] Darwin scholars, members of the "Darwin industry," also delve into other material still tucked away in the Darwin archives of the Cambridge University library. Further, as available material by Darwin continues to increase in volume, so, inevitably, do

[1] The Sketch of 1842 and the Essay of 1844 were published in Darwin and Wallace 1985, pp. 241–254. For the "big species book" see Darwin 1973. For the *Notebooks*, see Darwin 1987. The *Marginalia* can be found in *Darwin* 1990. *The Correspondence of Charles Darwin* is being published by Cambridge University Press. The first volume, covering the years 1821–1836, appeared in 1985 under the editorship of Frederick Burkhardt and Sydney Smith; vol. 11, for the year 1863, appeared in 1999, the senior editors being Burkhardt and Duncan M. Porter.

commentaries on it.[2] In this situation, it is difficult to step backward and consider briefly Darwin's stance and his impact in some of the areas we are concerned with. Acknowledging our debt to numerous editors and commentators, we will consider, first, the genesis of Darwin's theory, then the methodology he followed in its presentation when he eventually made it public, its implications for taxonomy, and for the place of man in nature. Most of these topics are, or have been, controversial. We will have to take a stand on them to the best of our knowledge and ability.

The Genesis of a Theory

In discussing Cuvier's insistence on the permanence of species, we referred to Richard Owen's Hunterian lecture of May 9, 1837, in which he announced that the "doctrine... of the transmutation of Species" was ". . . crushed in an instant when disrobed of the figurative expressions in which [it is] often enveloped; and examined by the light of a severe logic" (Owen 1837 in Owen 1992, p. 192; see Chapter 5). At about the same time – certainly by July, and probably even earlier – a new acquaintance of Owen, Charles Darwin, who had recently returned from a five-year voyage on Her Majesty's ship *Beagle*, was starting to write down in a series of notebooks his reflections on just such a theory, and on some speculations associated with it.[3] Indeed, the very specimens (of extinct mammals) that he had turned over to Owen for his analysis would provide a part of the data in which his theory would be grounded.

How did young Darwin, a member of the establishment if there ever was one, come to have dealings with what was then so heterodox a view? He was a scion of two eminent families, the Wedgwoods as well as the Darwins, and an accepted younger member of a prestigious Cambridge network that included figures like J. S. Henslow and Adam Sedgwick. He was also becoming a close friend of the eminent geologist Charles Lyell, whose work he had been studying with enthusiasm from the start

[2] In a single chapter on Darwin, it would be impossible to do justice to the secondary literature. Kohn 1985 represents a fair sample, and its bibliography is an excellent guide to the literature up to that date. We will have occasion to refer to some sources in the course of this chapter.

[3] References to the *Notebooks* are given by the letter designating the particular notebook, and the number in Darwin's pagination – e.g., B 32.

of the voyage. As we have seen, Lyell himself, like Owen in his Hunterian lectures, firmly rejected such theories as he had encountered of what we call evolution.

True, there were evolutionists in the London that Darwin had returned to. As Adrian Desmond has shown in his *Politics of Evolution*, radical thinkers concerned in particular to reform the hidebound medical and surgical professions were loudly proclaiming the virtues of Lamarck and of Geoffroy (whom they took to be a fellow transmutationist) (Desmond 1989). One of their heroes was Robert Grant, who had moved from Edinburgh to the newly founded University of London. During Darwin's two years as a medical student in Edinburgh, Grant had been his first guide in invertebrate zoology; he was a professed Lamarckian, and Darwin must have heard him proclaim Lamarckian ideas. However, Darwin avoided Grant in London; and, although he retained throughout his life a belief in the effect of use and disuse, he seems never to have been attracted to a thoroughly Lamarckian theory (or Lamarckian theory as portrayed by Lyell in *Principles of Geology*). Darwin did own a copy of Volume One of Lamarck's *Philosophie zoologique*, which he annotated copiously – but on a sheet attached to the back cover he wrote: "Very poor and useless Book" (Darwin 1990, p. 478). Of course, he also knew his own grandfather's poetical version of transformism. Indeed, what is usually called his first transmutation notebook is labeled "Zoonomia," a title used by the senior Erasmus, and it appears that young Darwin was in part following his grandfather's speculations in those first recorded musings.[4] However, it does not seem likely that the circle of scientists Darwin associated with in London in the late 1830s took Grandfather Erasmus's poetical ventures seriously as a contribution to science.

So how did the younger Darwin come to embark on so difficult and hazardous an intellectual journey? The Franciscan account, supported by Darwin's own autobiographical comments, was a simple one: As a student in Cambridge, he had found Archdeacon William Paley's argument especially attractive. You stumble on a watch somewhere; from its ingenious composition you infer a maker. Organisms are even more ingenious machines; you infer God as their maker. This is the old argument from design, which we discussed in the last chapter. In its insistence that organisms are designed machines, it is oddly reminiscent of Descartes's

---

[4] The heading "Zoonomia" occurs on B 1. Our interpretation here follows closely, if only cursorily, the argument of Hodge 1985.

argument for his *bête-machine*. Darwin, too, so the story goes, saw organisms as machines, but God was not needed to make them, in his view. Natural Selection could do the job.

A nice story, but it seems that Darwin's road to his developed theory was a little more circuitous than that. First, it is important to note that during the voyage, Darwin often referred to himself primarily as a geologist. He was, of course, wherever he went, an inspired and devoted naturalist, in the traditional sense of natural history: a tireless observer of a great variety of natural phenomena, from invertebrates through mammoths, as well as coral reefs and earthquakes. And he always retained the interest in generative processes, especially among invertebrates, that had been encouraged through his outings with Grant in Edinburgh. At a more comprehensive level, he had also been deeply influenced by his early reading of Alexander von Humboldt's work in physical geography, which promised an account of the phenomena of nature more all-embracing than the traditional natural history had usually aspired to be.[5] All these influences are not to be denied. Here, however, we want to remark only that his geologizing walking tour with Sedgwick, not long before he left for his new adventure, had made a great impression. And then the first volume of Lyell's *Principles of Geology*, which Captain Fitzroy of the *Beagle* presented to him before they sailed, captured his attention and directed a part of his efforts as observer and collector to the geological features of the places he visited. The receipt of the next two volumes during the voyage served to strengthen his interest and allegiance (Lyell 1830–1833).

Lyell, unlike Lamarck, willingly acknowledged the plain facts of the extinction of species, but refused to accept a transmutational beginning for their successors. Instead, he held that new species were created as perfectly adapted to the places left vacant in the order of nature. That thesis, Darwin found, was contradicted by many of the situations he observed in his travels. Thus, as Hodge has convincingly argued, Darwin came to the notion of transformism because, although firmly committed to Lyell's geological principles, he found Lyell's explanation of the origin of species inadequate (Hodge 1982). Introduced forms often proved better adapted to a given situation than their native cousins. And often closely related species appeared in very different habitats, and so on. So the criterion of adaptation – strictly, of instantaneous,

---

[5] Sloan (2001) argues that von Humboldt exercised the deepest influence on Darwin of any of his early mentors.

perfect adaptation – failed to account for the existence or persistence of many species that Darwin observed as he moved back and forth in South America and finally around the world. For example, he liked to talk about a woodpecker that lived where there were no trees to peck (Darwin 1859, p. 184).

Thus Darwin's early speculations – as the careful analyses of Kohn, Hodge, and Sloan, for example, have shown – were precisely not adaptational in their orientation (Kohn 1980; Hodge and Kohn 1985; Sloan 1985; Hodge 1985). When he does move to a stress on adaptation, moreover, he will of course soon move – as is well known – from the idea of perfect adaptation to that of relative adaptation: another difference from Lyell, as well as the natural theologians. He also noticed the resemblance between extinct and present forms, an observation neutral on the question of adaptation. He does, indeed, refer to adaptation early in the B notebook. He writes:

> . . . we see <<living beings>>. The young of living beings, become permanently changed or subject to variety, according <<to>> circumstance, – seeds of plants sown in rich soil, many kinds, are produced, though individuals produced by buds are constant, hence we see generation here seems a means to vary, or adaptation.
>
> *Notebook* B, 3–5

Thus we have adaptation as achieved by generation – but certainly not the instantaneous adaptation-at-creation envisaged by Lyell.

What, then, were Darwin's first steps toward a theory of transmutation? There has been a great deal of speculation as well as mustering of evidence on this question; we can only offer here a brief sketch of what seems to us the most plausible account (Hodge 1990). There is a passage in Darwin's "Ornithological Notes" of 1836 that shows him considering the possibility of changing species. Of the difference in varieties (not yet species!) of mockingbirds and tortoises on the Galapagos and of foxes in the Falklands, he writes: "such facts [would *inserted here*] undermine the stability of Species" (Barlow, ed. 1963, p. 262; quoted in Hodge 1990, p. 273). In his interpretation of this passage, Hodge conjectures that Darwin may be thinking of Lyell's reading of Lamarck, in which varieties are on their way to becoming species (Hodge 1990, pp. 259–262). He would be comparing this with cases on the mainland such as the Ant-birds and Oven-birds: congeneric species that turned up in very different habitats, or fossil and extant mammalian species with striking similarities. By the following March, John Gould had found the

Galapagos "varieties" to be in fact good species (including the finches since called "Darwin's," which Darwin himself had failed to distinguish or to label properly island by island). Owen had examined the mammalian specimens. Darwin could therefore go much further and acknowledge (if only to himself and in the privacy of his notebooks) that it was indeed completed transmutations, not their mere possibility, that he had been observing. It is ironic that Owen, at this time so firm an opponent of any kind of transformism, should provide important evidence in the development of just such a theory.

The first form that Darwin's transformist reasoning appears to have taken was again borrowed from Lyell, if only in disagreement with him. That was the theory of species senescence put forward by Giovanni Battista Brocchi, which the British geologist had described and (respectfully) dismissed (see *Notebook* B 35). Although this notion was soon abandoned, it did in a way set Darwin on a path he was to follow well into the next year, 1838, and to his "sexual" theory, in which he considered the fact of sexual reproduction to provide variety in the resulting offspring, and so to facilitate novelty and eventually the "birth" of new species (see Kohn 1980; Hodge 1982; 1990).

What is the interest of these early musings? David Kohn, in his paper "Theories to Work By," suggests that they gave Darwin training in formulating theories (Kohn 1980). More significant, perhaps, is the stress on generation. In agreement with Kohn's analysis, Hodge speaks of Darwin as "a life-long generation theorist" (Hodge 1985). Ever since his excursion into invertebrate biology with Grant in Edinburgh, Darwin had thought about problems of generation and inheritance. Even in his mature theory, the laws of heredity were evoked as necessary to the efficacy, or indeed the very existence, of natural selection.

One thing we can say for the Franciscan account of Darwin's development: As an undergraduate, he had been satisfied, even pleased, with Paley's argument for design, and he no longer was so. Paley, in his *Natural Theology*, had objected both to Erasmus Darwin's reliance on "generation" to produce new species and to David Hume's rather humorous suggestion, in the *Dialogues Concerning Natural Religion*, that the process of generation might be the source of the order we find around us. Hume, of course, was playing with thoughts about the origin of the universe itself, not simply of the multitudinous species that inhabit this portion of it. But the idea of generation as a process productive, not only of conservation, but of novelty, could well have appealed to the new, travel-seasoned Darwin. Although this may be rather playful speculation on our part, it would be delightful to think that the most superlative piece of philosophical literature in the English-language tradition might have had some influence on the author of the *Origin of Species* (see Hodge 1985, p. 211).

If, in 1838, Darwin did have something like a "theory to work by," however, it did need Malthus, and the concept of a struggle for existence, to trigger the notion of selection (always along with the laws of variation and generation and some effect of use and disuse) and so the theory we know as Darwinian.[6] What Malthus contributed, of course, to put it simply, was the recognition that population increase is likely to outrun food supply (the former increasing geometrically and the latter only arithmetically). Hence the struggle. This may not have been quite the instantaneous Aha! experience Darwin would later recall its being; but certainly in the months following his "reading Malthus for amusement," a new turn was taken in his thinking. It was only then that he would introduce and elaborate the analogy with artificial selection, with nature supporting slightly more viable small variations as the breeder selects the slighter more desirable variants to breed: an analogy that would lead him to decades-long inquiries of breeders as well as of naturalists and to many experiments of his own.

Still, it would be twenty years from his reading of Malthus to the time when Darwin would make his theory public. He wrote a sketch of it in 1842, and in 1844 in an essay to which he attached a letter to his wife Emma, instructing her how to find an editor for it in case of his death (see Darwin and Wallace 1958). It was also in 1844 that he confessed its existence to his friend Joseph Hooker at Kew (Darwin *Correspondence 3 [1844–1846]* 1987, p. 2.) But that was the year also in which, as we saw in Chapter 6, the notorious *Vestiges of Creation* burst upon the reading public. Darwin must have been horrified by the enraged response to it by older scientists such as Sedgwick. (See Chapter 6 for Sedgwick's reaction.) However careful he might be, any proposal of transmutation would surely be rejected by men he respected and admired. In 1846, Darwin began what turned out to be eight years' study of barnacles – work, he told Hooker, that fitted in beautifully with his

---

[6] Jean Gayon has stressed the distinction between the *hypothesis* and the *theory* of natural selection, a distinction that Darwin made explicitly only in the *Variation of Animals and Plants under Domestication* (Gayon 1998; Darwin 1868). While we recognize its importance, we can deal here only with the theory in its general sense, which includes the hypothesis.

species theory (*Correspondence 4 [1847–1850]* 1988, p. 140). In 1855, he invited a few friends to Down House (his country residence in Kent) to discuss the theory (Browne 1996, pp. 538–540), and after that he finally started writing his "big book" (unfinished, finally published in 1973 under the title *Natural Selection*) (Darwin 1973). That task was interrupted in 1858, when he received from Alfred Russel Wallace the manuscript of an essay he wanted Darwin to help him publish. It was an essay stating a theory of natural selection – not really exactly like Darwin's own, but close to it. Darwin was appalled. His friends persuaded him to have it presented at the Linnean Society along with an essay of his own, and a transcript of a letter he had written the year before describing his theory to the American botanist, Asa Gray (Darwin and Wallace 1958). That would establish his priority. In the meantime, he would write an "abstract" of the theory, anticipating more briefly (490 pages!) what the big book would have said. Its full title is: *On the Origin of Species by Means of Natural Selection, or the Preservation of Favoured Races in the Struggle for Life*. It was published on November 24, 1859, and people have been talking about it ever since (Darwin 1859).

Darwin put his case carefully, gently, and with much attention to "difficulties on the theory." Either intentionally or just instinctively, he pointed out, breeders selected the best available progeny for further breeding. Indeed, in Darwin's time, English breeders were doing this with startling success. Why should not something similar happen in nature? Animals multiply; unchecked, they would multiply in uncountable hordes. But the food supply in any habitat is limited (Malthus's argument). In the ensuing struggle, the somewhat better adapted will survive and leave progeny (the "favoured races") and the others will die out. That process is called Natural Selection. As Darwin himself put it:

> Whatever the cause may be of each slight difference in the offspring from their parents – and a cause for each must exist – it is the steady accumulation, through natural selection, of such differences, when beneficial to the individual, that gives rise to all the more important modifications of structure, by which the innumerable beings on the face of this earth are enabled to struggle with each other, and the best adapted to survive.
>
> Darwin 1859, p. 10

Of course it is difficult to envisage, say, a bird hatching out of a reptile's egg, or a highly complex organ such as the eye coming to be in bits and pieces. But it all happens very gradually, Darwin insists, through "numerous, successive, slight modifications" (Darwin 1859, p. 187). In considering the problem of supposing that the process of natural selection could gradually produce a complex organ such as the eye, he admits that this is a difficulty "insuperable by our imagination" (Darwin 1859, p. 187). Yet reason, if not imagination, can overcome it. And there are a number of kinds of evidence that suggest some form of transmutation, if not natural selection in particular. (We shall return to these shortly.)

That is a very crude sketch of what Darwin called his "one long argument" (Darwin 1859, p. 459). What we want to look into here, a little more closely, but nevertheless, and necessarily, sketchily, are some of the philosophical implications of Darwin's theory: its epistemic presuppositions and some of its consequences for the nature of biological explanation, as well as of the place of man in nature.

Epistemic Presuppositions

Darwin had first read Herschel's Preliminary *Discourse on the Study of Natural Philosophy* in 1831 (Darwin 1958, p. 67–8). Meantime, during the voyage, he had become convinced of the correctness of Lyell's uniformitarian approach in geology rather than that of his much more conservative mentor, Sedgwick. Lyell's uniformitarianism itself relied on Herschel's conception of scientific methodology. What was needed for good science, according to Herschel, hence also Lyell, and in turn Darwin, was a three-step procedure. First, one must find a cause efficacious today, so that one could study it in action, a *vera causa*. This is Lyell's actualism (see Chapter 6). Second, such a cause must be shown to be competent to effect the phenomena to be explained. Finally, it must be shown to be in fact – or with high probability – responsible for these phenomena. It is these three steps that do indeed govern the organization of the *Origin*.[7] The first four chapters show that (1) given the variation in nature (Chapter II) analogous to the variation that occurs

---

[7] Michael Ruse has supported his view that Darwin's methodology was both Whewellian and Herschelian by claiming that Darwin switched in the later chapters of the *Origin of Species* from Herschel's *vera causa* approach to Whewell's method of a consilience of inductions (without appeal to an existing cause). A recent presentation of his view, followed by Jonathan Hodge's defense of his adherence to the *vera causa* story, is to be found in Creath and Maienschein, eds., 2000 (Ruse 2000, pp. 3–26; Hodge 2000, pp. 27–47). Hodge points out, for example, that Whewell had not yet stated his theory of consilience in Whewell 1837, which is all Darwin could have known when first

in domesticated species, and has been controlled in a direction chosen by human beings (Chapter I), and (2) given the (Malthusian) struggle for existence that follows from the inevitable multiplication of members of the same species or of different species in the same habitat beyond the numbers their environments permit (Chapter III), (3) a process of natural selection must follow (Chapter IV). Thus, the existence of natural selection is deduced from phenomena whose existence the reader cannot but acknowledge. Chapter V deals with the (unknown) laws of variation that must underlie this process. Chapters VI through IX confront difficulties to be faced by the theory. Thus, together with the analogy provided in Chapter I, they defend its competence. Chapters X through XIII exhibit a wide range of data better explicable by a theory of descent with modification than by the traditional view of separate creations. This part of the book is aimed at going beyond hypothesis by arguing that natural selection is *actually* responsible for the appearance of new species.

In Chapter IX, the paucity of intermediate forms in the geological record had been presented as a difficulty; in the next chapter, however, that record, of the geological succession of organic beings, is presented as positive evidence for Darwin's theory. The two paleontological chapters conclude:

> If then the geological record be so imperfect as I believe it to be, and it may at least be asserted that the record cannot be proved to be much more perfect, the main objections to the theory of natural selection are greatly diminished or disappear. On the other hand, all the chief laws of paleontology plainly proclaim, as it seems to me, that species have been produced by ordinary generation: old forms having been supplanted by new and improved forms of life, produced by the laws of variation still acting round us, and preserved by Natural Selection.
>
> Darwin 1859, p. 345

Further support for the theory comes from the facts of geographical distribution (Chapters VI and XII), from the "Mutual Affinities of Organic Beings" – that is, from their classification, which, on the theory of descent, really does reflect a natural system, from morphology, from embryology, and from the existence of rudimentary organs (Chapter XIII). In all these areas, Darwin argues, his theory makes

---

working out his theory. But this is not decisive. See our n. 32, Chapter 6. The classic analysis of Darwin's method in the *Origin* is Hodge 1977.

intelligible matters that would remain mysterious under the usual hypothesis of separate creations. Thus it is probable that descent with modification, combined with natural selection, has been responsible for the coming into being, the persistence, and in many cases the extinction of the countless forms of living beings that have inhabited, and still inhabit, this earth. Finally, Chapter XIV provides a recapitulation of the argument.

In short, Darwin has argued that natural selection exists, that there is a good deal it could accomplish, and that an imposing array of facts in very diverse areas are probably due to its operation. In all this, of course, unknown laws of variation and of generation are at work: slight variations and their inheritance are the materials on which natural selection operates. Take, for example, the conclusion of the two chapters on Geographical Distribution. Darwin has assembled an imposing range of cases in which resemblance and difference between different species could reasonably be explained through migration and consequent modification. He concludes:

> On my theory these several relations throughout time and space are intelligible; for whether we look to the forms of life which have changed during successive ages within the same quarter of the world, or to those which have changed after having migrated into distant quarters, in both cases the forms within each class have been connected by the same bond of ordinary generation; and the more nearly any two forms are related in blood, the nearer they will generally stand to each other in time and space; in both cases the laws of variation have been the same, and modifications have been accumulated by the same power of natural selection.
>
> Darwin 1859, p. 410

Herschel's *vera causa* explanation is descended, via Thomas Reid, from Newton. But consider how different it is from Buffon's Newtonianism, which works precisely by outlawing the search for causes. Darwin's Herschel-like methodology also differed from other conceptions circulating in his own time – notably Whewell's notion of the consilience of inductions, which required no present-day causality, or any causality, or, a little later, from John Stuart Mill's canons of induction. If the later chapters of the *Origin* look like consilience, they are explicitly explanatory in view of the power of natural selection to produce the changes in question. True, Darwin contrasts the probability of descent with that of separate creations, and finds descent the more likely alternative. But such descent is said, over and over, to be produced through the action of natural selection on a series of slight, slightly improving, modifications. The real, present, perceptible power of selection must be acknowledged at the head of the argument.

If Darwin is presenting a classic *vera causa* argument, however, he is doing so recurrently in a form peculiar to himself, in oddly indirect and negative arguments. As Janet Browne puts it:

> He stood each problem on its head until it was not so much of a problem at all: sterility was not nearly so universal or complete as people supposed, and was merely a by-product of other changes; fossils were so infrequently preserved that of course there would be gaps in the record; every intermediate stage had some other purpose that we can only guess at today; animal habits could be inherited and acted on by natural selection.
>
> Browne 1996, p. 438[8]

His move, at least in the *Origin*, is to apologize recurrently for the apparent absurdity of his theory and to urge his readers to be CAUTIOUS in rejecting it because of its seeming improbability.[9] Although these adjurations are especially frequent in the "difficulty" chapters, they occur throughout, even in Chapter I on "Selection by Man." Here, after recounting his pigeon story – according to which all the varieties of the "Fancy" were probably descended from the common rock pigeon – he asks whether naturalists who admit such descent in domesticated species might not "learn a lesson of caution, when they deride the idea of species in a state of nature being lineal descendants of other species?" (Darwin 1859, p. 29). You think me rash to hold such a theory, he seems to be saying, but you shouldn't rashly throw it out. This odd turn of argument continues throughout the text.

Let us look at one instance of it in a little more detail. Take the case, often held against Darwin from that day to this, of the supposed gradual origin of "organs of extreme perfection and complication, in particular, the eye." He starts with an apology:

> To suppose that the eye, with all its inimitable contrivances for adjusting the focus to different distances, for admitting different amounts of light,

---

[8] Browne adds that this device is constant in Darwin's work; it seems to us especially conspicuous in the *Origin* (Browne 1996, p. 438).

[9] See, e.g., Darwin 1859, pp. 29, 103, 174, 190–192, 193, 195–197, 198–205, 306, 354, 407, 460.

and for the correction of spherical and chromatic aberration, could have been formed by natural selection, seems, I freely confess, absurd in the highest degree.

He then enumerates the conditions under which this staggering difficulty could be overcome: Reason tells him, he says:

> ...that if numerous gradations from a perfect and complex eye to one very imperfect and simple, each grade being useful to its possessor, can be shown to exist; if further, the eye does vary ever so slightly, and the variations be inherited, which is certainly the case...

(Note "variation" and "heredity": the two prerequisites for selection to take place) –

> ...and if any variation or modification in the organ be ever useful to an animal under changing conditions of life, then the difficulty of believing that a perfect and complex eye could be formed by natural selection, though insuperable by our imagination, can hardly be considered real.

Darwin admits that he does not know how a nerve becomes sensitive to light, but he suspects that any nerve that is sensitive at all can be so modified. How can we tell if such a gradual development has in fact happened? We cannot trace back its lineal ancestors, but sometimes we can look at "collateral descendants from the parent form" to see what gradations might have happened. Admittedly, we do not find much evidence in the Vertebrata (one of Cuvier's four *embranchements*), but we can do better in the Articulata. Here we do find a series with increasing complexity, from a nerve coated with pigment to a fully formed eye. Darwin continues:

> He who will go thus far, if he find on finishing this treatise that large bodies of facts, otherwise inexplicable, can be explained by the theory of descent, ought not to hesitate to go further, and to admit that a structure even as perfect as the eye of an eagle might be formed by natural selection, although in this case he does not know any of the transitional grades. His reason ought to conquer his imagination; though I have felt the difficulty far too keenly to be surprised at any degree of hesitation in extending the principle of natural selection to such startling lengths.

There follows the standard comparison of the eye to a telescope, which here does duty for Paley's watch: We know it to have been developed through human ingenuity, and take it that something similar obtains for the eye. But here is a challenge to the reader:

> ... may not this inference be presumptuous? Have we any right to assume that the Creator works by intellectual powers like those of man?

Along with the inevitable passage from Sir Francis Bacon, Darwin has set as epigram to this work a remark of Whewell in his Bridgewater treatise to the effect that the Deity works, not by repeated interventions, but by establishing general laws. That is the sentiment he is clearly appealing to here. There follows an eloquent account of how natural selection – as a substitute for the Creator – might produce that eagle's eye. We start with a layer, and then layers, of transparent tissue, sensitive to light, and slowly changing shape. Now comes natural selection:

> Further we must suppose that there is a power always intently watching each slight alteration in the transparent layers; and carefully selecting each alteration which, under varied circumstances, may in any way, or in any degree, tend to produce a distincter image. We must suppose such new state of the instrument to be multiplied by the million; and each to be preserved till a better be produced, and then the old ones to be destroyed. In living bodies, variation will cause the slight alterations, generation will multiply them almost infinitely, and natural selection will pick out with unerring skill each improvement. Let this process go on for millions on millions of years; and during each year on millions of individuals of many kinds; and may we not believe that a living optical instrument might thus be formed as superior to one of glass, as the works of the Creator are to those of man?

After this paean to natural selection, comes the admission that this theory could indeed be falsified:

> If it would be demonstrated that any complex organ existed, which *could not possibly have been formed* by numerous, successive, slight modifications, my theory would absolutely break down. But I can find out no such case (our italics).

How to prove an impossibility? We don't know the transitions – but here comes the warning:

> We should be extremely cautious in concluding that an organ could not have been formed by transitional gradations of some kind.

Then follows a discussion of organs that have been known to change function, including the swim bladder of fishes, and another case from the dear old barnacles. Conceivably, natural selection could have done all this. However, despite the caution not to abandon the new theory, it must be admitted that there are still difficulties:

> Although we must be extremely cautious in concluding that any organ could not possibly have been produced by successive transitional grada-tion, yet, undoubtedly grave cases of difficulty occur, some of which will be discussed in my future work.

These include, for example, neuter insects and electric organs. Of the latter, Darwin remarks, "We are far too ignorant to argue that no tran-sition of any kind is possible." Nor should we be disturbed by cases of evolutionary convergence, as we would call it. Darwin is "inclined to believe," as he puts it:

> ... that in nearly the same way as two men have sometimes independently hit on the very same invention, so natural selection, working for the good of each being and taking advantage of analogous variations, has some-times modified in very nearly the same manner two parts in two organic beings, which owe but little of their structure in common to inheritance from the same ancestor.

The section concludes with another reminder of our ignorance and an-other hymn to natural selection to compensate for it:

> Although in many cases it is most difficult to conjecture by what transi-tions an organ could have arrived at its present state; yet, considering that the proportion of living and known forms to the extinct and unknown is very small, I have been astonished how rarely an organ can be named, towards which no transitional grade is known to lead. The truth of this remark is indeed shown by that old canon in natural history of "*Natura non facit saltum.*" We meet with this admission in the writings of almost every experienced naturalist; or, as Milne Edwards has well expressed it, nature is prodigal in variety, but niggard in innovation. Why, on the the-ory of Creation, should this be so? Why should all the parts and organs of many independent beings, each supposed to have been separately created for its proper place in nature, be so invariably linked together by gradu-ated steps? Why should not Nature have taken a leap from structure to structure? On the theory of natural selection, we can clearly understand why she should not; for natural selection can act only by taking advan-tage of slight successive variations; she can never take a leap, but must advance by the shortest and slowest steps.
>
> Darwin 1859, pp. 186–194

What a strange mixture of ingredients in this argument! In his *Autobiography*, Darwin declares that he had worked "on true Baconian principles, and without any theory collected facts on a wholesale scale" (Darwin 1958, p. 119).[10] That is how a good British scientist was meant to talk in those days. But with that standard attitude, one expects caution in generalizing, let alone in speculating. Here is Darwin confessing his assent to a theory that seems "absurd in the highest degree," and warning his reader to be cautious – in rejecting it. Our very ignorance assists us in maintaining that caution. But then again, in the closing paragraph, ignorance alternates with knowledge (the transitions naturalists do find) in bolstering the theory. And then, over against the acknowledgment of immense difficulties faced by the theory (despite which one should be cautious about abandoning it), there are the positively dithyrambic passages in which natural selection is credited with almost miraculous powers. A century later, an evolutionist like Theodosius Dobzhansky was to remark without embarrassment, "Whatever we see, selection sees much more." Yet selection's discoverer had to build all kinds of hedges around such a statement. Why?

Part of the answer, at least, resides in the nature of Darwin's readership, of which he was well aware.[11] Whewell (who had been president of the Geological Society when young Darwin was its secretary), Sedgwick, even Lyell, not to mention Owen: Many of Darwin's readers among contemporary scientists would be highly skeptical of any attempt to undermine the stability of species. This would be in part because of the difficulty of practicing natural history if species didn't stay put. But it was also so, of course, because of the perceived threat to the special sanctity of humankind. That question Darwin avoided dealing with directly in the *Origin*. His *Descent of Man* would appear only some years later, in 1871. But the worry was there – especially after the furor over *Vestiges* in the 1840s. In those circumstances, an innovative thinker had better be apologetic, almost self-deprecating, though also firm in his approach. At the same time, it is clear that Darwin could not wholly contain his enthusiasm. From way back in the Notebook days, he had been convinced that he had "an excellently true theory"

[10] Compare, however, Darwin's remark to Lyell in 1860: "Without the making of theories I am convinced there would be no observation" (*Correspondence* 8 [1860] 1993, p. 233).

[11] See again the epigram to Darwin 1859, quoting Whewell.

(*Notebook* E, 119). After twenty years of experiment, observation, and fact-collecting, he was bound to feel exultation at what appeared to him its immense explanatory power. Admittedly, not all his readers saw it quite that way. Even those converted to some kind of transformism were inclined to hesitate at accepting natural selection. T. H. Huxley himself, Darwin's "bulldog," was never wholly convinced by that aspect of the theory.[12] But the point here is just to acknowledge the peculiar mixture of attitudes and arguments that characterize Darwin's "abstract."

Natural Selection as a Causal Agency: The Question of Teleology

Whatever his readers thought of it, however, it is clear that Darwin himself, when he refers to "my theory," has in mind not only descent with modification, but descent with modification mediated by natural selection. When we say that new species arise through natural selection, what kind of causality is intended? Nowadays, most of us think of natural selection as a collection of small changes, both internal and external to the organisms in question, such that some organisms will leave more offspring than others. Natural selection just is differential reproduction. If we expand that equation a little, we say that there are changes, both genetic and environmental, that combine to bring about a change in population structure. This is straight when-then, linear causality. So when we read Darwin's statement that "the law of Conditions of Existence" is a higher law than that of "Unity of Type," we are inclined to take the former "law" as expressing a set of necessary and sufficient conditions of the kind we are used to (Darwin 1859, p. 206).

It would be a mistake, however, to suppose that when Darwin refers to "the expression of conditions of existence, so often insisted on by the illustrious Cuvier," he has wholly eliminated the teleological connotation of that phrase. In fact, Darwin's own stance with respect to teleology, or final cause, appears to be irredeemably ambiguous. Let us look briefly at some of the evidence.[13]

In the *Notebooks*, which represent chiefly the years from 1837 to 1839, there are numerous passages in which the young theorist calls on final causes as contributors to the explanations he is constructing. At

---

[12] See, e.g., Desmond 1994, pp. 269–271. We will return to Huxley in the next chapter.
[13] For a thorough treatment of this question in the context of Darwin's religious beliefs, see Kohn 1989.

E 48, for example, he remarks: "My theory gives great final cause."[14] Moreover, when he first introduces his wedging metaphor, he writes:

> One may say there is a force like 100,000 wedges trying force <into> every kind of structure <of> in the oeconomy of Nature, or rather forcing gaps by thrusting out weaker ones. << The final cause of all this wedging, must be to sort out proper structure and adapt it to change.>>
>
> Notebook D, 135[15]

Yet in the very same period, in commenting on a work by one John McCullough published the previous year, he writes: "The final cause of innumerable eggs is explained by Malthus [Is it anomaly in me to talk of final causes] consider these barren Virgins!" – alluding, of course, to the old Baconian adage (Notebooks, Mac 58).[16] At one and the same moment, therefore, he seems to be both for and against teleology.

The same ambiguity appears two decades later. In 1855, Darwin discovers the principle of divergence, which reinforces selection, allowing it to work, not only in cases of geographic isolation, like that of island populations, but even sympatrically – that is, among individuals sharing the same environment. Of this new principle, in comparing the quantity of life in a fertile meadow as compared with a hillside blooming with heather, he notes, ". . . this is not final cause, but mere result from struggle." He adds: "I must think out this last proposition" (DAR 205.3: 167). Yet in the very next year, commenting on Huxley's attack on a teleologically minded naturalist, Darwin wrote to Hooker: ". . . to deny all reasoning from adaptation and so-called final causes, seems to me preposterous" (*Correspondence* 6 [1856–7] 1990, p. 147). Moreover, in the "Big Book," which he was writing between 1856 and 1858, Darwin explicitly invokes Cuvier's guiding principle in its teleological form:

> Finally, although within the same class species having a nearly similar structure may be adapted to the most diverse habits, I believe that each single species has had its whole structure formed through natural selection, either in ancient time for the good of its progenitors, or more

---

[14] According to its editor, David Kohn, this notebook was probably opened in October 1838 and completed by July 1839.

[15] The wedging metaphor was dropped after the first edition of the *Origin*.

[16] See also Darwin's marginalia in his copy of Whewell 1837 (Darwin 1990, pp. 867–8). In contrast to his later view that the Cuvier and Geoffroy principles are complementary rather than opposed, Whewell here fiercely defended the teleological side of the debate; Darwin expresses doubts.

recently for its own individual good . . . This conclusion seems to me to accord with the famous principle enunciated by Cuvier "celui des conditions d'existence, de la convergence des parties, de leur coordination pour le rôle que l'animal doit jouer dans la nature."

Darwin 1973, pp. 383–4[17]

" . . . ought to play," " . . . is to play": that is just the kind of teleological language Cuvier cherished and Geoffroy abhorred. But, looking back, it accords well with the *Notebooks* sense of "final causes." In the same sense, in 1874, we find Darwin's American supporter, Asa Gray, remarking, in an article in *Nature*, that Darwin has brought biology "back to teleology." Darwin replies in a letter, "What you say about teleology pleases me especially" (Gray 1963, p. 237; F. Darwin, ed. 1887, American ed., Vol. II, p. 267).

This would suggest that Darwin is taking "conditions of existence" in something like Cuvier's sense – in other words, as a synonym for "final cause." There is other evidence as well. By the time Darwin is writing the *Big Book*, and then, in 1858–9, the *Origin*, the old Cuvier–Geoffroy antagonism has become not so much a rivalry as an accepted duality. As Darwin puts it, it is generally agreed that "all organic beings have been formed on two great laws . . ." His own theory, he had noted earlier, "unites these two grand classes of views" (DAR 72: 161, March, 1847). And what were these two grand classes of views? As Whewell declared in 1855 (in an anonymous work called *Of the Plurality of Worlds*), leading comparative anatomists recognize that their investigations must be guided by "Unity of Composition, as well as the principle of Final Cause" (Whewell 1855, p. 211).[18] Darwin commented: "How exactly this agrees with descent and selection for [final cause of (deleted by Darwin)] advantageous structures" (DAR 205. 5:143). (Note that Darwin both refers to final cause and deletes it.!) Further, on a sheet attached to his copy of Owen's *On the Nature*

---

[17] The reference to Cuvier is repeated at the close of the chapter (p. 386).

[18] Note the change of view by Whewell compared with his position indicated in note 16. In Whewell 1840, Whewell had written that "whenever laws . . . of general analogies of one frame with another . . . are discovered, we can only consider them as the means of producing that adaptation which we so much admire . . . Our discovery of laws cannot contradict our persuasion of ends; our morphology cannot prejudice our teleology" (Whewell 1840 I, IX, 6, ¶14). (When Asa Gray remarked that Darwin had "wedded morphology to teleology," he was probably ringing changes on Whewell's remark [Gray 1963, p. 137].)

of Limbs (published in 1849), he wrote: "Final causes not sole governing principle" [final causes doing service for conditions of existence, as Cuvier would have wished] and "Some think falsely (I argue that conformity of plan is opposed to idea of design . . ." (*sic*) (Darwin 1990, p. 655).[19]

From all this, it is evident that by "conditions of existence," Darwin meant something different from our contemporary, non-finalistic, reading. For him, that phrase still held something of its original teleological intent. Not that he simply borrowed Cuvier's principle. He changed it from a static to a historical principle, so that for Darwin it was the "good" of this individual or of its progenitors that was involved, whereas for Cuvier, it was only this organism, whether living or extinct, whose "conditions of existence" were in question. For Darwin, it was the power of natural selection (combined, of course, with the laws of variation and the laws of generation) that had brought about the conditions of existence the naturalist was observing. For Cuvier, the origin of those conditions would have been either, in Lyell's phrase, "a mystery of mysteries," or would entail some pious hand-waving in the direction of the Deity (a move, be it said, that Cuvier made less often than his British admirers). Nevertheless, we cannot say – as many said and some still say – that Darwin abandoned teleology. He may have wanted to, sometimes, but he didn't. "For the good of . . ." discourse is constant throughout the *Origin* and elsewhere.

True, any divinely ordained teleology has certainly been abandoned. Natural selection is the cause. Nor is Darwin's teleology Aristotelian, as Cuvier would have been happy to acknowledge his variety of final cause to be. Aristotelian final causes require a fixed end-point. In Darwinian nature, on the contrary, it is past and present dynamic that dictates the immediate future. Moreover, chance has a more positive role in Darwin's world than in Aristotle's. Admittedly, even the Stagyrite recognizes that, however stable his universe and the kinds it contains, some things do happen "by chance." But not to worry! On the whole, in Aristotelian nature, things – and kinds of things – stay put just as they ought to do. In Darwin's world, in contrast, numerous slight variations just happen to be available when the environment changes. These make possible the gradually emerging new adaptations that lead to the origin of new varieties, of (what is the same thing, only more so) new species. Presumably,

---

[19] In notes on Baden Powell, Darwin also refers to Owen's "splendid sentences" proclaiming the inadequacy of design alone to explain the unity of form (DAR 71:45).

there are laws of variation, but since we don't know them, we can only say that the variations that do occur are random with respect to the needs of the organism. If they don't occur, extinction follows. If they do, thanks also to the (then) unknown laws of generation, the favorable variations will persist, and "evolution" will follow. In distinction from the stable Aristotelian (or Cuvierian) cosmos, with its fixed end-points and its ever-circling order, chance enters into a Darwinian world in two senses. There is the random occurrence of the variations that permit the origin of species, and there is the fundamental contingency of nature so envisaged: whatever happens, it might have been otherwise. As Stephen Jay Gould often emphasized, it is that radical sense of contingency that most centrally characterizes the Darwinian revolution in our intellectual heritage.

Where, then, is teleology in Darwin's view of things? Perhaps we should that say he retained whatever suggestion of teleology was inherent in the very concept of adaptation. That is a vague and weak answer, but it is the best we can do, and even so, we must also acknowledge his moments of doubt about any teleology at all.

Darwinian Species

In the *Origin*, summing up his argument so far in the chapter on "Variation in Nature" Darwin declares:

> From these remarks it will be seen that I look at the term species, as one arbitrarily given for the sake of convenience to a set of Individuals closely resembling each other, and that it does not essentially differ from the term variety, which is given to less distinct and more fluctuating forms. The term variety, again, in comparison with mere individual differences, is also applied arbitrarily, and for mere convenience sake.
>
> Darwin 1859, p. 52

Surely these are fighting words. If Darwin's tone is usually gentle, or gentlemanly, often even apologetic, here he seems to be reducing a work entitled "On the Origin of Species..." to a declaration that his supposed subject-matter does not exist at all. If the term species is arbitrary, what are naturalists doing when they make distinctions between specimens that seem "related" or "unrelated," and when they worry about whether certain forms represent "good species" or are merely varieties? What, indeed, was Darwin himself doing during the *Beagle* years, or afterward, when he asked specialists in London to look into the sorting of his specimens? What did he think John Gould was doing when he found that the mockingbirds – and finches – of the Galapagos Islands were distinct species rather than just varieties? Darwin himself had made many similar distinctions, as, for example, between two "species" of *Rhea*. And what about his long and laborious investigations of the barnacles, which, he told Hooker, confirmed his faith in his "species theory"? Either he is really undermining it – finding that "species" is only a term arbitrarily applied – or he is in a situation like that of Epimenides the Cretan, who said, "All Cretans are liars."

Yet that cannot really be how Darwin meant us to take his seemingly skeptical remarks. It was not his custom to denigrate the work of other naturalists. On the contrary, he sought their advice, along with that of breeders, and of any one who might give him information bearing on his own work in an amazing number of areas. And he thought his own species work would help, not hinder, them. "Systematists," he tells us in the concluding chapter of the *Origin*, "will be able to pursue their labours as at present," but they will no longer be burdened with "the vain search for the undiscovered and undiscoverable essence of the term species" (Darwin 1859, pp. 484–485). In Darwin's time, a definition would characterize a species as of necessity eternally just thus-and-so. As latter-day evolutionists would put it, systematists need no longer worry about finding a fixed definition for the category – that is, the *concept* of species. They can just go on sorting groups of organisms – that is, taxa, the entities themselves, rather than our concepts of them – as they have done all along.[20] There was no threat to Linnaeus's binomial nomenclature, for example. It seemed to work as well after as before Darwin's proposed innovation.[21]

Admittedly, it does seem a little odd to go on classifying one-knows-not-what. Indeed, controversies about the meaning of the term "species" have persisted long after Darwin's day. We shall return to them in Chapter 10. What Darwin did definitively do, however, was to alter radically the significance of classification. Earlier naturalists had been worried by the difficulty of constructing a natural, rather than a merely artificial, system. Linnaeus himself recognized that the distinctions he

[20] See Beatty 1985, pp. 265–281, which provides an interesting resolution of the problem, and compare it with Beatty's earlier, more skeptical essay (Beatty 1982).

[21] Some recent writers have denied that Linnaean classification is acceptable in a fully Darwinian system. We cannot go into this topic here, but see Chapter 10 for the relevant background.

used were to some extent arbitrary. Yet he was sure there was a natural system that the Creator had ordained. How was one to know when one had found it? Buffon, as we saw, made the concept of species historical, but in a limited way: Each species consisted of ancestors and descendants related through generation, but each species just was what it was, distinct from others, except in some cases of "degeneration." In fact, in reaction against Linnaeus, Buffon distrusted classification altogether. Cuvier, of course, was a great classifier, and even Geoffroy was so to some extent, although he was more interested in constant features across species than in sorting out species themselves. And Blumenbach and his followers at Göttingen were looking for a natural system built on the ruins of the Great Chain of Being. But in all these cases it is difficult to say what is meant by two species being, say, more closely related than some others. For Darwin, on the contrary, being related has the same meaning for collections of other organisms that it has for human beings. For groups of human beings, it is possible, on principle, to construct a family tree. After Darwin, the whole multitudinous and variegated biota on and over and below this earth forms a family, not nearer to or farther from some ideal type, but related to one another through generation, just as our families are – though, of course, in a much more complex pattern, of which we can perceive only scattered bits. After all the vain efforts of earlier classifiers to find the basis for a natural, rather than an artificial, system, Darwin has provided, it seems fair to say, the first grounded, realistic basis for systematists' work. As he put it at the close of the chapter on Natural Selection:

> As buds give rise by growth to fresh buds, and these, if vigorous, branch out and overtop on all sides many a feebler branch, so by generation I believe it has been with the great Tree of Life, which fills with its dead and broken branches the crust of the earth, and covers the surface with its ever branching and beautiful ramifications.
>
> Darwin 1859, p. 206

Admittedly, this change in the import of taxonomists' work was not instantly acceptable to every one. Lyell accepted it only reluctantly, and Owen not at all. And, as we mentioned earlier, many people were shy of accepting natural selection as a major means of species change. But transformism as such, the notion that species descend from other somewhat altered species, took hold surprisingly quickly. As early as 1866, for example, the BAAS appeared to find it acceptable.

One obvious result was a change in the meaning of the term "homology." Geoffroy had dealt at length with what he called "analogies." Owen rechristened them "homologies," and distinguished them from analogies. By homologies he meant structures built on the same plan, examples of the same "archetype." Analogies, in contrast, are similarities of function without the same underlying identity of structure. What Darwin did, as he himself put it, was to make the archetype into a real (formerly) living ancestor, so that homology becomes a historical concept: Homologous structures are signs of common descent (Darwin 1990, p. 655).[22]

Back in the 1830s, in the first transmutation notebook, Darwin had written in his copy of the Cuvier–Geoffroy quarrel, "unity [of form] due to descent." And twenty-two years later, in the *Origin*, when he refers to their respective principles, he says again: "... unity of type is explained by unity of descent" (Darwin 1990, p. 301; 1859, p. 206). Analogies, on the contrary, provide examples of what we call convergent evolution: They do the same work, but their ancestries are different. Thus, a bird's wing and a bat's wing are homologous: They both exhibit the structures common to vertebrate limbs. But a bird's wing and a butterfly's wing, though they are both instruments of flight, are only analogous; they do not share a common structure, and that means they do not share a common ancestor – unless, of course, one is going back to the beginning of life (before there were wings of any sort) – and takes life to be monophyletic – that is, to have a unique origin from which all later beings are descended. Inquiries about homology do not usually go that far! Indeed, if they did, they would be useless. Darwin himself was careful not to speculate about the origin of life; he did remark that there might be several kinds at the beginning of the story, or maybe even one (e.g., Darwin 1859, p. 484). What concerns us here is simply the point that, thanks to Darwin, we think of homology as a historical concept. It is difficult for us to think of it in pre-Darwinian, non-historical terms.

Darwinian Anthropology

Until now we have been dealing chiefly with relationships between philosophy and biology – the subject matter with which natural historians or their successors, comparative anatomists, concerned themselves. We

---

[22] This is a remark on Owen's *On the Nature of Limbs*. Compare Darwin 1859, p. 35.

have largely skirted the question of our own place as a species in that natural scene. Given the reception of Darwin's work, and, indeed, the nature of his own speculations (chiefly in the *Notebooks*), we must deal with that question, if only briefly. Back in the early notebook days, Darwin had filled two separate notebooks with remarks, the first one labeled "This Book full of Metaphysics on Morals and Speculations on Expression," the second described as handling "Metaphysics and Expression" (*Notebooks* M and N, *passim*). They deal extensively with the presumed kinship of human beings with other animals, particularly as evident in the expression of emotions. That is a subject Darwin will return to explicitly much later, in 1872 (Darwin 1872). Both here, and in other notebooks as well, he confessed the materialistic (but not atheistic!) implications of his new theory. Thus in *Notebook C*, for example, wondering whether thought could be hereditary, and concluding that only the structure of the brain could be so, he continues: "love of the Deity effect of organization, oh you Materialist!" And further: "Why is thought being a secretion of brain, more wonderful than gravity being a property of matter?" (*Notebook* C 166). In *Notebook* M, in fact, Darwin offers a suggestion of the sort his critics reviled and journals like *Punch* loved to caricature. He writes:

> Origin of man now proved. – Metaphysics must flourish. – He who understands baboon <will> would do more for metaphysics than Locke.
>
> *Notebook* M, 84, Darwin 1987

Notoriously, that is just the kind of remark Bishop Wilberforce would ascribe to Darwin in Oxford in the spring of 1860, when Huxley and Hooker would rise to his defense.

In the *Origin*, however, there is only the slightest hint of such a view. In the closing chapter, Darwin risks a prediction, if in the most general terms. After reminding his readers of the length of time, "incomprehensible to us," in which the changes he has been describing must have taken place, he continues:

> In the distant future I see open fields for far more important researches. Psychology will be based on a new foundation, that of the necessary acquirement of each mental power and capacity by gradation. Light will be thrown on the origin of man and his history.
>
> Darwin 1859, p. 488

For Darwin, that light was already clear enough, although he would not offer it to the public until 1871 in *The Descent of Man* and in the next year in *The Expression of the Emotions in Man and Animals* (Darwin 1871; 1872). "Man still bears in his bodily frame the indelible stamp of his lowly origin," he wrote (Darwin 1871, p. 405). That is the chief lesson of both of those works. Their author still concedes that his reasoning will be "highly distasteful to many persons" (Darwin 1871, p. 404). He admits, too, that it is in part highly speculative, but he has followed the evidence as best he could. There is in the *Descent of Man* still something of the apologetic air of the "difficulties" arguments in the *Origin*, but it is much milder. In 1859 Darwin was already a respected member of the inner circle of British scientists. By 1871, he was internationally known, the founder of a much cited theory: of "der Darwinschen Theorie," of "la théorie darwinienne," even of Darwinism, or "Darwinismus." He had others, notably Huxley at home (with his popular exposition of "Man's Place in Nature") and Ernst Haeckel in Germany, to back him up. Even Lyell had finally come around (Lyell 1863). By this time, also, incidentally, descent with modification has taken on the name we know it by: Darwin refers to "the general principle of evolution" (Darwin 1871, p. 390).

The *Descent* has two parts, the first arguing for the ancestry of *Homo sapiens* in some climbing, tailed, ape-like creature, the second a lengthy investigation of sexual selection. Most difficult to provide is an account of the origin of our "higher" intellectual and moral gifts. Darwin is convinced, however, that, starting from the social instincts that we share with many other animals, natural selection could lead to the kind of cerebralization that has permitted the rise of the human arts and sciences. And from those social instincts also, he argues, our moral sense could have issued, partly through selection – since those who help others will make their tribe or their clan fitter to survive. Moreover, once we invented language, the praise and blame of others would contribute to the rise, and maintenance, of the golden rule. Finally, as Darwin sees it, moral sense is transfigured into a positively Kantian conscience. (Darwin 1871, Chapter III, pp. 70–106).[23]

If Darwin's intent is to exhibit our origin from "lower" animals, he achieves this in part by elevating them, especially in terms of the life of the emotions, almost to our level. He recounts anecdotes showing how animals – dogs or baboons, for instance – exhibit undeniable signs of love or jealousy. He also finds active intelligence in varying degrees

---

[23] In the text, Darwin in fact begins with Kant and works downward to an incipient, instinctive moral sense. The sage of Königsberg would not have been amused.

throughout the animal kingdom: Ants, for example, are much more intelligent than beetles (Darwin 1871, p. 145). Indeed, ten years later, in his work on earthworms, Darwin will celebrate the intelligence of those lowly turners of the soil in picking just the right leaf, and turning it just the right way, to get it into their burrows (Darwin 1881). Darwin was not alone in what seems to us this rather subjective reading of animal behavior. But it does assist him in his effort to demonstrate our place within the natural world.

Darwin now admits, however, that in the *Origin*, he had perhaps attributed too much to natural selection, and this in two respects. First, he confesses:

> ...I probably attributed too much to natural selection or the survival of the fittest. I have altered the fifth edition of the *Origin* so as to confine my remarks to adaptive changes of structure. I had formerly not sufficiently considered the existence of many structures which appear to be, as far as we can judge, neither beneficial nor injurious; and this I believe to be one of the greatest oversights as yet detected in my work.
>
> Darwin 1871, p. 152

As he had earlier moved from a belief in perfect adaptation to that of relative adaptation, he now admits that there are structures that seem to have no special adaptive origin. His excuse, he says, is that he had wanted to show two things: descent with modification instead of special creation, and natural selection as the principal carrier of that process. In *The Descent of Man*, he is still confident of the power of natural selection, but in his eagerness to overthrow the dogma of separate creations, he acknowledges that he may have exaggerated its force.

Second – and that is the concern of half the book – Darwin declares that he should have given, and now does give, more attention to the widespread phenomenon of sexual selection. Male birds, with the conspicuous patterns they display, may be more visible to predators, but that risk is outweighed by their attractiveness to the females who choose to mate with them. Throughout nature we find such occurrences. This is not directly a struggle to survive, but it is certainly a struggle to leave more offspring than the less attractive fellow next door. In the twentieth century, there was extensive debate as to whether sexual selection was a kind of natural selection or, as Darwin himself presented it, an alternative to it. It is certainly natural and it is certainly selective, but it does often appear to run counter to the promotion of its performer's own survival. On either reading, it constitutes a process well worth investigation in a Darwinian biota. There are studies of sexual selection in the late twentieth century that rely explicitly on the principles Darwin had set forth in 1871.[24]

The work on the expression of the emotions was to have formed a part of the previous work, but was too long to be contained in it. This was a theme that had long fascinated Darwin and that has been elaborated by many comparative psychologists since then. Darwin has shown how processes such as the human frown or smile could have arisen from facial contortions characteristic of other mammals. He also spends a good deal of time on showing that facial expressions or other spontaneous gestures in human beings are universal within our species. In both of these books – *Descent* and *Emotions* – Darwin has set forth in great detail the view he had so long been developing of man as an animal – if a rather special one – among other animals.[25]

A mind as far-ranging as Darwin's cannot, of course, be said to have relied on any one stimulus for its development. Darwin's interests and the data he collected have an almost unbelievable scope (always, be it said, from the point of view of an English gentleman in Victoria's reign). But, granting that, we may close this brief account, as Darwin himself closed the *Descent*, with a reference to the young traveler getting his first sight of the Fuegians in their island home. As we noted earlier, he acknowledges that his main conclusion, "that man is descended from some lowly-organised form," "will be highly distasteful to many persons." But, he continues:

> . . . there can hardly be a doubt that we are descended from barbarians. The astonishment I felt on first seeing a party of Fuegians on a wild and broken shore will never be forgotten by me, for the reflection at once rushed into my mind – such were our ancestors. These men were absolutely naked and bedaubed with paint, their long hair entangled, their

[24] See, e.g., Ryan 1985. At the same time it should be noted that Darwin is thoroughly a man of his time and place in his attitude to those not fortunate enough to be British; see, for example, the remarks he quotes with approval on the "careless, squalid, unaspiring Irishman" versus the "frugal, foreseeing, self-respecting, ambitious Scot" (Darwin 1871, p. 174).

[25] In concentrating on the *Origin* and the *Descent*, we have necessarily neglected other important works. For example, *The Variation of Animals and Plants Under Domestication* (Darwin 1868), as we mentioned earlier, introduces the distinction between the hypothesis and theory of natural selection (see Gayon 1998); it is also the work in which Darwin announced the doctrine of pangenesis, his (unsuccessful) theory of reproduction.

mouths frothed with excitement, and their expression was wild, startled, and distraught. They possessed hardly any arts, and like wild animals lived on what they could catch; they had no government, and were merciless to every one not of their own small tribe. He who has seen a savage in his native land will not feel much shame, if forced to acknowledge that the blood of some more humble creature flows in his veins. For my own part I would as soon be descended from that heroic little monkey, who braved his dreaded enemy in order to save the life of his keeper; or from that old baboon, who, descending from the mountains, carried away in triumph his young comrade from a crowd of astonished dogs – as from a savage who delights to torture his enemies, offers up bloody sacrifices, practises infanticide without remorse, treats his wives like slaves, knows no decency, and is haunted by the grossest superstitions.

Darwin 1871, pp. 404–5

"Such were our ancestors!" A scion of the Darwins and the Wedgwoods, faced with those screaming savages, can surely go one step further and look for our ancestry in some not-yet-human source. Darwin concludes:

...we must acknowledge, as it seems to me, that man with all his noble qualities, with sympathy which feels for the most debased, with benevolence which extends not only to other men but to the humblest living creature, with his god-like intellect which has penetrated into the movements and constitution of the solar system – with all these exalted powers – Man still bears in his bodily frame the indelible stamp of his lowly origin.

Darwin 1871, p. 405

CHAPTER 8

✦

Evolution and Heredity from Darwin to the Rise of Genetics

Introduction

In the *Descent of Man*, Darwin admitted that, in his enthusiasm for what appeared to him a major discovery, he had perhaps overstressed the importance of natural selection. He now sees that there may be changes in the biota that are merely "chemical" – we would say, neutral – but, as he has always recognized, only adaptations are subject to natural selection. Still, Darwin's theory was, in his view, basically a theory of natural selection. Yet, paradoxically, as we noticed earlier, although the *Origin* persuaded many of the fact of descent with modification, it did not convince so many of natural selection as the chief agency of change. The Darwinism, or "Darwinismus," of the *Origin's* first fifty years is certainly evolutionary, but it is scarcely recognizable as Darwin's theory. With respect to this period, Peter Bowler has written both of "the eclipse of Darwinism" (an expression introduced by Julian Huxley in his *Evolution: The Modern Synthesis* [Huxley 1942]) and of a "non-Darwinian revolution" (Bowler 1983; 1988).

However, the situation is more complicated than that. Before we move on to the evolutionary synthesis of the 1930s to 1950s and the philosophical questions associated with it, we need to identify some strands in that complex story. Again, as with Darwin, we will have to select a few points from a very rich subject-matter. Indeed, we are already restricting our account of the nineteenth century severely by focusing on evolution – let alone Darwinian evolution – and neglecting such topics as vitalism or the growth of the cell theory. We will refer briefly to the latter in this chapter and to the former when we come to more recent debates about the reduction of biology to physics and chemistry. However, as we warned at the start, we are not attempting

221

a survey. At the same time, even limiting our account to the evolutionary context, and to the associated topics of variation and heredity, we have to try to sort out, without too much oversimplification, some themes and some figures in a confusing, and even confused, cluster of events.

The Survival of Selection

First, we should note that although there was much that was not strictly Darwinian in late nineteenth-century evolutionary literature, there were also continuing efforts, both empirical and theoretical, to vindicate the theory of natural selection. Even this story has its complications, however, and we will have to make at least one detour in telling it.

Darwin's cousin, Francis Galton, started out to investigate questions opened up by Darwin, particularly with respect to laws of inheritance, but found himself going in a direction tangential to that pursued by his famous kinsman. His work, moreover, influenced equally two antagonistic groups of workers at the turn of the century: the biometricians, chiefly Karl Pearson and W. F. R. Weldon, and their enemies. These included the new Mendelians, under the leadership of William Bateson, and other mutationists (including Bateson himself in the last years of the nineteenth century, before the "rediscovery" of Mendel), all of whom insisted on the importance of sudden changes, rather than Darwin's slight "continuous variations," as the motor of species change. The biometricians were the most conspicuous defenders in their time of the study of variation as Darwin seemed originally to have understood it, but we will have to come to them via Galton. We will therefore return to them, in that context, and in this section will consider two other topics that show the persistence of interest in selection: one, its indirect confirmation through study in the field, and the other, an attempt to entrench it, theoretically, as all-powerful in evolution.

The first is the study of mimicry, a development that Darwin himself lived to celebrate and to incorporate in later editions of the *Origin*. In 1862, W. H. Bates published his "Contributions to an insect fauna of the Amazon valley," in which he described a number of species of edible butterflies of the family of the Pieridae that closely mimicked unpalatable, or even poisonous, members of the Heliconidae (Bates 1862). Even in their geographical variation, the mimics closely followed the models. Such variation, Bates argued, must be caused by "natural selection, the selecting agents being insectivorous animals, which gradually destroy those sports or varieties that are not sufficiently like to deceive them" (Bates 1862, p. 512; quoted in Mayr 1982, p. 522). Bates's discovery was soon confirmed and extended. Then, in 1879, Fritz Müller, a German naturalist working in Brazil, discovered that mimicry may occur not only between inedible and edible species, but between two equally unpalatable or poisonous species. If, in a given region, predators learn to avoid all the butterflies with a given appearance, both species are protected by their mutual mimicry (Müller 1879). Both Batesian and Müllerian mimicry served as striking indirect confirmations of the reality of natural selection.

On the theoretical side, a conspicuous defender of the significance of natural selection as the major force directing evolution was August Weismann, Professor of Zoology at the University of Freiburg im Breisgau. Darwin had conceded that natural selection operates under the constraints provided by the laws of variation and of generation. His own theory of heredity, developed under the title of pangenesis, was less than successful, although, as we shall see, it did influence later workers, in particular, Hugo de Vries, one of the co-rediscoverers of Mendel in 1900 (see de Vries 1910). Indeed, in inventing the term "pangen" or "pangene" (based on Darwin's term), De Vries became, through the mediation of the Danish botanist W. L. Johannsen, the source for the fateful coinage "gene" (Johannsen 1913).

With phenomena such as those of mimicry, one is inclined to forget the question of the hereditary basis for whatever traits come to be selected, and the power of heredity seems to be blunted by the existence, which Darwin accepted, of heritable effects of use and disuse. Weismann took a much firmer stand on both these questions. He had come to his theory of heredity through work in the fast-developing field of cytology, which included in particular the study of the chromosomes. He was confident that these structures, with the complex transformations it was being discovered they went through, were the bearers of heredity. On this foundation, he was able, unlike Darwin, to deny the effect of use and disuse, and hence to banish altogether the inheritance of acquired characters. Not only was there no positive evidence for such a process, he argued, but the very possibility of it was excluded by the character of the hereditary material, or germ plasm, which he declared to be, in principle, segregated from the rest of the body. Heredity, he insisted, was based on something in the nucleus of the germ cells that was transmitted across generations. In each generation, the soma has to develop from that transmitted germ (Weismann 1892).

How then, does evolution – the modification that occurs with descent – happen? To begin with, sexual reproduction, involving as it does the crossing of chromosomal ingredients, permits, and produces, variability. But that will make some individuals better adapted than others to some new or modified environment, and hence available for natural selection, the prime mover in the evolutionary process. Thus, Weismann argues, for example, in an 1885 lecture on "The Significance of Sexual Reproduction in the Theory of Natural Selection," that it is unnecessary to assume, as some have done, any "internal force" in the evolution of an organism such as the whale (Weismann 1892, pp. 251–297). Within the very general mammalian pattern, everything about the animal is clearly explained through its gradual adaptation to aquatic life. Given the absence of Lamarckian effects, we need, in addition to the continuity of the germ-plasm, only the power (the "*Allmacht*") of natural selection to explain the origin of the forms now inhabiting our globe. Weismann's position was variously called "ultra-Darwinism" or "neo-Darwinism." At first glance at least, it seems a striking anticipation of the mid-twentieth century Synthesis, combining as it does a theory of inheritance through the nucleus of the germ-cells with a stress on the creative power of natural selection.

However, there are complications. We must mention two in particular. Darwin had spoken of the laws, not only of generation, but of variation – laws unknown to him, but necessarily assumed as the basis for the small individual variations on which selection works. Weismann suggests that such variations could chiefly be traced back to the unicellular ancestors of more complex forms, where germ and soma had not yet been separated. Then sexual reproduction would maintain the already available divergences, and, by the mixing of germ plasms, yield germinal material ready to be selected as environments changed. In later writing, meeting various objections, he modified his theory by suggesting that the sacrosanct germ-plasm itself might sometimes be altered by external influences.[1]

Second, although those speculations did not threaten the power of natural selection, there was another aspect of Weismann's thought that did seem to undercut his pan-selectionism. Primarily through work on the development of sphingid caterpillars, he came to accept the doctrine of recapitulation: Ernst Haeckel's thesis that ontogeny recapitulates

---

[1] See the careful study of Weismann's views of variation as derived from changes either within or beyond the germ plasm in Winther 2001.

phylogeny, with new stages being added in linear sequence to a series of earlier forms. Thus, allegedly, for example, the gill-slits of our embryos would be the equivalent of the gills of modern adult fishes, and so on. Unlike Fritz Müller, however, who in his 1864 *Für Darwin* had proposed recapitulation as compatible with selection (through a range of variations in developmental rates), Weismann acknowledged that such a process would have to be generated by some force other than selection, even if selection could limit or control it (Weismann 1904, vol. 2, pp. 151–19; Gould 1977, p. 101).[2]

However, while acknowledging these additional, and in the second case, even seemingly contradictory, aspects of Weismann's work, we wanted to refer to him here, as we said earlier, chiefly to indicate that even in the late nineteenth century there were prominent biologists who supported, not only some kind of evolutionary theory, but an evolutionary theory in which classic Darwinian selection played a prominent role. Yet if recapitulation is invoked by Weismann, and even by Müller, it does not play the leading part in their work that was given it by its most enthusiastic popularizer, Ernst Haeckel.

The Reign of the Biogenetic Law

It was Haeckel, in fact, who coined the terms "ontogeny" and "phylogeny" and named his thesis that the first recapitulates the second the "biogenetic law." (He also coined the term "ecology," although that then not yet existent discipline seems to have little to do with the theories he developed.) For him, the story of descent with modification was largely internal: It was the terminal addition of further adult stages, and the consequent condensation of earlier forms into a tighter time series, that chiefly characterized the evolutionary process. In contrast to this picture, the great biologist K. F. von Baer – while, however, rejecting any form of transformism and embracing Cuvier's inalterable *embranchements* – had argued earlier in the century against just such a concept of a linear sequence of adult stages, insisting that animals developed from a more generalized embryonic stage to more specialized adult stages. Thus the gill slits in our embryos would mirror the same early stage in fish development, whereas their adults and

---

[2] There is a detailed account of this aspect of Weismann's work in Stephen Jay Gould's now classic *Ontogeny and Phylogeny* (Gould 1977, pp. 102–109; see also Gould 2002, pp. 208–210).

ours would differ sharply. In Darwin's reference to embryology, he had followed what came to be known as "von Baer's laws."[3] Yet Haeckel took as his leading principle a thesis strikingly contrary to this, with no suspicion, apparently, that he was proposing anything but strict Darwinism.

Indeed, Haeckel visited England in 1866, and sought out the great master Darwin to kneel (metaphorically) at his feet. Since neither seems to have been fluent in the other's language, it must have been a strange occasion. Haeckel also visited Down House again in later years, and despite language barriers seems to have considered himself a loyal Darwinian. Yet Haeckel in fact stated and, through the success of his writings, helped spread far and wide a view of evolution in which, under the guidance of his "law," the search for phylogenies was everything and natural selection was of little weight. This fashion held sway for some decades: Note, for example, that both Weldon and Bateson, later antagonists in the early days of Mendelism, had started by working with the embryologist Francis Balfour at Cambridge, who was interested in problems of phylogeny, and Bateson had gone on to investigate the question whether *Balanoglossos* was, though unsegmented, an ancestor of the chordates (Provine 1971, pp. 37–8). That was the kind of problem that appealed to evolutionary biologists under the dominance of a Haeckelian perspective.

Was this Darwinism, or, as Bowler would insist, rather a kind of pseudo-Darwinism? In answer to this question, we must note that both the background and the tenor of Haeckel's evolution differ markedly from Darwin's. True, early in the long years of meditating privately, or close to privately, on his theory, Darwin had talked in his notebooks of the brain secreting thought, and had exclaimed against himself: "Oh, you materialist!" Yet he was always careful to keep in the background that aspect of his thought. It was of course the implications for the special status of *Homo sapiens* that enraged critics such as Sedgwick or the notorious Bishop Wilberforce. But, as we observed earlier, except for a very small hint, Darwin himself discreetly omitted any reference in the *Origin* to such dangerous extrapolations (see Chapter 7). He knew his own gentlemanly, and ladylike, society too well for that.

---

[3] It seems that in the nineteenth century, every one liked to find "laws" in biology, although the degree to which they were thought to be "nomic" or necessitating varied considerably.

In Germany, however, there seems to have been more room for a movement in the direction of a programmatic materialism. Of course, there, too, there was also opposition on the part of theologians, idealistic philosophers, and even scientists. Thus Ludwig Feuerbach, for example, a reformed Hegelian and perhaps the prime mover in initiating German materialism, lived privately, unable to hold an academic position, and Fritz Müller had emigrated to Brazil in angry reaction to clerical opinion. But, at the same time, perhaps for the very aura of revolutionary enthusiasm that surrounded them, the materialists also received great popular acclaim. For example, Ludwig Büchner's *Kraft und Stoff* (*Force and Matter*), first published in 1855, had its nineteenth German edition in 1898. Although Büchner went through periods of depression, and doubts about his own philosophical position, the popularity *of Kraft und Stoff*, as well as his lectures on Darwin, allow him to serve as a good example of the general materialist trend in mid-nineteenth century German thought. In *Kraft und Stoff*, he had stressed the demotion of Hegelian spirit, and with it of a unique or non-material human spirit. Thus, in the preface, he wrote:

> Starting with the knowledge of the fixed relationship between force and matter as an indestructible foundation, the empirical-philosophical view of nature has to come to results that decisively ban every kind of supranaturalism and idealism from the explanation of natural events. Its explanations must be conceived completely independently of the assistance of any external power existing outside things.
>
> Büchner 1855, quoted in Gregory 1977, p. 106

That was in 1855. But then, with the translation of the *Origin* into German in 1860, Büchner and his readers were ready for the naturalization of man that was implicit, though not explicit, in Darwin's argument. It was that message – the acknowledgment of man's place in nature – that they specially welcomed. On the other hand, Büchner and his fellow materialists were in general skeptical of the power of natural selection to direct evolution, and wished rather to stress the progress they thought inherent in the advance of life. Obviously, this went with their left-wing social and political views, and their hope for liberation from the superstition and ignorance of the priest-ridden past. So, even short of Haeckel, the "German Darwin," Büchner and those like him would fall, in Bowler's terms, into the class of "pseudo-Darwinians."

Haeckel himself, to be sure, disclaimed the materialist label; there was, he said, a spiritual side to his conception (Gould 1977, p. 421,

n. 6).[4] In fact, Büchner himself, in his later years, preferred to call himself a "monist." But Haeckel, like the notorious author of *Kraft und Stoff*, was certainly fostering a comprehensive monism that seemed to many of his readers to support a materialist metaphysic. So there was an appeal to a different, and broader, readership. Here was a grand new vision, to be substituted for tired old traditional beliefs and superstitions.

It may have been, in part, that aspect of Haeckel's writings that appealed so much to T. H. Huxley. Though already known in the battles that followed the publication of the *Origin* as Darwin's bulldog, Huxley was far from a convinced Darwinian, *sensu strictu*. At first he had disliked the idea of descent with modification altogether – he wanted species to last a long, long time and each to go back a long, long way, and he never came to appreciate natural selection as a principal agent of transformation. But he clearly welcomed Darwin as a force against the establishment – an establishment he perceived as being firmly Anglican and attuned to the discourse of natural theology (Desmond 1994; 1997a, b). As we noted in Chapters 6 and 7, scientific interests were becoming relatively institutionalized throughout this period, but it was still largely through club-like associations of what Martin Rudwick aptly calls "the gentlemen of England" that this was happening. It was still extremely difficult for a poor man like Huxley to make a living from science, a situation he bitterly resented.[5] Now here was Haeckel, suggesting a new and much less gentlemanly cosmology to support Darwin's transmutationist theory.

Further, Haeckel's "Darwinismus" grew out of traditions different from the British. There were traces there of *Naturphilosophie* – the German bent to seek grand overall unities in the phenomena of nature, and a heavy involvement in a prestigious tradition of work in morphology. It was a much less ecological approach than Darwin's, which saw even anatomical differences (as with his beloved barnacles, for example) in the context of the demands of a particular environment. To be sure, the morphological emphasis may also have contributed to Huxley's enthusiasm for Haeckel's work when he read *Generelle morphologie*, published the year Haeckel first came to England – 1866 (Haeckel 1866).

---

[4] Indeed, Haeckel can be claimed (or indicted!) as a forerunner of Nazi "mysticism."

[5] For insight into the complexities of scientific professionalization in nineteenth-century Britain, see Desmond 2001, Waller 2001, and Alberti 2001, all in *Journal for the History of Biology* 54 (1) (Spring 2001).

Although of course Darwin himself was a great anatomist – Huxley had first encountered him through the barnacles book, when Darwin gave him a copy, which the younger, and much poorer, man could not have afforded to buy – Darwin's stress on "conditions of existence" would have been peripheral to Huxley's more strictly anatomical interests. It is not clear whether Darwin himself ever noticed the difference between his own point of view and Haeckel's; his German was not up to reading much of the recapitulationist's work. It is certainly the case, however, that the more internalist focus of Haeckel's speculations, stemming at least in part from the dominance of morphological interests in German biology, did give a direction to his thought very different from an evolutionism governed by an interest in slight variations subject to selection in slightly different environments.

Thus what resulted from Haeckel's idiosyncratic perspective was a picture that differed sharply, both literally and metaphorically, from Darwin's. Consider the single diagram in the *Origin*. Despite his eloquent reference to the "tree of life," what Darwin presents is a sketch of a bushy rather than tree-like structure, where some lines persist, some branch, some die out, and there is no single direction taken by the whole biota. Haeckel's most famous diagram strikingly presents one great single trunk, with little branches coming off it, to be sure, but leading unambiguously from monad to man (see Simpson 1953).[6]

There are two implications of this difference in perspective that need to be mentioned. First, obviously, the Darwinian picture gives no special place to mankind, while Haeckel's tree leads, as many pre-evolutionary scenarios had done, straight to man as the pinnacle of nature. That aspect, incidentally, would not have appealed to Huxley; it was the new and humbler reading of "man's place in nature" that attracted him in Darwin's work, despite his disagreement on much of the master's theory. Second, and more significantly, Haeckel's single-minded, internalist version encouraged, as Darwin's stress on the incompleteness of the fossil record failed to do, a search for phylogenies: a persistent inquiry about just what was descended from what. Such searches characterized much of late nineteenth-century and early twentieth-century evolutionary thought. Who was the first ancestor of the vertebrates? How did we get from *Eohippus* to *Equus*? The American paleontologist Othniel C. Marsh presented the equid story in a famous, nicely linear, diagram that appealed to the sense of progress associated with such phylogenetic

---

[6] Haeckel's diagram is reproduced in Gould 1977, p. 171 (fig. 20).

stories (see Simpson 1953, p. 259–60). As we noted earlier, it was still this kind of inquiry that Bateson pursued early in his career.

Further, the search for phylogenies took place conspicuously under the guidance of the biogenetic law. After all, a phylogeny just *is* a sequence of ontogenies, and so each ontogeny is a bead in a string that sums up to a phylogeny. It should be easy enough to sort out. You add bits on the end, condense the sequence in the middle, and that's that.

But then we have to ask: How do new ontogenies happen? Here, Haeckel himself, and, increasingly, many paleontologists after him, fell back on so-called Lamarckian explanations – that is, explanations in terms of acquired characters and their inheritance.[7] Darwin had still allowed some influence of use and disuse, but in the case of Haeckel and those who followed him, the inheritance of acquired characters assumed new prominence and respectability, at the expense of natural selection. Two leading American paleontologists of this period, Alpheus Hyatt as well as Marsh, were outspoken Lamarckians.

In apparent contrast to this emphasis, many students of the fossil record took an opposite direction, and found, or claimed to find, in the phylogenies they were tracing, evidence of "orthogenesis" – that is, an intrinsic tendency to develop, from one generation to another, and one species to another, in a given direction, without the stress on adaptive relations characteristic of either classic Darwinian or Lamarckian accounts. Oddly enough, the first popularizer of this view, Theodor Eimer, started out accepting the inheritance of acquired characteristics, and hence a stress on the effect of the environment, but came to combine this with an emphasis on inner, seemingly directed, tendencies. Paleontologists such as Marsh and Hyatt also managed to blend Lamarckian with orthogenetic lines of thought. How did two seemingly contrary points of view appear to so many serious scientific workers to fit so well together?

In Eimer's case, his animosity to Weismann may have had something to do with it. As against the segregation of the germ plasm, you insist on external influences, and against the alleged power of selection, you stress the seemingly internally impelled directness of phylogenetic development. Thus, in his work on butterflies, *Die Orthogenesis der Schmetterlinge*, Eimer declared: "...the causes of definitely directed

---

[7] The inheritance of acquired characters is, of course, only one of Lamarck's basic principles, but "Lamarckian" is commonly used as a (usually pejorative) term referring only to that aspect of his thought.

evolution are contained . . . in the effects produced by outward circumstances and influences upon the constitution of a given organism" (Eimer
1897, p. 22, quoted in Bowler, 1983, 152). Yet Eimer believed orthogenesis, for all its external initiation, to be "a universal law," holding
"not only for the markings" – as he claimed to have shown in the lep-
idopteran case – "but also for the other morphological characters of
animals, and also for those of plants" (Eimer 1897, p. 21; Bowler 1983,
p. 152). Even the cherished Darwinian case of Batesian mimicry had
resulted, he argued, from laws of coloration expressed in parallel in a
number of species.

However, if in Eimer's case the combination of Lamarckism and orthogenesis may have had some relation to his dislike of Weismann, that
would not explain the position of other workers such as the American
paleontologists. Perhaps what they share, as against cytological interest in the germ plasm, and particularly in the germ cell's nucleus, is an
interest in the whole organism, responding spontaneously to its environment, and then perhaps carrying such inner activity through a sequence
of generations. Particularly in the case of an invertebrate paleontologist
such as Hyatt, the habit of looking at whole organisms might encourage
such an attitude.

Whatever its *raison d'être*, orthogenesis in a number of varieties
was an influential position throughout a number of decades. Early in
the twentieth century, for example, the senior paleontologist of the
American Museum of Natural History, Henry Fairfield Osborn, espoused a theory of what he called "aristogenesis": the coming to be
of the best as the motive force in evolutionary history. As late as 1926,
L. Berg published a work called *Nomogenesis*, announcing the laws that
govern, internally, the course of evolution. If Darwin had conquered
the world for descent with modification, he certainly had not effected
a similar victory for his theory of natural selection. Well into the twentieth century, evolution was conceived, even if only vaguely, in terms
of the biogenetic law; development and evolution – later considered
alternatives – were thought to be the same thing writ small or large.

Francis Galton

Darwin's younger cousin, by training a mathematician, was attracted
to the argument of the *Origin*, but from the first, it seems, with special
attention to the problem of the laws of heredity that underlie the process of transmutation. He was especially interested in the hypothesis of

pangenesis, which Darwin proposed in *Variation of Animals and Plants Under Domestication* (1866). At first, his interest was favorable, but experiments in which he tried to confirm its existence served instead to falsify it. He injected blood from rabbits of one color into those of another hue, and found that there was no consequent difference in the color of their offspring. So Darwin's notion that gemmules carrying hereditary characters flowed from all the cells of the body into the germ cells failed his experimental test.

At first, also, Galton seems to have been impressed by the power of selection to mold at least domesticated forms – and he thought, as he said in an 1865 article on "Hereditary Talent and Character," that with due care and encouragement the same plasticity could be achieved in the direction of improving human abilities and achievements, both mentally and morally (Galton 1865). Indeed, eugenics, in particular what came to be called positive eugenics, was Galton's life-long obsession. In the essay on "Hereditary Talent and Character" he had exclaimed: "What an extraordinary effect might be produced in our race, if its object was to unite in marriage those who possessed the finest and most suitable natures, mental, moral, and physical!" (Galton 1865, 163). And he proceeds to imagine a Utopia – or, as he puts it, "a Laputa, if you will," in which public examinations testing "every important quality of mind and body" would fasten on ten exceptional young men of twenty-five and ten young ladies of twenty who would be suitable to mate with them. The "Trustee" in charge would announce their selection to the "deeply-blushing young men" (!), and would conclude his official announcement as follows:

> ... It appears that marriages between you and these ten ladies ... would offer the probability of unusual happiness to yourselves, and, what is of paramount interest to the State, would probably result in an extraor-dinarily talented issue. Under these circumstances, if any or all of these marriages should be agreed upon, the Sovereign herself will give away the brides, at a high and solemn ceremony, six months hence, in Westminster Abbey. We, on our part, are prepared, in each case, to assign [£]5,000, as a wedding present, and to defray the cost of maintaining and educating your children, out of the ample funds entrusted to our disposal by the State.
>
> Galton 1865, p. 163

Utopia, indeed! But something like this was Galton's lifelong ideal. Our moral and mental development – particularly, of course, that of the English! – had outrun our physical evolution – hence our sense of original sin – and we needed to do some active arranging to help us catch up.[8]

So much for Galton and eugenics – an embarrassing connection, but one that dogged Darwinism for a long time and that we cannot wholly ignore (see Chapter 11). What chiefly interests us here, however, is Galton's considerable influence on Victorian (and Edwardian) biological practice, which went, paradoxically, in two contradictory directions. On the one hand, it was Galton who first introduced statistical methods into theories of heredity, and hence of evolution, and so paved the way for the work of the biometricians, like Weldon, who would attempt to produce by statistical means direct confirmation of the existence of selection in natural populations. On the other hand, with the "law of ancestral heredity" that he himself developed by his use of statistical methods, Galton came to negate any creative force of selection, or any effect of cumulative tiny variations, in evolution, and so to encourage the focus on discontinuity that permitted the initiation of work in the opposing direction of Mendelian, or, more generally of mutationist, theorizing and experimentation. It is always unfair, in the case of a complex thinker, to separate isolated strands in his thinking, but as we were constrained to do with Weismann, so here, we must single out themes in Galton's work that appear to bear especially on the twin topics of the persistence of selection and the neglect of, or opposition to, that emphasis in the decades from the 1860s to the rise of the evolutionary synthesis in the 1920s and '30s.

First, then, let us look, if sketchily, at Galton's introduction of statistical methods into evolutionary theory. In his early work, such as the paper from which we have been quoting, he reported information from various collective biographies, something like what would later be the National Dictionary of Biography, and from other sources, such as the list of Lord Chancellors, or of Cambridge graduates, and so on, and counted the number of distinguished kinsmen in these lists. Thus he shows an interest in relatively large collections, but this is hardly, in any sophisticated sense, statistics. By 1875, however, when he published an article called "Statistics by Intercomparison, with Remarks on the Law of Error" (Galton 1875), he had begun to rely on the work of the

---

[8] According to Waller 2001, there was a gap in Galton's writings on eugenics, but since he returned to the cause with sustained enthusiasm, it seems fair to call it a life-long interest.

Belgian statistician A. Quetelet, in particular with respect to the law of deviation. And by the time he published "Typical Laws of Heredity" (Galton, 1877), Galton had developed a device of his own, which he called the "quincunx," which enabled him to trace the effect of the "law of deviation" over generations. This was a most significant step in the history of evolutionary theory.

Celebrators of Darwinism often acclaim the master as the initiator of "population thinking": up to Darwin's time, they say, all had been "typology" or "essentialism" and biology had been unable to move forward. Then came the *Origin*, and – in the spirit of "God said, let Newton be, and all was light" – the life sciences were rescued from their dark ages and the new, good population thinking reigned instead. But, as we noted earlier, Darwin was not fully a population thinker; he was concerned with small variations in individuals and the accumulation of such variations in changing environments so that first new varieties and then new species would arise. The criticism of the engineer Fleeming Jenkin, which he took very seriously, had challenged him for this very reason.

Jenkin raised three objections against Darwin: (1) the prospect that new variations would be swamped if inheritance were blending; (2) the question of the size of variations: minute or "sport-like"; and (3) the problem of their quantity: were there only a few, or many?[9] Darwin tried to reply at least to the third of these questions, but never satisfactorily. It is true, of course, that he had abandoned the thought of species as inalterably fixed, but he had rejected them, not in favor of large aggregates – that is, of populations, statistically treated, as an alternative – but of attention to the lives and characters of numerous individuals. It is Galton, with his mathematical training and his reliance on Quetelet, who turned to populations as the unit to be considered in the history of life.[10] His influence in this respect will be evident when we turn to the work of the biometricians.

In the meantime, we must consider the anti-selectionist direction taken by Galton as his thought developed. It may be summed up in what Pearson would baptize "the law of ancestral heredity." First, Galton

[9] For an illuminating analysis of Jenkin's objections and Darwin's response, see Gayon 1998, pp. 83–102.

[10] Darwin referred to Quetelet on the normal curve in human height in an essay in *Nature* in 1873, pp. 177–182, and his intent in that paper with respect to rudimentary organs is explained by his son George in a letter to *Nature* published in the same volume, reprinted in Darwin 1977, pp. 292–3.

believed that each individual carried, at least in "latent" form, characters derived from all its ancestors, in decreasing quantities, but still for many generations back. Each individual, he believed, has half its characteristics from its parents and the other half from the sum total of earlier ancestors. That half is again divided in the same way, and so on backward. In decreasing proportions, something flows into each of us from all our ancestors. That is what is meant by calling his conception "ancestral heredity." Second, when he applied the law of deviation, Galton found that groups, or populations, of individuals always tend to approximate a constant mean. In other words, when you look statistically at the history of a population, you get a steady, Gaussian curve. Thus, if two tall people mate, their children will still be tall, but, on average, they will not be as tall as their parents, and the same for couples who deviate from the mean in being unusually short. The offspring, as he put it, are "meaner" than their parents. This tendency Galton first called reversion, later, regression. Now from this it follows, third, that small variations of the kind Darwin envisaged will not after all give rise to new species. As he came to put it, Galton believed there were "positions of organic stability" that could be unsettled only by a relatively abrupt move to another such state. If you think of Henri Bergson's metaphor of evolution as a snowball that swells as it advances – though, of course, Bergson was thinking in terms of evolutionary inner forces – for Galton, you would have to substitute a polygon that somehow or other bumps from one face to another. Thus selection can only conserve, not originate. If there is a point of equilibrium itself a little off center, selection can keep it there. But chiefly selection is a source of normalization, and in no case of novelty. As Galton put it in an article called "Typical Laws of Heredity," published in 1875–6:

> ... the ordinary genealogical course of a race consists in a constant outgrowth from its center, a constant dying away at its margins, and a tendency of the scanty remnants of all exceptional stock to revert to that mediocrity, whence the majority of their ancestors originally sprang.
>
> Galton 1875–6, p. 298

This seems a gloomy outlook, both for Darwinian selection and for Galton's own beloved eugenics. In 1894, in a paper in *Mind* (Galton 1894), he relates how he had come to this view, referring to recent work of his own: *Natural Inheritance* (Galton 1889), *Finger Prints* (Galton 1892a), and the Preface to a new edition of *Hereditary Genius*. And in that same essay, which is entitled "Discontinuity in Evolution," he goes so far as to celebrate the appearance of William Bateson's outspokenly anti-Darwinian *Materials for the Study of Variation, Treated with Special Regard to Discontinuity in the Origin of Species* (Bateson 1894). This was six years before the "rediscovery" of Mendel and twelve years before Bateson himself would suggest the name "genetics" for the newly developing science of heredity. It was enough to put Galton squarely on the side of the anti-gradualists, and a camp stridently opposed to that of his biometrician friends, and in a sense, disciples, Weldon and Pearson.

The Biometricians

In 1898, Karl Pearson, Goldschmidt Professor of applied mathematics and mechanics at University College, London, presented a modified version of Galton's law, now the "Galton–Pearson law," which allowed alteration in the population mean and thus permitted evolutionary change. Originally, Pearson had had no special interest in biological questions, but in 1891 he had made the acquaintance of Weldon, who had moved from Cambridge to take up a chair at London (though in 1899 he moved to a professorship at Oxford). Weldon was a keen believer in Darwinian selection, and found in Pearson's mastery of statistics a guide to his empirical work, at first on pelagic shrimp, but later, and more successfully, on the common shore crab. With Pearson's technical guidance and from his own careful study of two sets of 1,000 specimens, Weldon concluded that when the water was turbid, crabs with wider carapaces were selected against, due to accumulation of debris in their gills. That sounds like a valid inference from carefully established data to the existence and efficacy of natural selection, and hence to a causal conclusion.[11]

What is most interesting philosophically about Weldon's conclusions, however, is their interpretation in terms of Pearson's conception of scientific methodology. When editing, and completing, the manuscript of the late W. K. Clifford, Pearson had read Ernst Mach on mechanics, and had become a convinced programmatic phenomenalist. The scientist must recognize, he insisted, that all he is doing is collecting and summarizing his sensory experience. He has no right to go beyond sense perception to infer hidden forces or substances or any such (literally!) nonsense. Thus when he speaks of "cause," all this means, scientifically, is invariable expectation. For example, if I am used to the sequence CDEFG, and I

---

[11] For an account of biometry, see Gayon 1998, pp. 198–252.

meet C, I expect DEFG to follow (Pearson 1937/1892, pp. 42–3).[12] Pearson refers in this connection to J. S. Mill, but his formulation sounds oddly like Hume's constant conjunction, with the idea of necessary connection reduced to a little less than necessary. To insist that causal inferences may be made only on the basis of sound statistical data is one thing; to declare that statistical correlations and the expectations based on them are in fact all that statements of causal connections amount to is another matter. If the expectation of lung cancer stops people from smoking, well and good; but we doubt if pulmonary physicians dealing with the lungs of afflicted patients see it entirely that way.

The fact remains, however, that this was indeed Pearson's phenomenalist position, and it didn't seem to prevent Weldon from doing careful, and, we would have thought, naively realistic, empirical work under the other's auspices. Phenomenalism may of course, have influenced the empirically minded Weldon in keeping him from speculations about underlying entities that might have produced his results, as Sloan suggests in his careful analysis of the biometricians' work (Sloan 2000). In any case, whatever its philosophical presuppositions, the Pearson–Weldon collaboration did produce impressive results; it was not, as Bateson was to insist, just a stubborn expression of ignorance once the work of Mendel had been brought to prominence (Provine 1971, p. 71).

At the same time, it must be admitted that a quarrel between Pearson and Bateson (due to the latter's actions in the Evolution Committee of the Royal Society) did have its practical effect (Provine 1971, pp. 49–51, 54–55). In its wake, Pearson and Weldon, with the collaboration of Galton (whom Pearson admired sufficiently to write his biography and to assume the Chair of Eugenics endowed by Galton on his death in 1911), founded a new journal, *Biometrika*, designed to champion their gradualist position against the emergent Mendelians in the early years of the twentieth century. To them and their disciples, it appeared that Mendel's laws could scarcely be universal; if you work statistically on large populations you scarcely expect – or find – neat ratios such as those exhibited by Mendel's peas. Both Pearson and Weldon did try, in different ways (Weldon in unpublished MSS) to deal with the new fashion. Their cause proved to be an unsuccessful one; yet their work may well have contributed to the eventual emergence of a statistical Darwinism in the third decade of the century, chiefly through the work

---

[12] The section in Pearson is called "The Brain as a Central Telephone Exchange."

of R. A. Fisher, who published papers in 1918 that clearly reconciled Mendelism with the gradualist Darwinian position.

Discontinuity and the Rise of Genetics

In 1894, as we saw in connection with Galton's approval of his work, William Bateson had published his *Materials for the Study of Variation*, in which he emphatically supported the significance of discontinuous variation rather than the small "individual variations" championed by Darwin and investigated by workers like Weldon. Meantime, in 1889, the Dutch botanist Hugo de Vries had published his *Intracelluläre Pangenesis* (de Vries 1899), in which he put forward his own version of a discontinuist theory, based, with significant differences, squarely on Darwin's theory of pangenesis. Unlike Bateson, de Vries insisted on the material existence of units of heredity, which he called "pangens." "To my mind," he wrote:

> Darwin's provisional hypothesis of pangenesis consists of the following two propositions:
> 1. In every germ-cell (egg-cell, pollen-grain, bud, etc.) the individual hereditary qualities of the whole organism are represented by definite material particles. These multiply by division and are transmitted during cell-division from the mother-cell to the daughter-cells.
> 2. In addition, all the cells of the body, at different stages of their development, throw off such particles; these flow into the germ-cells, and transmit to them the qualities of the organism, which they are possibly lacking (the "transportation-hypothesis").
>
> de Vries 1910, p. 5

De Vries accepted the first of these hypotheses, but not the second. That is why he called his units "pangens" rather than "gemmules" – Darwin's term – which seemed to him to connote the second, mistaken hypothesis. The pangens, de Vries argued, were "morphological structures, each built up of numerous molecules" (de Vries 1910, p. 70). These structures were contained within a given cell (hence "intracellular"). They varied independently, and it was on them, de Vries maintained, that heredity – and changes in heredity – depend.

Further – and this would be fundamental to his developed "mutation theory" – pangens vary in two different ways (de Vries 1901–3; 1910, pp. 74–5). If they change in number or distribution, small variations, like those stressed by Darwin, would follow, but these, de Vries believed, could have no major evolutionary consequences. On the other hand, qualitative alterations in the pangens, particularly when the organism was under stress, would produce the larger and more significant changes "on which depends the gradually increasing differentiation of the entire animal and vegetable world" (de Vries 1910, p. 74). So hereditary units such as those Darwin had postulated would mediate evolutionary change of the very kind Darwin had been proud to deny. Darwin's theory, he had boasted, supported the maxim *"natura non facit saltum"* ("Nature makes no leaps"). For de Vries, leaps are the very stuff of nature's history.

In 1899, Bateson invited de Vries to attend an international conference of the Royal Horticultural Society, a group in which Bateson was active, and he also paid a visit to Bateson at Cambridge. Bateson was delighted to meet another supporter of discontinuity (Provine 1971, p. 66). The two became friends, at least for a while, although later it seems differences in their views drove them apart. As we have seen, de Vries emphasized from the start a material basis for heredity; Bateson would be skeptical of such a position, and wanted to find some less particulate foundation for the phenomena of inheritance. De Vries, too, though the first in print as rediscoverer of Mendel, became less and less confident of the universality of Mendelian laws, while Bateson became a more and more fervent support of Mendelism.

However, we are still in 1899. The next spring, de Vries was to publish the first of the papers that announced the "rediscovery" of the work of Gregor Mendel. In the *Comptes Rendus* of the French Academy of Sciences, he published an article on "the law of disjunction of hybrids," in which he failed to mention Mendel's name, although he did adopt from Mendel's paper of 1865 the terms dominant and recessive (de Vries 1900a). De Vries also sent a German version, which did mention Mendel, to the *Berichte der deutschen botanischen Gesellschaft* (H. de Vries 1900b). A German botanist, Carl Correns, who had been working on similar questions, seeing de Vries's French paper, hastily prepared an essay of his own, giving Mendel his due; this was published in the *Berichte* in May (Correns 1900). In addition, an Austrian named Erich von Tschermak published in the same volume a summary of his dissertation, which had been issued in full in an Austrian publication (Tschermak 1900). Although it has been questioned whether Tschermak was really thinking in Mendelian terms, these three have generally been considered to be the "rediscoverers," and it is "Mendel's laws of heredity" that they are thought to have rediscovered.

Although this story was long taken for gospel, and is still a textbook "truth," its accuracy has since been questioned, notably by Robert Olby in his by now classic paper, "Mendel No Mendelian?" and by other students of the period, as well as of Mendel's own famous essay, "*Versuche über Pflanzenhybriden*" (Olby 1985, 234–258; Mendel 1865). The standard story was that Mendel stated three laws of heredity, which the botanists of 1900 first rescued from their unfortunate obscurity (Olby 1985, 234–239). What – we do not want to say "in fact," since historical facts are almost always more complicated and less clear-cut than they appear – but what, to the best of our knowledge, was Mendel interested in and what did he discover in the course of pursuing that interest?

Like many others of his day – and as late as Bateson's friends in the Royal Horticultural Society – Mendel was interested in hybrids and hybridization, and especially in the stirring question as to whether hybrids could be so bred as to give rise to new species. This was not a question raised through the agency of Darwin and the *Origin* (which Mendel, if he knew it at all, certainly did not make central to his argument in connection with his work on *Pisum*). It had been pursued by such predecessors of Mendel as J. G. Koelreuter and C. F. von Gaertner, and was of interest to his distinguished contemporary Karl Nägeli. It is the problem of hybridization, among practical breeders as well as those with more theoretical scientific interests, that most strikingly captured the attention of biologists, especially of botanists, in this period. Indeed, both de Vries and Correns refer in the titles of their papers to rules governing hybrids – and Tschermak to crossing in *Pisum sativum*.

What, then *did* Mendel discover about hybridization? What he discovered, although it fell short of reaching to the origin of new species, was "a generally applicable law of the formation and development of hybrids" (Mendel 1865, p. 2). Suppose you have two plants which you have found to breed true for alternatives to a given character, such as tall or dwarf stems. You find that when you cross tall with dwarf, all the offspring (the F1 generation) are tall, but when you cross the hybrids with one another, you obtain three tall and one dwarf (the F2 generation). Now inter-cross the dwarfs among the F2 generation, and they will be dwarfs like their grandparents, and will go on breeding true as long as you keep self-crossing them. The same goes for one-third of the tall-stemmed plants in the F2 generation. The remainder however, when crossed, will again produce the characteristic 3 to 1 tall-to-dwarf ratio. From this pattern it is clear that tall and dwarf are constant, segregating characters, which persist presumably through any number of generations. It is clear, moreover, that tall was "dominant" over dwarf, as Mendel put it, and dwarf "recessive" to tall, and that both of these phenomena occurred in definite ratios, as Mendel's professor at Vienna, an authority on combinatory analysis, would have led him to expect. There were characters, or factors, as Mendel called them, which recurred in these neat proportions. That gives him an algebraic law for the development of hybrids: "development" understood mathematically rather than embryologically.

All this is expounded purely empirically in the first part of Mendel's famous paper. In a second section, Mendel proceeds to speak of the "reproductive cells of hybrids," and the way in which constant forms are produced. It is here that we expect to find a statement of what has since been called classical transmission genetics, with meiosis and then a choice between AA, Aa, or aa at fertilization. But notice that Mendel always speaks of A, Aa, or a. His formula for the F2 generation is A + 2Aa + a. Thus he never doubles his factors in speaking of the pure lines, or suggests that there is here a pair that has to be separated before reproduction. Indeed, he remarks that "only the differing elements are mutually exclusive."[13]

A few more steps were needed to produce what we think of as Mendelian heredity. It was Bateson, the self-appointed chief advocate of Mendelism in Britain, who coined the terms "homozygote" and "heterozygote," as well as "allelomorph" (soon abbreviated to allele), and thus provided the revised conception fundamental for classical genetics. The pure-bred tall plant now has two determinants for A and the equivalent dwarf has aa, with two hybrid Aa's in between. A law of hybridization has been expanded so as to offer space for a full theory of heredity. Yet at the same time, it must be remembered, Bateson resisted the belief that the Mendelian "factors" correspond to definite, locatable entities in the nucleus, a belief already insisted on by de Vries and developed, with de Vries's approval, as the chromosome theory. Even much later, when he was driven to admit the success of that theory, Bateson did so only grudgingly. Apparently, this was too materialistic for him, and he favored some kind of vibratory theory as a basis for heredity (Provine 1961, p. 61).

---

[13] This is Robert Olby's translation, in Olby 1985, p. 251; see p. 258 for the original: "Wobei nur die differierenden sich gegenseitig ausschliessen." Cf. Stern and Sherwood 1966, p. 43.

Yet, as we have already noted, it was also Bateson who would suggest to the Third International Conference on Hybridization in 1906 that their subject matter should be called "genetics." In his Presidential Address, he pointed out to his colleagues that with the Mendelian revelation, the mystery breeders used to face had been replaced by order. But the newly emerging discipline, he told them, was still nameless and could be described "only . . . by cumbrous and often misleading paraphrases." He continued:

> To meet this difficulty I suggest for the consideration of this Congress the term Genetics, which sufficiently indicates that our labours are devoted to the elucidation of the phenomena of heredity and variation. In other words, to the physiology of Descent, with implied bearing on the theoretical problems of the evolutionist and systematist, and application to the practical problems of breeders, whether of animals or plants.
>
> Bateson 1907, p. 91

Retroactively, then, the conference issued its report under the title: "Report of the Third Conference of Genetics, Hybridisation (the cross-breeding of genera or species), the Cross-breeding of Varieties, and General Plant-breeding."

Another chapter in our story. In 1903, the Danish botanist W. L. Johannsen had published studies he had undertaken on "pure lines" in the common bean, *Phaseolus vulgaris*. The work was conceived in line with Galton's use of statistics, and was indeed dedicated to Galton. What Johannsen found, however, was not regression to a mean for the total population, but constancy of what he called "pure lines," where only slight, "continuous variation" was observed, but where larger gaps obtained between the different lines. This reinforced the distinction, dear to the first Mendelians, between relatively insignificant variations and the "mutations" that would have to be present to initiate new species (Johannsen 1903). It seems to be in that spirit and context that Johannsen, in 1909, introduced the concepts of "genotype" and "phenotype." Here "genotype" represents an underlying, theoretically stable and experimental testable, hereditary structure, whereas the phenotype constitutes the observable character of the organism, tractable to biometrical measurement (Johannsen 1909).

So do we now have the conceptual apparatus necessary to equip classical transmission genetics, and to serve as a foundation for the eventual discovery of the double helix and the genetic code? It may look that way. There are those germ cells (whose segregation had already been proclaimed by Weismann). What remains but advances in biochemistry and eventually molecular biology to get it all sorted out? Much work had been done since Schleiden and Schwann, in the 1830s and '40s, had done their pioneering work in what was to become cell theory and eventually the increasingly complex and prestigious field of cytology. After all, even Mendel had written about the reproductive germ cells, and it must be clear that there were structures in those cells, perhaps in particular in the nucleus, on which the phenomena of heredity were based. Yet matters had not yet been so definitely decided. That there were Mendelianly segregating factors of heredity was plain, but did they correspond to definite tracts of some substance, probably nuclear, and probably in the chromatin? It was Johannsen himself who introduced the term "gene," but he did so with no predilection in favor of locating the genes in some definite physical substrate. Starting from the pangen of de Vries (and Darwin), he proposed to drop the first syllable and speak of "the gene" or "the genes." But he did not, like de Vries with the pangen, identify his "gene" with any material particle. As he put it:

> The word 'gene' is . . . completely free of any hypothesis. It expresses only the fact that the characteristics of the organism are determined in the make-up of the gametes by particular 'conditions,' 'factors,' 'units,' or 'elements.' These are at least partially separable, and thus to some extent independent – in short, exactly what we wish to call genes.
>
> Johannsen 1913 in Burian 2000, p. 1130b[14]

Whether these were, in some functional but less than locatable way, "factors" in heredity, or whether there were actually things called genes located in the cell, and in particular in the chromosomes, thus appeared to some workers in the new genetics still quite debatable.

In parallel to the work of geneticists, however, cytologists had been studying the chromatin of the cell since the 1880s, and, theoretically at least, Weismann had already located his germ plasm in what would later be called the chromosomes. An important step forward in this development was the experimental work of Theodor Boveri on sea urchins and of Walter Sutton on grasshoppers, culminating in their publications of 1902–1904 (Darden 1991, pp. 72–79). They confirmed that the chromosomes were morphologically individuated entities, which persisted from one cell division to the next, and that they were indeed the carriers

---

[14] Translated by Burian from the 1913 edition of Johannsen 1909, in Burian 2000, p. 1130b.

of individual hereditary characters. As Lindley Darden has pointed out in her analysis of this history, there were striking parallels between their conclusions about the chromosomes and those of geneticists about the relationships between allelomorphs in Mendelian populations (Darden 1991, p. 85). But there were also difficulties. For example: How could a limited number of chromosomes carry so many Mendelizing traits? Surely great clusters of them would always have to be carried together. Not only Bateson, but many geneticists were still skeptical, including a young American experimentalist named Thomas Hunt Morgan, who had found embryology too speculative a discipline and had turned to the fruit fly *Drosophila* as a suitable organism for the study of genetics. It was only after some years of work that Morgan found in the "sex-restricted" mutant "white eye" convincing evidence for the chromosomal location of the hereditary material (Morgan 1911). From then on, he and his students in the "fly room" at Columbia University presided over the definitive establishment of classic transmission genetics (Morgan 1916). We might say it took *Drosophila*, with the happily enlarged chromosomes of its salivary glands and its amenability to breeding and interbreeding in population cages in Manhattan, to complete the revolution initiated half a century earlier in a monastery garden in Moravia.

Still, few seeming revolutions in science remain unqualified. As late as 1954, L. J. Stadler had distinguished between an "operational gene" – "the smallest unit of the gene string associated with a specific genic effect" – and a "hypothetical gene," an actual material particle with a specifiable location on a chromosome (Stadler 1954).[15] If, with the advent of molecular genetics, the "hypothetical gene" ceases to be hypothetical, it must be remembered, on the other hand, that such localizability is not as simple as some might have wished, and that it sometimes seems more reasonable to describe genes operationally, in terms, for example, of their accessibility to selection (see Beurton, Falk, and Rheinberger, eds. 2000). However, that is another, and later, story than the one we have been sketching here; we will return to it at the end of Chapter 9.

Finally, to return to our period, the first couple of decades of the twentieth century: we must mention one more feature in the new development

---

[15] See the account in Burian 2000. We are grateful to Dr. Burian for bibliographical and historical guidance in connection with these developments, through his 2000 article as well as personal communication.

of Mendelism. In 1908, a British mathematician, G. H. Hardy, and a German physician, W. Weinberg, independently discovered that in an ideal interbreeding population, Mendelian factors would reach an equilibrium, and, other things being equal, remain there (Hardy 1908; Weinberg 1908). That insight was implicit in Mendel's own discoveries, but making it explicit in a mathematical formula provided a base line from which evolutionary speculation, and experiment, could begin. Ideally, dominant and recessive factors are balanced in a ratio of 1 to 2 to 1, and selection can be measured in terms of the imbalance of that otherwise stable equilibrium. Of course, things are never ideal in fact, but thinking of it this way allowed geneticists, and some years later also what were then still their Darwinian opponents, to envisage a null-point situation from which to inquire into the agencies, whether selection, mutation or random drift, that might unsettle that equilibrium and so induce changes in population structure. In the absence of such disequilibriating factors, a largely randomly interbreeding population would not change the frequency of alternative alleles. In a way, this insight is reminiscent of Galton with his normal curve, to which organisms tend to revert, but what has altered the situation fundamentally is the reliance, in Mendelian terms, on the factors present in the parental genome only. There is no longer the "ancestral" ingredient, threatening regression, and eliminable only by unlikely saltations to some alternative stable position.

To allude to the Hardy–Weinberg law is to anticipate the way in which Mendelian genetics would be used two decades later to support, rather than to oppose, Darwinian evolutionary theory. At the time, however, evolutionary theory seemed to be going one way and the new science of genetics another. Nevertheless, there were several ways, not only the Hardy–Weinberg discovery, in which the new discipline of genetics was preparing the way for an eventual synthesis of the two traditions. Jean Gayon, in his *Darwinism's Struggle for Survival*, has given a most illuminating analysis of the consequences Mendelism had, or would eventually have, for the Darwinian theory; we conclude this chapter by enumerating the effects that Gayon specifies (Gayon 1998, pp. 289–309).

First, as we noted in connection with the Hardy–Weinberg ratio, the Mendelian approach to heredity – as distinct from the original, more limited question of hybridization – was to liberate heredity from its Galtonian, and indeed, popular, connection with descent. Remember that on Galton's model, each individual – and all the individuals in each population – received half its characteristics from its parents, but the rest in decreasing amounts from more remote ancestors. A long line of descent was what mattered. In fact, breeders often held that the longer a trait had been persisting, the greater its effect. And there was always the threat of regression in the background. In Mendelian terms, in contrast, it was the genotype of the parents, or, strictly, their contribution to the next generation, that mattered; never mind questions of genealogy, "good blood," and the like.

Second, Gayon points out, "Mendelian genetics swept away the notion of heredity as a 'force' or 'tendency' that varies in intensity" (Gayon 1998, p. 289). The genes in question, whether dominant or recessive, are either present in the gene pool of a given generation or they are not. True, dominance and recessivity turn out not to be absolutely all or none affairs, as some have sometimes thought, but the presence or absence of a given gene in a given genome, and statistically, in the gene pool of a population, is not subject to "force-like" considerations.

Third, it was accordingly now possible to accept the changes induced by selection as relatively stable, and so subject to experimental study. It could be asked how varying factors such as mutation, drift, or selection itself would affect the structure of populations. Here we meet again with a difference in philosophical approaches. Pearson, as we have seen, was an ardent Machian, and even if the empirical work of his friend Weldon seemed to issue in what look like causal statements, it was the observation of large numbers of cases in the field that were involved, not theoretical hypotheses subjected to experimental treatment. You looked in a great many cases to see if you could find selection happening and reported what you found. With the acceptance of the principles of genetics, in contrast, biologists acquired a basis for their work in a clear, causally deterministic theory. Starting from a stable base, they could investigate experimentally possible consequences of their theoretical position in circumstances they could not only observe but manipulate.

The importance of these interrelated effects of the rise and acceptance of Mendelism will be apparent when we turn, in the next chapter, to the Modern Evolutionary Synthesis of the twentieth century, both its achievements and the challenges to it that have arisen in recent decades.

CHAPTER 9

✦

The Modern Evolutionary Synthesis
and Its Discontents

What Is the Modern Evolutionary Synthesis?

The Modern Evolutionary Synthesis has served as the dominant interpretive framework that has guided professional evolutionary biology during the central decades of the twentieth century. Just what was being synthesized? In the 1920s and 1930s, there occurred a synthesis between Mendelian genetics and Darwinism, whose champions had earlier been at each others' throats (see Chapter 8). The central figures in this effort were R. A. Fisher, J. B. S. Haldane, and Sewall Wright. However, the phrase "modern evolutionary synthesis" refers not only to the synthesis between Mendelism and Darwinism, but to a sustained effort extending from the late 1930s through the 1940s to use "population genetics," in the sense worked out by these pioneers, to unify – synthesize – a wide array of biological disciplines. That is how Julian Huxley, grandson of Thomas Huxley, used the phrase in his 1942 book *Evolution: The Modern Synthesis*, which gave the Synthesis its name. Huxley listed "ecology, genetics, paleontology, geographical distribution [biogeography], embryology, systematics, [and] comparative anatomy" as "converging upon a Darwinian center," thereby rescuing Darwinism from its turn-of-the century "eclipse" (Huxley's phrase) and, in Huxley's opinion, bringing into being for the first time a truly modern, theoretically based, biology (Huxley 1942; 1943, p. 25[1]). Key figures in the Modern Synthesis in Huxley's sense were Theodosius Dobzhansky, Ernst Mayr, George Gaylord Simpson, and G. Ledyard Stebbins. All these figures wrote books in a series published by Columbia University Press between 1937

---

[1] Page references to Huxley's *Evolution: The Modern Synthesis* are to the 1943 American edition.

and 1952. In these so-called "canonical" works of the Synthesis, which were based on the annual Jesup lectures at Columbia, population genetics was brought to bear on topics that included the nature and origin of species, their geographical distribution, and the relationship between species and *taxa* above the species level.[2]

So entrenched is the Modern Synthesis in evolutionary thinking that anyone who criticizes its key doctrines takes up a heavy burden of proof. This has not, of course, deterred numerous challengers from having their say; in fact, hardly any aspect of the Modern Synthesis has remained uncontroversial during the last thirty years or so. Disputes raged in the 1970s, for example, about how to bring behavior, especially cooperative behavior, into the Darwinian fold. The 1980s witnessed widely publicized expressions of doubt on the part of some paleontologists about Simpson's assurance that "macroevolution" – evolution above the species level – is merely the result of accumulated "microevolutionary" change at the level of populations. And perhaps most significantly for the future of evolutionary studies, the 1990s saw the consolidation of a long-simmering challenge from developmental biology, which had been left out of the Synthesis at its formation, but has in recent years become increasingly germane to evolutionary studies. Many of these challenges to the Synthesis – or at any rate to the way it was originally formulated – reflect the growing influence of molecular biology, especially molecular genetics, a field that did not even exist when the Synthesis was first formed, which came from quite different intellectual roots (in biochemistry), and which, after James Watson and Francis Crick's revelations about the structure and function of the DNA molecule, has continued to go from triumph to triumph. Throughout this chapter, we will mention how aspects of molecular genetics, including, most recently, developmental genetics, have affected the Synthesis.

The work of defending, expanding, challenging, and, perhaps, replacing the Modern Synthesis has tended to bring out the philosopher in many evolutionary biologists. It has also attracted to the scene professional philosophers, whose efforts to clarify key concepts in contemporary evolutionary theory have resulted in substantive contributions to the field and have stimulated the formation of a new specialty within professional philosophy of science – the philosophy of

---

[2] It has recently been shown that when Dobzhansky gave the first lectures in this series they were not yet called Jesup lectures. See Cain 2002.

biology. The concerns of both philosophizing biologists and biologically informed philosophers have been based on three sorts of issue. First, there are conceptual issues about the meaning, or "analysis," of key terms in modern evolutionary science: "fitness," "adaptation," "unit of selection," "gene," and so forth. Second, questions have arisen about the sorts of inferential patterns on which the Synthesis depends: Are there universal laws in evolutionary biology, as there presumably are in physics? Is evolutionary theory predictive? If it is not, how can it be explanatory? Third, there are questions having to do with how the Modern Synthesis – and, recently, the stress on evolution and development ("evo-devo") – has affected the central themes of philosophy of biology we have been tracing in this book, such as teleology, reductionism, and classification. This chapter will be concerned in the main with the first two sorts of issues.

The Making of the Modern Synthesis

If the Modern Evolutionary Synthesis has a theoretical core, it seems fair enough to say that this core consists in the impressive efforts of mathematical population geneticists in the 1920s and '30s to reconcile Darwinism with Mendelism. Early twentieth-century Mendelians, as we have already learned in Chapter 8, were committed to viewing evolutionary change as neither gradual nor the result of natural selection. De Vries, one of the rediscoverers of the Mendelian laws, wrote "The theory of mutation assumes that new species and varieties are produced from existing forms by certain leaps" (De Vries 1905, p. vii). However, in a series of works beginning in 1918, the Darwinian eugenicist and statistician R. A. Fisher showed that the probability of a genetic mutation's being diffused throughout a population is inversely proportional to the effect of that mutation (Fisher 1918; Fisher 1930). Large mutations – mutations producing what the geneticist Richard Goldschmidt (1940) called "hopeful monsters" – have in fact very little hope. Fisher also argued that small mutations are likely to spread through a population under even small selection pressures, and that the rate at which natural selection propels slightly fitter genotypes through an indiscriminately interbreeding ("panmictic") population is proportional to the amount of genetic variation available (or, more technically, to the statistical *variance* of the *additive* genetic variation). Once it is lodged in a population, an allele (an alternative form of a gene) is likely to stay there, even when there is selection against it, as Haldane especially was concerned to show. For in diploid organisms – organisms in which there is room for two alleles at each single locus on the chromosome, one from each parent – recessive alleles are exposed to selection, whether for or against, only when they show up in two copies at a single locus on the chromosomes. Tucked away in single doses, they are very hard to eliminate completely from a population, and may in fact later be selected for when circumstances change.

There are variant versions of population genetics and of the sorts of Darwinism – genetical Darwinism – that can be built on it. Especially significant is the long-standing tension between Fisher's and Wright's versions. Although not sharp enough to be called contradictory, their opposed models of how natural selection normally works have been carried forward into different versions of the Modern Synthesis. On any version of the Synthesis, however, whether it be Fisher's, Wright's, or anyone else's, the relationship between genes and traits is assumed to be many-to-many. There is typically no single gene for a single trait (although malfunction in a single gene can certainly knock out the expression of a trait in an individual organism, as it often does in genetic diseases). On the contrary, many genes presumptively go into making a single trait and, conversely, many traits often depend on the same genes. "The value of any gene depends in general on the array of other genes with which it is associated," wrote Wright. In the guinea pig populations on which he worked throughout his long life, for example, Wright eventually found "more than fifty loci...that affect coat color" (Wright 1982, p. 625). Moreover, on any and all versions of genetic Darwinism, organisms (and the genes they contain) are construed as members of populations, and changes in them over multiple generations are treated statistically and probabilistically. In this connection, as we mentioned in the last chapter, theoretical population geneticists use the Hardy–Weinberg formula as a base line from which to calculate the probabilities of changing gene frequencies in populations under this or that set of circumstances. The formula is a binomial expansion of the Mendelian laws. It tells you that the population-wide distribution of a variant gene, an "allele" or "allelomorph," at any given time t will be $p^2AA + 2pq\,Aa + q^2aa$, if p is the probability of a dominant gene A, and q that of recessive a (in a diploid chromosomal system). The formula enables you to show how alleles will be distributed in a population unless and until they are disturbed by a dose of one or more "evolutionary forces" – mutation pressure, selection pressure, gene flow, genetic drift, and factors such as population density and the frequency of an allele in a population at a given time.

In spite of much that they held in common, however, Wright crossed swords with Fisher when he challenged the latter's assumption that mutation pressure and natural selection are, on the whole, the only forces at work in evolution, and that in consequence large panmictic populations present the most favorable conditions for "mass" natural selection to create adaptations. "I would not deny the possibility of very slow evolutionary advance through [Fisher's] mechanism," Wright wrote in an early review of Fisher's *The Genetical Theory of Natural Selection*,

> but it has seemed to me that there is another mechanism which would be more effective in preventing the system of gene frequencies from settling into a state of equilibrium . . . If the population is not too large, the effects of random sampling of gametes in each generation bring about a random drift of the gene frequencies about their mean positions of equilibrium . . . . If the population is not too small, this random drifting leads inevitably to fixation of one or the other allelomorph, loss of variance, and degeneration. At a certain intermediate size of population, however, . . . there will be a continuous kaleidoscopic shifting of the prevailing gene combinations, not adaptive in itself, but providing an opportunity for the occasional appearance of new adaptive combinations that would never be reached by a direct selection process . . . A favorable condition would be that of a large population broken up into imperfectly isolated local strains . . . Complete isolation originates new species differing from the most part in nonadaptive respects, but is capable of initiating an adaptive radiation.
>
> Wright 1931, in Wright 1986, p. 159

For Wright, the difficulty with Fisher's view is not simply that "this is an exceedingly slow process" (Wright 1982, in Wright 1986, p. 628). The problem is that for at least two reasons, gains in an adaptive direction are offset by losses in another. For one thing, adapted organisms introduce significant changes into their own environments, undermining the selective value of the very traits that make for fitness in the first place. (This was later known as "The Red Queen Hypothesis," after the character in *Alice in Wonderland* who must run very fast just to stay in place [Van Valen 1973; see Wright 1982, in Wright 1986, pp. 628–9]). For this reason, Wright argued, "a species occupying a small field under the influence of severe selection is likely to be left in a pit and to become extinct, the victim of extreme specialization to conditions which have ceased" (Wright 1932, in Wright 1986, p. 167). A second source of difficulty is that in any finite population, but especially in a small one, some less than optimally adapted alleles will inevitably accumulate through mutation pressure and chance, or what Wright called "genetic drift." The smaller the breeding population, in fact, the more likely it is that "the tendency toward complete fixation of genes, practically irrespective of selection, will lead in the end to extinction" (Wright 1932, in Wright 1986, p. 165).

Wright's proposal was to think of genetic drift as part of nature's solution rather than as posing a problem. Noting that in the wild a great many species do not in fact live in large panmictic populations such as those assumed by Fisher, but instead in the small breeding populations that Wright called "demes," Wright hypothesized (in the passage quoted extensively here and in many other places; see Wright 1932) that natural selection is at its most effective when an allele gets a toe hold in a small population (a "deme") purely by chance ("genetic drift"). If they subsequently prove to be adapted, such alleles will spread to other local populations of the same species by way of the "gene flow" that results when emigrants move to other demes. These populations will be selected and will spread their characteristics throughout the range of the species.

Wright's work was no less theoretical, mathematical, or probabilistic than Fisher's. Genetic drift, for example, is simply the biological instantiation of a law of probability theory according to which small samples often fail to exhibit the expected distribution that will be approached as sample size grows. Nonetheless, Wright's model was considerably more empirical than Fisher's; it was inspired by, and sought to explain, well-known practices of animal breeders, who had long known that concerted selection in a single direction makes for unfit animals, and who, in consequence, had long known that they should divide their breeding populations in the way Wright now ascribed to nature (Provine 1986). Wright called his model the "shifting balance" theory of genetic evolution. He did so because the interaction of evolutionary factors or forces was presumed to shift somewhat unpredictably between mutation, migration, selection, and drift, even though on the whole it tended to settle into the most favorable, or balanced, situation for *continued* adaptive response. In this way, Wright added greatly to the number of factors that must be taken into account in any realistic evolutionary situation, as well as to the number of likely or possible scenarios that evolutionary change can in principle exhibit.

Wright's influence was conspicuous in Dobzhansky's 1937 *Genetics and the Origin of Species*, the earliest and the most seminal of the four "canonical" works of the Synthesis.[3] Dobzhansky's opposition to Fisher was no less pronounced than Wright's; he called him an "extreme selectionist" (Dobzhansky 1937, p. 151; Dobzhansky 1941, p. 188). Dobzhansky's great theme, however, was how species are actually formed in nature by way of Wright's arsenal of forces, combined with genetically based, and not just geographical, mechanisms for isolating populations. Dobzhansky's method for dealing with this subject was to bring specimens of different species and sub-species of fruit flies (*Drosophila*) to the laboratory, where variations and changes in their chromosomes were carefully noted by microscopy. He then used population genetical models to portray the evolutionary dynamics of the natural populations from which his specimens had come. In his three-cornered approach, Dobzhansky masterfully blended the field research practices that he had brought with him from Russia (which in the 1920s had a very active and influential genetics community; see Adams 1994), the laboratory techniques he had learned in T. H. Morgan's famous "fly room" lab at Columbia, and population genetical analysis. In the last respect, he had considerable help from Wright (Provine 1986, pp. 327–365). (Dobzhansky confesses that he just whistled when he saw a difficult equation; of Wright, he remarked, "Papa knows best" [Dobzhansky 1962–3, p. 399].)

In *Genetics and the Origin of Species*, Dobzhansky treated many of the same concepts with which Darwin had dealt in the great work to which Dobzhansky's own title alludes: variation, heritability, natural selection, adaptation, and, finally, speciation. He spent most of his time on variation because, as he explained, "the most serious objection raised

---

[3] Dobzhansky 1937 was seminal because the other architects of the Synthesis acknowledged the power of his lead and applied it to their areas. "When Dobzhansky gave the Jesup Lectures at Columbia University in 1936," Mayr wrote, "it was an intellectual honeymoon for me" (Mayr 1980d, pp. 419; 420–21). Simpson testified that "the book profoundly changed my whole outlook and started me thinking along lines...less...traditional in paleontology" (quoted by Mayr in Mayr 1980c, p. 456). Stebbins (whose own Jesup lectures we must unfortunately neglect) wrote, "The present author was stimulated to apply the theory [of Dobzhansky's book] to plants" (Stebbins 1980, p. 174). For his part, Dobzhansky disclaimed some of the credit. "The reason why that book had whatever success it had was that...it was the first general book presenting what is nowadays called...'the synthetic theory of evolution'" (Dobzhansky 1962–3, pp. 397–400).

against [Darwin's theory of natural selection] is that it takes for granted the existence, but does not explain the origin, of the heritable variation with which selection can work" (Dobzhansky 1937, p. 149). Genetics now removes this objection. For one thing, variation is not directly dependent on mutations; the process of meiotic division is constantly creating new, potentially adaptive genetic combinations, a fact that becomes both obvious and significant once one gives up "one gene-one trait" thinking. Dobzhansky's laboratory work on specimens of wild *Drosophila* populations showed that rearrangements of the chromosome, as well as the duplication of whole chromosomes ("polyploidy"), can also provide sources of variation. As a result "an enormous, and hitherto scarcely suspected, wealth of genic diversity has been detected in wild populations of *Drosophila*" (Dobzhansky 1941, p. 62).

Genetic variation is not only plentiful in natural populations. Dobzhansky went on to assert that it is not bunched up at one end or another of a statistical curve, as it was for "classical population geneticists," such as H. G. Muller, but is distributed throughout a population (Dobzhansky 1970, pp. 197–199; see Muller 1948). Since Dobzhansky's initial efforts, the amount of genetic variation diffused throughout natural populations has, in fact, proven to be very large; a population of a given species will often contain ten or more alleles for any locus on the chromosome. This stress on the availability and wide diffusion of variation, considered as the material or fuel on which natural selection works – the *balance* as against the *classical* theory – is characteristic of the tradition in population genetics that Dobzhansky brought from Russia and diffused in the United States. It is a salient theme in the work of his student Richard C. Lewontin, for example, who has pursued Dobzhansky's early conviction that the amount of genetic variation within a natural population is typically higher and more spread out than the amount of variation between populations (Lewontin 1974). This discovery has important implications. It implies, for example, that terms such as "race" are biologically far more problematic than had hitherto been assumed. It also implies that eugenics (to which Fisher was deeply attached) is bad biology, depending as it does on the false assumption that one can locate exceptionally bad and exceptionally good traits, and genes "for them," at the extreme ends of a "normal" statistical distribution (Beatty 1994).

As to heritability, Dobzhansky recognized that more than genes are heritable; for example, he did not entirely dismiss the long-standing claims of embryologists for cytoplasmic inheritance (Dobzhansky 1937, pp. 72, 296). Nonetheless, he argued that compared with genetic inheritance, these phenomena have almost no effect on evolutionary change (Dobzhansky 1937, p. 72). Dobzhansky even *defined* evolution as "a change in the genetic composition of populations" – a bold and, in retrospect, somewhat reckless, conceptual commitment, since this definition had the immediate consequence, which Dobzhansky explicitly embraced, that population genetics by itself constitutes a *complete* theory of evolution (Dobzhansky 1937, p. 11). On the other hand, Dobzhansky did see that natural selection – the statistically differential retention of heritable genetic variation in a population – has an empirical, not a definitional, relationship to adaptation. Natural selection is an explanation, and in fact the best explanation, of adaptation (Dobzhansky 1937, p. 150). Dobzhansky also saw that a contingent relationship obtains between adaptive natural selection and evolution, since at least in early editions of *Genetics and the Origin of Species* he wholeheartedly commended Wright's shifting balance theory, and so accorded a certain independence in evolutionary dynamics to genetic drift and other population-genetical "forces."

Having set all this forth, Dobzhansky came to his main point about the nature and origin of species. A species, he argued, is "a stage in a process" in which "a once actually or potentially interbreeding array of forms becomes segregated into two or more separate arrays which are physiologically incapable of interbreeding" (Dobzhansky 1937, p. 312). A species is a real thing in nature – a set of populations reproductively isolated from others by more than merely contingent circumstances, such as geographical isolation. A species is not primarily, then, a taxonomic rank. If it is so, it is because it is founded on a real entity in nature (a claim to which Darwin was attracted, but about which he was not always confident [Beatty 1982; 1985]). As real entities, species come into existence and go out of existence at definite points in time and space. They come into existence, Dobzhansky argued, when the interbreeding populations of which they are made up have found a way not just to occupy, but to keep themselves on, an adaptive peak by blocking gene flow between themselves and other populations, while at the same time they maintain sufficient genetic variation within themselves to adapt to an environment that inevitably changes, often by the organisms' exploitation of its resources. Typically, Dobzhansky said, geographical isolation is a prelude to speciation. But "the formation of isolating mechanisms" properly so called involves genetic change, "mostly not single mutational steps, but the building up of systems of complementary genes"

(Dobzhansky 1937, p. 256; see Dobzhansky 1941, p. 282, for a clearer statement). Giving a realistic interpretation to Wright's metaphor of an "adaptive landscape," Dobzhansky concluded that "the genotype of a species is an integrated system adapted to the ecological niche in which the species lives" (Dobzhansky 1937, p. 308).

We will have more to say about the species concept and its relationship to systematics in the next chapter. For the present, we may turn briefly to Mayr's and Simpson's contributions to the making of the Modern Synthesis. Mayr's 1942 *Systematics and the Origin of Species* builds on Dobzhansky in the way Dobzhansky built on Wright. Mayr, an avian systematist working at the Museum of Natural History in New York, was as enthusiastic about Dobzhansky as Dobzhansky had been about Wright when he first heard him speak. "When Dobzhansky gave the Jesup lectures at Columbia University in 1936," Mayr later wrote, "it was an intellectual honeymoon for me. He came down to the museum and I was able to demonstrate to him the magnificent geographic variation of South Sea Island birds. . . . I was delighted with the book that came out of his lectures" (Mayr 1980d, in Mayr and Provine 1980, pp. 419–420). For his part, Dobzhansky cited some of Mayr's observations in his book (Dobzhansky 1937, p. 54).

Still, differences in emphasis appeared when Mayr published his Jesup lectures in 1942. Whereas Dobzhansky had written from the perspective of changing gene frequencies, Mayr concentrated on what was observable at the phenotypic level. In part for this reason, Mayr was not taken with Dobzhansky's definition of a species as a stage of evolution; for him, species are simply geographically distributed populations that, for one reason or another, have been reproductively isolated from others. (See Chapter 10 for more on Mayr's "biological species concept.") More importantly, Mayr ascribed a greater causal weight to geographical isolation in the formation of species. He vigorously defended a particular version of "allopatric" speciation – "peripatric speciation" – according to which a new species almost always begins with a very small "founder population" that exists at the edge of its species range. Isolated, and frequently stressed, these populations are prone to "genetic revolutions," which, in the statistically rare case that they prove to be adaptively successful, can radiate outward, frequently into the range of a soon-to-be-displaced parent species (Mayr 1954; Mayr 1970, p. 534; Mayr 1988, p. 446). For Mayr, a "genetic revolution" happens when inbreeding in a small, isolated population uncovers many recessives, some harmful or even lethal. This in turn disturbs the genetic balance or homeostasis that in well-adapted organisms extends across the genome. Extinction is often the result. In a small number of cases, however, a new species takes hold (Mayr 1988, p. 446).

In his 1944 *Tempo and Mode in Evolution*, the paleontologist George Gaylord Simpson further consolidated the Modern Synthesis by arguing that evolution above the species level is simply an extrapolation over time of what natural selection, and the forces associated with it, create at or below the species level. Dobzhansky and Mayr had already anticipated this claim on purely theoretical grounds. If species are formed in the ways that Dobzhansky and Mayr say they are, it seems reasonable to infer that higher taxa are simply the result of ever-accumulating genetic distance between species, some caused by adaptive change between speciation events ("anagenesis"), some concentrated at the branching events when new species are born ("cladogenesis"). Dobzhansky said he was "reluctantly forced" to put an equal sign between microevolution and macroevolution – reluctantly because this was still very much a "working hypothesis" (Dobzhansky 1937, p. 12).[4] Simpson was more confident. While he acknowledged that his aim was to establish only that there is nothing about macroevolutionary phenomena that is inconsistent with the microevolutionary processes acknowledged by the Synthesis, Simpson was in fact quite convinced that everything that happens in phylogeny is the result of gradual natural selection at the level of populations, albeit working at different "tempos and modes," some of which, especially those connected with speciation, are quite rapid ("quantum evolution"). There are, on this view, no grounds for postulating a separate mechanism for large-scale evolutionary change, such as the sorts of "systemic macromutations" defended by Goldschmidt, whose *Material Basis of Evolution* the founders of the Modern Synthesis all loved to

---

[4] This is not to imply that Dobzhansky was not trying hard to argue that population genetics controls macroevolution. The issue had already been framed in Russia in a debate between S. Chetverikov and Dobzhansky's teacher, I. A. Filipchenko. If it were written in Russian rather than English, Dobzhansky's book would have to be seen as intervening on the side of Chetverikoff. See Adams 1980; Burian 1994. The same issue was also being debated in Germany. Bernard Rensch's 1947 *Neuere Probleme der Abstammungslehre*, (Stuttgart, Enke), which was translated into English in 1960 on Dobzhansky's recommendation (Rensch 1960), defended the continuationist view in Germany. The Synthesis, although it was most intensively and institutionally pursued in the United States, was an international accomplishment. We regret not being able to spend more time on the development of the Synthesis in other countries.

hate (Goldschmidt 1940). As a result, the concepts of the emerging Synthesis are implicitly said to constitute a more or less complete theory of evolution in their own right. Nothing more is needed. (This confident extrapolation was to come under fire in 1972, when Niles Eldredge and Stephen Jay Gould published their theory of "punctuated equilibrium" [Eldredge and Gould 1972]).

During the evolutionary debates of the 1980s, it was claimed by Gould and others that after an initial period of "pluralism" about evolutionary forces that was dominated by Wright, the Synthesis had "hardened" into an "adaptationist program" according to which natural selection alone is more or less responsible for adaptation and for evolution (Gould 1983). There is a good deal of truth to this claim. In the late 1940s and early 1950s, evolutionary biologists such as A. J. Cain and E. B. Ford were demonstrating that phenomena hitherto seen as the result of genetic drift and other non-adaptive forces – patterns in the shells of the snail *Cepaea nemoralis*, for example – are seasonally polymorphic adaptations (for *Cepaea*, see Cain and Sheppard 1950; Cain 1954). During the same period, Fisher and Ford challenged Wright about whether drift played any role in the evolution of a moth called *Panaxaia dominula* (Fisher and Ford 1947). This controversy forced Wright to take pains to deny that he had ever said that genetic drift (which in his 1942 book Huxley had called "the Sewall Wright effect," and had commended as a relatively independent cause of evolution) is sufficient by itself to produce speciation. Drift, Wright insisted, is, and had always been, intimately connected with interdemic selection in the origin and spread of adaptations and of species (Wright 1948).

During the same period, Mayr and Dobzhansky, too, were becoming more adaptationist. Mayr confessed in the 1970s that

> In one respect I myself became a stronger selectionist after 1942, for I realized that [the selective] neutrality of alleles could not explain cases of polymorphism that had remained stable for fifty, eighty or more years, or that had stayed the same over large geographical distances. If such morphs were truly neutral, accidents of sampling should produce large fluctuations, while their absence would seem to indicate the presence of balancing selective forces.
>
> Mayr 1988, p. 528

The "balancing selective forces" to which Mayr refers here allude to Dobzhansky's theory of "balancing selection," which was conceived as an adaptationist alternative to Wright's "shifting balance" theory. The balance that is struck among evolutionary forces, according to Dobzhansky, is actually a result, in the large number of lineages fortunate enough to possess them, of adaptations that preserve, rather than merely utilize, genetic variation, with the result that variation is retained in gene pools that can later be favored by selection in a changing environment. This theory had a certain resonance in the debates of the day; by making adaptive natural selection central, Dobzhansky reduced the impression left by Wright that drift and other factors are independent forces. Still, Dobzhansky's interest in this topic was not new. When he came to Morgan's fly room in 1927, he brought with him the hypothesis of the Russian geneticist Sergei Chetverikoff that "species soak up heterozygous mutations like a sponge" (Chetverikoft 1926; see Adams 1980). They can be presumed to do so because natural selection often favors the heterozygote as a sort of hedged bet. By studying cases in which natural selection favors the heterozygote ("heterosis"), Dobzhansky came to believe that heterozygote superiority ("hybrid vigor"), and perhaps even the evolved diploid structure of chromosomes on which it depends, are themselves adaptations for retaining variation that might prove valuable in changing environments (Dobzhansky 1970, pp. 197–199).

Conceptually Clarifying the Modern Synthesis

In a rather grand sense, most, if not all, the figures we have been considering had philosophies, in the sense of privately held metaphysical interpretations of the scientific theory that they shared. Thus Wright, when asked to write about his metaphysical interpretation of evolution, proved himself to be a monistic idealist (Wright 1964). Simpson, by contrast, was a rather dogmatic materialist. And Dobzhansky, for his part, was an orthodox Christian who worried about how a good God could come up with balancing selection, which sentenced to death a predictable number of homozygotes for a recessive lethal allele, presumably in order to preserve the balance that could lead to further evolutionary advance.

It was not in such matters, however, that philosophical reflection on the Modern Synthesis centered. Private interpretations, where they were known at all, were off the table in journals such as *Evolution*, which was founded in 1947 by Mayr and others who were hoping to project the Synthesis as a mature, value-neutral science. Just because those who forged the Synthesis were seeking a professional status that had eluded the Darwinisms of an earlier day – compromised as they had been by Spencer's "Social Darwinism," Galton's and Fisher's eugenics, and other such value-laden, ideological concepts – the Synthesis came under philosophical scrutiny of a different sort. Efforts to bring its various component disciplines – genetics, paleontology, systematics, and so forth – together to forge something approaching a single discipline implied that the Modern Synthesis was solid enough to count as empirical science based on well-defined theoretical core. How to think of that core became a matter of concern. And with the rise of biochemistry, evolutionists also worried about the relationship of their subject to others threatening its status. (For the relevant literature on the second topic see Chapter 10.)

Historically, the Synthesis could be validated, at least in part, by showing that this version of natural selection confirmed and clarified Darwin's own insights.[5] In this connection, for one thing, the makers of the Modern Synthesis could point to the fact that Darwin's ambiguity about whether species are real units in nature or chimeras that merely reflect back one's own arbitrary taxonomic categories had now been resolved in favor of realism about species, and in favor of the possibility of the "natural system of classification" that Darwin ardently sought (Darwin 1859, p. 485; Beatty 1982; 1985; see Chapter 10 for more on this topic). Then, too, the balance that Darwin was seeking between the elements of chance and necessity in the evolutionary process had now been struck. Natural selection had been misunderstood by physics-oriented luminaries such as Herschel as "the law of higgledly piggledy." On the other hand, Spencer's pseudo-Darwinism, which was unjustifiably well-known and well-respected in the waning decades of the nineteenth century, treated organisms as passive functions of the environments by which and to which they were molded, thereby stressing the element of necessity to the point of determinism. The situation had, if anything, grown worse during the contest between Mendelians and Darwinians. With their ill-conceived focus on large mutations as the origin of species, the early Mendelians stressed chance to the virtual exclusion of natural selection. The Darwinian biometricians were on the right track, but they had no theory of speciation or of other fundamental evolutionary processes (see Chapter 8). The Modern Synthesis, by contrast, hit just the right note (Eldredge and Grene 1992, pp. 46–49).

---

[5] This is particularly true of Mayr, who has reconstructed the *Origin* in a way that makes it anticipate the Modern Synthesis (Mayr 1991).

Mutation and recombination are the chance element in evolution. But they are simply the first step in a two-step process. The second step is natural selection of the variations thus produced, which in turn gives rise to adaptations. Ranging as it does over a vast sea of genetic variation, natural selection is a creative, shaping force rather than an executioner of the unfit, as it had appeared to earlier Darwinians (and still appeared to Muller) as well as to mutationists such as de Vries. In consequence, the beings that natural selection creates are not passive pawns of their environments, but exhibit all the abilities to respond actively and creatively to the challenges presented by their environments that great biologists from Aristotle onward could not help recognizing. To be sure, the necessitarian element is a bit more prominent in the legacy of Fisher, and the chance element in the legacy of Wright and Dobzhansky. Nonetheless, in spite of the persistence of the tension between these two orientations – and its continuous eruption into controversy – the Modern Synthesis could plausibly present itself as the triumph, after several near-death experiences, of the Darwinian tradition.

Another important conceptual implication of the Modern Synthesis emerges from the balance between chance and necessity that we have just noted. The first occurrence of a mutation that enhances the reproductive prowess of an individual that happens to possess it does not, and cannot, count as an adaptation. For an adaptation is a trait that not only has the *effect* of enhancing fitness, or that is useful to its possessor in fighting life's battles, but that has come to be by way of cumulative, directional selection over many generations, *just because it has this effect* (Williams 1966; Lewontin 1978; Burian 1983; Brandon 1981a; 1990, pp. 40–44). Thus, by definition, natural selection of a concerted kind is a necessary condition for a particular member of a populations having this or that particular adaptation. Whether it is a sufficient, as well as a necessary, condition is a more difficult question. It is so only if one assumes as a general rule that natural selection is the overwhelmingly dominant force at work in evolution, that it always brings forth traits that are optimal for reproductive success, and that one is entitled to use criteria of "engineering efficiency" or "reverse engineering" to establish that this or that trait is the best solution to a problem posed by the environment (Burian 1983, p. 302; see also Burian 1992a). However, these general assumptions are full of dangers, in spite of the fact that they have been widely adopted in "hardened" versions of the Synthesis. (Compare the changes in successive editions of Dobzhansky's authoritative book.) This approach tends, for example, to neglect cases in which concerted natural selection moving in a single direction undermines adaptedness, and its method of reverse engineering presumes, rather than establishes, that organisms are, or are very much like, designed entities rather than products of evolutionary *bricolage* or tinkering (Gould and Lewontin 1979; for the contrary view, see Jacob 1977).

We come now to another conceptual issue in the Modern Synthesis. In spite of the increasing acrimony of the controversies we have briefly recounted, the growing theoretical integrity and influence of the Synthesis soon made it a pressing task to determine whether an old accusation that had dogged Darwinism ever since Darwin had been persuaded to use the phrase "survival of the fittest" could be laid to rest. The "survival of the fittest" was a doubly unfortunate coinage. On the one hand, it was redolent of Spencer's sense of organisms as passive pawns of very demanding environments. On the other, it seemed to be an empty tautology, and so not to be explanatory at all. Of course the fit survive, since the fit are by definition the survivors! The need to rebut this old charge was made especially urgent by the fact that, responding to the growing prestige of the Modern Synthesis, no less a personage than Sir Karl Popper revived the charge of tautology by claiming that natural selection, while valuable as a metaphysical "conjecture," was empirically empty since it was non-falsifiable (Popper 1972). (That Sir Karl should have thought so is in part explained by his misunderstanding of the new Darwinism; overstressing the chance element, he thought that "each small step [leading to an adaptation] is the result of a purely accidental mutation" [Popper 1972, pp. 269–70].)

Beginning in the 1970s, professional philosophers of science helped the new Darwinians respond to this charge. So convincing were these responses that Popper soon withdrew them (Popper 1978). Philosophers of science differed somewhat, however, in their approaches to the problem. One tack was to think of the new Darwinism as a full-fledged theory in the empiricist mode, resting on laws of nature established by testable hypotheses in ways that conformed more or less to logical empiricist or to Popper's own falsificationist norms. Those who took this view could then respond to the charge of tautology by claiming that at the level of axioms there is nothing wrong with a little tautology. The term "fitness" is an undefined, primitive term in the theory, or is defined by its relationship to other primitives, such as "force" or "mass" in Newton's laws (Rosenberg 1983).[6] The link to empirical data comes

---

[6] That fitness is an undefined term is also the view of the philosopher of biology Elliott Sober (Sober 1984; 1993). Unlike positivists, however, Sober distinguishes between

much further down in the continuum between theory and data, and the entire theory is to be judged, like any other axiomized theory, in terms of its theoretical parsimony and elegance, its predictive capacity, and (on the logical empiricist assumption that explanations are simply retrodictive predictions) its consequent explanatory prowess.

However, a quite different sort of reply to the tautology objection soon gained favor. Fitness is indeed defined in contemporary evolutionary theory by reproductive output, as the objection assumes. But a well-known thought experiment soon made it clear that fitness is not defined by actual reproductive output, but instead by expected reproductive output in a given environment, which data about actual output can help establish by statistical analysis, but not define. If two identical twins are on a mountain top, the experiment goes, and one is struck dead by lightning, while the other goes on to produce a large and fecund family, both twins must be pronounced equally fit, even if one is luckier than the other (Scriven 1959). For fitness means relative adaptedness of one sub-population compared with another in a given environment, and the various "components" that confer relative adaptedness consist in the myriad small advantages by which selection gives a reproductive advantage to one phenotype rather than another. Expected fitness, or relative adaptedness, is in this sense a *dispositional* property of organisms, a propensity to reproduce that is not undermined by occasions, such as a bolt of lightning, when the expected result fails to happen in a statistically irrelevant case (Mills and Beatty 1979; Brandon 1978; 1996, pp. 3–29). On this view, there is no tautology at all. Differential reproduction is the result of ordinary when-then causality operating on populations through a host of underlying, mechanistic causes. The problem is explaining how diverse components of fitness come together to generate an expected number of offspring – often a very difficult task, especially if the inquirer is judicious enough not to infer too hastily from laboratory experiments based on engineering criteria (Burian 1992a).[7]

One's preference for either the first or the second way of deflecting the tautology objection is roughly correlated with one's attitude about whether evolutionary biology rests on laws of nature. Philosophers

"source" laws (molecular biology) and "consequence laws" (population genetics), and acknowledges that the conditions under which these two sorts of law meet, and the evolutionary "forces" that join them, are many and various – and not especially law-like.

[7] On some versions of the evolutionary theory, such as Fisher's, the tautology may be inherent and incurable. See Grene 1961.

of science who take the first approach – what's wrong with a little tautology? – have generally been at pains to show that genetical Darwinism can be axiomatized and formalized (Williams 1970; Rosenberg 1983). Some support might conceivably be found to support this view in theoretical population genetics, and even in the classic works of the Synthesis. *Defining* evolution as changes in gene frequencies in populations, like Dobzhansky, for example, encouraged the idea that the Hardy–Weinberg formula is a law of nature and, in consequence, encouraged the idea that the forces that provoke deviations from Hardy–Weinberg equilibrium can be measured with enough precision to make for a predictive science. However, there is a serious objection to this approach. The Hardy–Weinberg formula, and even the Mendelian "laws" on which it is based, do not fit the usual notion of what is required for a law of nature (Beatty 1981; Beatty 1995). For one thing, they are not spatio-temporally unrestricted. The Mendelian laws, and mathematical permutations of them such as the Hardy–Weinberg formula, describe the working of an admittedly large, but finite number of historical entities that come into existence at a certain time and go out at another. Further, the Mendelian laws and the Hardy–Weinberg formula do not stand up to counterfactual cases. The so-called laws of transmission genetics are full of exceptions. Accordingly, as in other matters of contingent history, it is plausible for an investigator working on a recalcitrant case to deny the law instead of concluding that an unknown factor *must be* interfering with the effect that it would otherwise produce. The failure of some populations, species, indeed of whole lineages to obey Mendel's laws would not have nearly the drastic effect of, say, the failure of a well-entrenched law in physics. Hence one must be skeptical about assimilating the conceptual structure of evolutionary biology to that of physics, or to physics-based models of science.

The philosopher Alexander Rosenberg's response to this difficulty is telling in what it reveals about just how compelling the objection is. Without changing his formalist, physics-oriented, law-governed standards of what will count as good science, Rosenberg has over time become ever more doubtful that population genetical Darwinism is in fact predictive enough, and by logical empiricist standards explanatory enough, to be said to rest directly well-confirmed scientific laws, and so to count as solid science in the same sense that physics is solid science (Rosenberg 1994). Another, alternative philosophical response is almost as revealing. The so-called "semantic view of theories," which has been advocated by a large number of philosophers of biology, attempts to maintain the high scientific status of evolutionary biology by revising the concept of natural law that is embedded in the so-called received or logical-empiricist account. The semantic view treats laws as component parts of definitions of kinds of systems (such as a Newtonian system or a Mendelian system), not as verifiable or falsifiable statements directly referring to the world (van Fraassen 1980; Giere 1979). Proponents of this view then assert that their definitions are brought to bear on real phenomena by way of models (as our planetary system can be successfully modeled as a Newtonian system). If a given chunk of reality fails to conform sufficiently to the model (usually measured by statistical methods), the model is not thereby falsified, but merely declared inappropriate to the case at hand. Disobedience to Mendel's laws carries, on this view, few consequences, unless its inapplicability becomes so widespread as to undermine its utility generally.[8]

One might take an even more assertive approach to the maturity of the Synthesis by denying from the start that the adequacy of evolutionary biology should be held hostage to *any* physics-based model of good science. Ernst Mayr, in his role as a philosopher of biology, has most strenuously defended this view. Mayr denies that evolutionary biology ever has had, ever will have, or ever should have (counterfactual sustaining) laws like those of physics (Mayr 1985; Mayr 1988, pp. 18–19). The so-called Mendelian laws, and the Hardy–Weinberg formula, which projects them to a populational level, are to Mayr's way of thinking interpretive concepts that are no better than their capacity to facilitate detailed inquiry, which in turn is utterly dependent on the artful judgment

---

[8] Among those who have used the semantic account of theories to preserve the lawfulness of the Mendelian laws and their population level expansion, the Hardy–Weinberg Equilibrium, are Beatty 1981, Lloyd 1988, and Thompson 1989. There are other ways than the semantic theory of recognizing that evolutionary biology has laws, even though the laws have exceptions. Brandon refers to "schematic laws" (Brandon 1978, in Brandon 1996, p. 27). Sandra Mitchell thinks of lawfulness more as an ontological than an epistemological notion; it refers to the degrees to which a system will rebound in the face of disturbances and perturbations (Mitchell 2000). By this standard, Newton's laws are very robust, even though we now know that they too have evolved very much as the Mendelian laws have. The Mendelian laws, by contrast, are less lawful, but still laws. For some philosophers of biology, the consequences of the semantic view are not severe enough. "The semantic position," writes Michael Ruse, "downplays the importance of laws in scientific activity . . . blurring the line between science and non-science . . . Significantly, one person (a practicing biologist, in *Nature* no less) has already argued that the semantic view opens the way to a justified belief in miracles" (Ruse 1988, p. 21). (The reference to "one person" is to Berry 1986.)

and experience of scientists in reconstructing past events and patterns in evolutionary history. Models or concepts are not laws at all, in either the received or the semantic senses. This sort of *ex post facto* reconstruction can be facilitated, of course, by bringing various scenarios to bear on the evidence. But this does not imply that adequate accounts are, either in principle or practice, predictions. If logical empiricism or its cousins imply that explanation is deduction from laws that just happen to refer to past rather than future events, then so much the worse for logical empiricism as a universally valid philosophy of science! If evolutionary explanations are necessarily retrospective reconstructions of particular events and sequences presented, after much research, in the form of narratives constructed by those whose close acquaintance with natural history makes them informed judges, then lack of predictive or extrapolative prowess is no vice (Mayr 1985; Mayr 1988; Lewontin 1991a). On this view, it might even be said that the maturation of evolutionary biology has given substance to Whewell's hope for "palaetiological" sciences (see Chapter 6).

A strong version of the "autonomy of evolutionary biology" (from physics and philosophy of physics) stance has grown up around Mayr's trenchant proclamations on these matters. One reason Mayr took this view goes back to his preference for doing evolutionary biology at the phenotypic level, which we have already mentioned. For Mayr, this meant that the Modern Synthesis is continuous with the tradition of natural history. After molecular geneticists, in their first flush of post-Watson and Crick discovery, began to intimate in the late 1950s that evolutionary biology would soon be absorbed into molecular biology, Mayr became even more resistant to any implication that biological disciplines such as biogeography or systematics, rooted as they are in natural history, had simply been waiting around until mathematical population geneticists finally made of their results something better than "stamp collecting," and resistant, too, to the implication that evolutionary natural history itself might go out of business altogether once population genetics had been "reduced" to molecular genetics (see Chapter 10).

To be sure, Mayr concedes that both population and molecular genetics have removed barriers to the Modern Synthesis by showing the consistency of Darwinism and Mendelism (see also Kitcher 1984). But whereas theoretical population geneticists, to Mayr's way of thinking, had been staying away from problems and concepts that are crucial to any realistic biological science, such as the nature of species and speciation, natural historians had long been making steady progress on just these issues. It was naturalists, Mayr argued, who first began to treat species as "collections of potentially or actually breeding populations distributed across a definite span of space and time" (Mayr's "biological species concept," see Chapter 10), who documented the leading role of geographical isolation in speciation, and who were already producing a systematics appropriate to an evolving, population-based world (Mayr 1959; 1980a). "Systematics," Mayr wrote, "contrary to widespread misconceptions, was not at all in a backward and static condition during the first third of the twentieth century. Population thinking came into genetics from systematics and not the reverse" (Mayr 1959, p. 3). Indeed, Mayr has asserted that the Modern Synthesis did not fully coalesce until a meeting in 1947, when representatives from a variety of substantive biological specialities agreed that what they already knew could be anchored in approaches to evolution set forth in the canonical works of Dobzhansky, Mayr, and Simpson (Mayr 1980b).

It will be useful for the reader to bear in mind that this or that philosophical solution to the tautology objection, to the status and utility of laws, and to the symmetry or asymmetry between explanation and prediction has often been adopted by evolutionary scientists in order to support preferred versions of the Modern Synthesis itself. Sociobiologists, for example, have inclined toward empiricist reconstructions; their critics have taken fairly strong stands on the ineliminably narrative, and in that sense historical, argumentative strategies of the evolutionary sciences (Gould 1989; Lewontin 1991a). At the same time, it is no less true that philosophers of biology have often erected their philosophical reconstructions on their prior commitment to this or that research program within the Synthesis. This two-way flow between science and meta-science is characteristic of contemporary debates about evolution.

The Units of Selection Controversy: The Challenge of Genic Selectionism

In 1966, an astute evolutionary biologist named George C. Williams published a book entitled *Adaptation and Natural Selection*. He argued vigorously against an unfortunate ecologist named V. C. Wynne Edwards, who had said that birds expand or contract the number of their young in response to the availability or scarcity of food "for the good of the species" (Wynne-Edwards 1962; Williams 1966). Williams claimed that although the phenomenon of regulating clutch size is real enough, any proposed explanation in terms of "the good of the species" is misguided. "Organic adaptations," Williams wrote, are by definition "mechanisms designed [exclusively] to promote the success of an individual organism," not groups – and certainly not species (Williams 1966, p. 96).

Williams took this to be an uncontroversial statement of "the current conception of natural selection, often termed 'neo-Darwinism'" (Williams 1996, p. 96).[9] In fact, he meant something special by it. He meant that the *only* scientifically respectable and logically valid formulation of the Modern Synthesis is one in which organisms, while they and their traits are *targets* of selection (Mayr 1982, p. 588), are too "temporary" to serve as *units* of selection – that is, as entities that are selected in the concerted, cumulative way that over multi-generational time makes for adaptations. The only entities that have sufficient temporal staying power to do this, Williams claimed, are genes, conceived (in post-Watson–Crick fashion) as chunks of DNA that survive repeated meiotic divisions.

To be sure, Williams had methodological reasons for putting things this way. Defending genetic Darwinism as mature science, he was working under the influence of logical empiricist ideals, according to which a theory is a symbolic machine for making predictions from the most parsimonious number of conceptual assumptions. These, it is presumed, will always be better if they work on the most basic entities involved: particles within atoms, for example, or atoms within molecules, or, for Williams, genes – sections of DNA that code for proteins – within the nucleus of cells. This analytical approach may have its biases (see Wimsatt 1980). But the implication is that these biases are far better than the mystifications of Wynne-Edwards.

There is nothing in Williams' gene's-eye view of genetic Darwinism, or what soon became known as "genic selectionism," that denies the fundamental two-stage account of natural selection built into the Synthesis. Genes are differentially retained over generations only because the phenotypes they "code for" (through the production of amino acids and then proteins, followed by the folding up of proteins into tissues) are better able than other phenotypes to deal successfully enough with their environments to leave, on the average, more offspring. For Williams, in fact, who was strongly in the adaptationist camp, only phenotypes

---

[9] We have not used the term 'neo-Darwinism' in referring to the Modern Synthesis, although this is commonly done, because the term was first, and properly, used to name Weismann's version of Darwinism at the end of the nineteenth century.

that optimize the replication rate of the genes that code for them will be preserved in the long run; he was among the first genetic Darwinians to take natural selection as both necessary and sufficient for the evolution of an adaptation properly so-called. Nonetheless, Williams was loath to say that the genes exist for the sake of the better functioning of the organisms that contain them. On the contrary, organisms exist for the sake of the genes. The critique of Wynne-Edwards then followed almost automatically. If genes are not there for the sake of organisms, they are certainly not there for the sake of the group – for the sake, that is, of a fleet herd of deer rather than a herd of individually fleet deer, to use what has become a standard example. *A fortiori*, they are not there for the sake of the whole species.

There was still another reason for preferring Williams' way of formulating the Synthesis. One of the chief points recommending his "parsimonious" reconstruction of the Modern Synthesis was that genic selectionism helped resolve one of Darwinism's nagging challenges: the widespread existence of cooperation among conspecifics, and even of self-sacrificing "altruism," in a world that is assumed *ex hypothesi* to reward competitive over cooperative behavior. Following the brilliant lead of William D. Hamilton, Williams proposed that fitness "is measured by the extent to which [an allele] contributes genes to later generations of the population of which it is a member" (Williams, 1966: 23). According to this idea – the idea of "inclusive fitness" (Williams 1966, p. 97; Hamilton 1964) – genes are fit not only if they elevate the reproductive success of the organism whose traits they directly code for, but of its kin, who after all carry the same alleles in the proportion to which they are genetically related. This idea, "kin selection," was originally suggested in an off-handed way by Haldane, who expressed it in well-known quip: "I will die for two brothers or eight cousins." In the hands of Hamilton, seconded by the parsimonious Williams, inclusive fitness and kin selection came to imply that from a gene's point of view, it matters not a whit what body it is in, so long as it stands a better chance of getting itself multiplied in the next generation. From this perspective, cooperative behavior would be sure to evolve whenever it possibly could, as it certainly seems to have done in the case of social insects.[10] This revelation enhanced the prestige of Darwinism by

---

[10] Using game theory – a set of mathematical tools that enable one to calculate, if not selection pressures, then at least payoffs for responding to them, which were beginning to spread far and wide among evolutionists and ecologists – Hamilton already had

alleviating one of the chief anomalies it had borne more or less since its inception.

The notions of inclusive fitness and kin selection proved important in solving problems about the evolution of behavior. They played a role, for example, in E. O. Wilson's controversial proposal for a Darwinian "sociobiology," in which cooperative behaviors that the Synthesis had earlier conceded to culture were now to be reabsorbed as adaptations into evolutionary biology (Wilson 1975; see Chapter 11). Kin selection and inclusive fitness also played a key role in the Oxford biologist Richard Dawkins' trenchant reformulation of William's genic selectionism in his so-called "selfish gene" hypothesis. Dawkins desired, as he confessed in *The Selfish Gene*, to "reassert the fundamental principles of Fisher, Haldane, and Wright, the founding fathers of 'neo-Darwinism' in the 1930s" in order to make of genetic Darwinism a better theory (Dawkins 1989, p. 273).[11] In order to do so, Dawkins agreed with the substance of Williams' argument: The phenotype, and the organism considered as a sum of traits, exists for the sake of getting the genes that code for phenotypic adaptations maximally represented in the next generation. Only genetic combinations that have accomplished this are still around to tell the tale. But Dawkins thought that positivist delicacy had prevented Williams' case for genic selectionism from being "full-throated" enough (Dawkins 1989, p. 12). Williams had recommended it primarily on the grounds of empiricist semantics, reductionist ideals, and theoretical parsimony. The same argumentative resources could be used, however – and in Dawkins' view should be used (especially in the light of advances in molecular biology – to claim that the genes that figure in genic selection – "selfish" genes – have a palpable, realistic

shown persuasively that cooperative, even self-sacrificing, behavior readily arises in haplodiploid species such as the social insects (Hamilton 1964). In such species, sisters share a higher proportion of their genes with one another than with their brothers. If sisters could deflect their reproductive efforts onto a single queen, and concentrate their energy instead on raising sisters and starving their brothers, their genes would be represented in higher numbers in successive generations. In the course of that process, the role-differentiated cooperative adaptations so obvious in social insects would have been both predictable and causally explained. Despite the fact that not all *Hymenoptera* are social, and that not all social species have hard-wired adaptations for cooperation, this is a very powerful vindication of kin selection's claim that cooperation is proportional to genetic relatedness.

[11] Dawkins confesses that Fisher – "the greatest biologist of the twentieth century" – is his hero, in part because Fisher was willing to assign context-independent fitness values to each allele (Dawkins 1989, pp. ix; 124).

causal force in creating and preserving the phenotypes that shelter them. Williams had asserted that organisms are too "temporary" to be units of cumulative selection. Dawkins, playing the ontologist, was more vehement about it. Organisms, he wrote, are "like clouds in the sky or dust storms in the desert," while genes "like diamonds, are forever" (Dawkins 1989, pp. 34–5). They are "immortal coils" of DNA that, starting with the speck of protein that gave the first "naked replicator" a marginal advantage in the primeval seas, create their adapted phenotypes in order to get themselves maximally represented in the next generation (Dawkins 1989, p. 21). Why? Because the very nature of these chunks of DNA is to make copies of themselves. They are what Dawkins called "replicators," not the only kind of replicators, to be sure, but in evolutionary biology the most important and influential kind. Because these replicators are successful in proportion as they code for bits and pieces of the phenotype that, under selection pressure from the external environment, mediate between them and the outside world, phenotypes are said to be the "vehicles" or "survival machines" of the "replicators," a distinction that the philosopher of biology David Hull put in a less prejudicial way by distinguishing between replicators and interactors (Hull 1980).

Just as Williams presented his version of genic selection as a way of articulating the Synthesis, not replacing or refuting it, so did Dawkins (Dawkins 1986, p. 273). And just as Williams had recommended his way as superior to the organism-centered versions of Mayr and Dobzhansky because of its ostensibly greater ability to explain cooperation and altruism, so did Dawkins; for Dawkins, selfish genes causally explain kin selection, and kin selection provides evidence that genes are in actuality selfish (in the sense of being inherently bent on self-replication). But Dawkins also touted his formulation as preferable because it explained additional facts that pre-molecular versions of genetic Darwinism did not, and could not, even know about. Simply by noting that replicators make more of themselves unless something stops them, the selfish-gene hypothesis can explain why so many alleles that code for proteins are selectively neutral (a fact that was first promoted as "non-Darwinian evolution" [King and Jukes 1969]);[12] why the genome is cluttered up with

---

[12] In 1968, Motoo Kimura reported that the rate at which amino acids are replaced in proteins as they evolve is more or less constant and, in consequence, probably not under the control of natural selection (which would presumably wax and wane with selection pressure). What Kimura called "the neutral theory of protein evolution"

"junk DNA," only a small percentage of DNA in each cell coding for proteins at all; and why genes come in huge families and copies that provide variation on which natural selection can experiment, while one or a few copies do the basic work of "running" the organism (Dawkins 1989, pp. 275–6).

Very interesting ideas. However, trouble soon broke out. Genic selectionism measures fitness as genetic contribution to successive generations averaged over all contexts in which the allele is in play. Evolutionary biologists and philosophers of biology who wish to retain as central the kind of selection pressure that the environment presents to the whole organism have a good deal of trouble with genic selectionism so construed. Unless the organism can deal successfully enough with its environment to reproduce, it cannot pass on its genes, no matter how "good" they are. But the consequent relationships among genes and traits are so complex that unless it can be shown otherwise, the entire genome should be regarded as the least unit of replication (Mayr 1963; Lewontin 1974; Wimsatt 1981). "Fitness at a single level ripped from its interactive context," Lewontin wrote, "is about as relevant to real problems of evolutionary genetics as the study of the psychology of individuals isolated from their social context ... The entire genome is the unit of selection" (Lewontin 1974, p. 318). Just as important, the relationship between genotypes and phenotypes is so context-dependent – so tied to particular, highly unstable environments and to the large number of interacting evolutionary forces at work in them – that it makes little sense to talk about the "average" contribution of each gene to fitness, as the Williams–Dawkins line requires (Brandon 1985; 1990).[13]

was provocatively rebaptized "non-Darwinian evolution" a year later by two molecular biologists, Thomas Jukes and Jack King (King and Jukes 1969). The constant non-adaptive ticking of amino acid substitution was then dubbed by Zuckerkandl a "molecular clock" that would give biology the prestige of physics and chemistry, with their atomic and other natural clocks (Zuckerkandl 1987). At first, Darwinians were dismayed by this finding; many presumed that it must be possible, at least in principle, to assign to a unique fitness value to each amino acid. It did not take long, however, for Darwinians to recognize that neutralism is non-Darwinian only if one assumes that evolution must occur at one level alone, presumably its lowest one, and that the only force driving it is natural selection. This way of handling neutralism by relegating it to doings in the nucleus is one of several sources of the shift toward hierarchical thinking that entered into the Synthesis, and philosophical reflection on it, in response to the reductionistic proclamations of molecular geneticists like Watson and Crick.

[13] As early as 1930, Wright had noticed and rejected Fisher's assumption "that each gene is assigned a constant value, measuring its contribution to the character of the

The effect of genes, like good wine, does not travel well. Admittedly, the organism is not an immortal replicator, or an entirely faithful replicator at all (with the exception of clones, which have lost their evolutionary importance compared with the advantages of sexual reproduction and meiotic division). But the individual genes that are passed on by organisms are, from the orthodox perspective, not in any better shape with respect to Dawkins' criteria of "longevity, fidelity, and fecundity."

Genic selectionism (usually of a reconstructed sort) has not gone without persuasive defenders (Sterelny and Kitcher 1988; Sterelny 2001).[14] Nonetheless, something of a consensus has formed among philosophers of biology against genic selectionism, at least as a general formulation of genetic Darwinism. Many of those taking this critical view have based their case on the theoretical coherence, and empirical existence, of group selection. The theoretical coherence comes from the fact that efforts at logically reconstructing population-genetical Darwinism have shown that, formally at least, natural selection can in principle range over any and all entities that show variation, heritability, and differential retention, and so cannot be theoretically restricted to genes, or even organisms (Lewontin 1970). This recognition gave rise to the idea that there might be a variety of *levels*, as well as units, of selection – that is, to the idea that selection might range over entities below the organismic level, such as cell lines, as well as groups above it (Brandon 1990). There might well be selection for genes – genic selection – at the level of the intracellular milieu, just as there is selection for adapted traits at the level of organisms within an environment. By the same token, there can be a level of selection for groups for properties that cannot be accounted for in terms of the traits of their constituent organisms. And there might even be selection for species-level properties, as we will see in the next

individual . . . in such a way that the sums of the contributions of all genes will equal as closely as possible the actual measures of the population" (Wright 1930; in Wright 1986, p. 84).

[14] Sterelny's version of gene selection does not treat genes as the only replicators or organisms as the only interactors. Neither does he discount the possibility, indeed the actual existence, of group selection, or even of species selection. However, placing great store on Dawkins' notion of the extended phenotype – a trait that affects fitness even though it does not belong to the organism's own body, such as the housing that the caddis fly cobbles together for itself out of debris – Sterelny argues that in such cases, and in the many cases where several species are interacting, genes are both the long-term beneficiaries and, at a certain level of generality, the principal causes of evolutionary change. See Sterelny 2001; Sterelny and Griffiths 1999, Chapter 3.

section. But all this is an empirical matter, not a matter for stipulative decision-making or for philosophical legislation.

An empirical case of group selection had long before been put forward by Lewontin and the Columbia geneticist Leslie Dunn (Lewontin and Dunn 1960). There exists, it seems, an allele – the t allele in the house mouse – that will kill every male in which it is homozygous. To make matters worse, there are more of these lethal homozygotes than one might expect, since the allele beats Mendel's laws: It gets itself over-represented in each generation by a process called "meiotic drive." These facts might have led to the extinction of house mice altogether were it not for the fact that they live in small demes. In some demes, the t allele is lopsidedly, even fatally, rampant. In others, it is not. The continued existence of the house mouse seems to be the result of an equilibrium between gene-level selection and interdemic selection. For Wright, the demic lifestyle of the mouse would illustrate what he meant by a "shifting balance." Dobzhansky and Lewontin might find ways to interpret it as a matter of balancing selection, and to that extent as an adaptation in which group-level processes, such as living in small communities, compensate for an unfortunate quirk of the genes. Williams, for his part, has had to admit that selection on irreducibly group-level traits is possible, and that it exists in the case of the house mouse, although he thought it an oddity (Williams 1966, pp. 117–119[15]). Most genic selectionists are inclined to discount the very existence of selection at the level of groups.[16] For them, a fleet herd of deer just *is* a herd of fleet deer, since *ex hypothesi* its fitness is the averaged sum of the inclusive fitness of each individual in the herd. In consequence, cooperation arises not from selection between more-or-less cooperative groups, but from behaviors (including cooperative behaviors) that aid the reproductive success of individuals – or, more fundamentally, from their genes. The two sides, it seemed, saw the same phenomenon in different ways.

The ecologist David Sloan Wilson, joined by the philosopher of biology Elliott Sober, has presented a case for group selection that would make it far more common than the sort of interdemic selection that Lewontin, and Wright before him, had in mind (Wilson, D. 1989; Sober and Wilson 1994; 1998; see also Wade 1978). Their argument begins by subtracting gratuitous philosophical assumptions. Fitness is

---

[15] Williams 1992 develops a less parsimonious account of genic selection than Williams 1966. See Chapter 10.

[16] The exceptions include Sterelny. See n. 14.

not identical with the way of measuring it. It is a function of the collection of observable, phenotypic traits that causally bias reproductive output in this or that direction. It is predicated of interactors, not replicators. (There is selection *of* genes, as Sober trenchantly put it, but there is selection *for* traits that make an organism fit [Sober 1984].) Some of these traits – dam building among beavers, for example – foster interaction among members of a population, involving them in a common fate and, on the whole, improving their individual fitnesses. Such "trait groups" do not require that organisms lose their individual identities, or that they be immune to cheaters and freeloaders – to beavers, for example, who shirk work. For an elegant mathematical demonstration shows that one rotten apple does not always spoil the barrel; a cooperating trait-group that contains a few cheaters will still outproduce one that does not (Sober 1993, pp. 98–102; see Sterelny and Griffiths 1999, pp. 160–165). Thus, for Wilson and Sober, the trait group is a unit that is selected *at* the level of groups. Moreover, group-level traits reliably inherited generation after generation are, *contra* Williams and Dawkins, adaptations that must be attributed to the comparative success of groups against other groups; the properties that produce the fitness are not reducible to properties of individuals apart from their relationship with the group. If Hamilton and his followers cannot see this, it must be because they do not calculate fitness correctly. The use of averaged gene frequencies biases the question in favor of individual level selection, and misses what is causally going on where individuals interact with environments as members of groups.

During the 1980s, we may conclude, the Modern Synthesis did not stay stable. It tended to bifurcate into genocentric formulations, which, with some help from the "selfish" properties of the molecular gene, looked back to Fisher's adaptationism, and, at the other extreme, into "expanded" versions that sought to recover Wright's convictions that natural selection can and does operate at various levels, and that resolute "pluralism" about interactions among various evolutionary forces must be maintained. Advocates of the second orientation do not deny that something like genic selection exists. But, properly identified, it occurs only within the milieu of the cell, where "selfish genes" serve as both interactors and replicators (Brandon 1988; Brandon 1990; Lloyd 1988; Gould and Lloyd 1999). Acknowledging the existence of genes as beneficiaries of the selection process cannot count, for those who hold this view, as a viable articulation of the Synthesis in the age of molecular genetics.

The Challenges of Punctuated Equilibrium and Species Selection

Forces tending to drive the Modern Synthesis apart have intensified as doubts have accumulated about how successful Simpson had been in treating evolution above the species level – macroevolution – simply as a long-term effect of microevolutionary forces. Simpson was aware of how important "putting an equal sign" (Dobzhansky 1937) between micro- and macroevolution was to efforts to present the Modern Synthesis as a general, complete, adequate theory of evolution. "If the two prove to be basically different," he wrote, "the innumerable studies of microevolution would become relatively unimportant and would have minor value in the study of evolution as a whole" (Simpson 1944, p. 97). He was aware that his task would be difficult if it meant that evolution by natural selection and associated mechanisms must move at a stately, uniform, Lyellian rate. For he admitted that "the face of the [fossil] record does really suggest normal discontinuity at all levels, most particularly at high levels" (Simpson 1944, p. 99). There are obvious breaks, moreover, in the geological record, such as the mass extinction and subsequent sudden Cambrian explosion of about 500 million years ago. Simpson's solution was to attribute speciation to rapid ("tachytelic") natural selection, usually operating in the mode that he called "quantum evolution" (after the analogy with quantum leaps in physics), which frequently involves genetic revolutions in something like Mayr's sense.

Nonetheless, Simpson thought that cladogenesis – what happens when lineages branch at speciation events – took place against a background of "anagenesis" – that is, the slower, cumulative, largely adaptive evolution in the "phyletic" mode that is presumably constantly going on, albeit at different rates, *between* branching events. For Simpson, as for the other founders of the Synthesis, speciation certainly involves adaptive natural selection. But the converse is not true; not all adaptation, indeed comparatively little of it, involves speciation. According to the Synthesis of the 1940s, speciation events occur as temporal and spatial dots in an ocean of ongoing adaptive change, which is largely responsible for evolutionary "trends" (such as Simpson's paradigm of the increased size of species of *Equus* from the Eocene to the Pleioscene) and for the "grades" on which traditional, Linnaean classifications are based.

When Eldredge and Gould proposed their thesis of "punctuated equilibrium" in 1972, then, their innovation was not to assert that speciation is rapid, or even "punctuated." Simpson himself noted that what

Eldredge and Gould meant by 'punctuation' is essentially what he "meant by 'quantum evolution' in *Tempo and Mode in Evolution*" (Simpson 1984, p. xxv). Strictly speaking, Eldredge and Gould did not even contest Simpson's claim that "nine tenths of the pertinent data of paleontology fall into patterns in the phyletic mode," since, as the by-then venerable Sewall Wright pointed out, Simpson himself had admitted that "there might be episodes of tachytely in the phyletic mode" (Wright 1982, in Wright 1986, p. 622; Simpson 1944, p. 203). Rather, Eldredge and Gould's innovation was to deny that much anagenetic evolution occurs between branching points. Evolution, they claimed, is concentrated in comparatively rapid events of speciation (Eldredge and Gould 1972; see also Gould and Eldredge 1977). These "instants," it should be noted, can be fairly long in absolute terms; they are short only relative to the total lifetime of the species measured by geological time (Eldredge and Gould 1972).

Eldredge and Gould's claim was about pattern, not causes, and it was offered on empirical grounds. In the fossil record of marine invertebrates, for example, morphological change appears typically to have occurred within 5,000 to 50,000 years of speciation in species that lasted anywhere from 5 to 10 million years (Eldredge and Gould 1972). To be sure, there is plenty of adaptation in the phyletic mode going on within that larger time. The point, however, is that it doesn't seem to be going anywhere with respect to phylogeny; it consists of "oscillations around some modal value for phenotypic features examined" (Eldredge 1989, p. 65). But, in addition to empirical support, punctuated equilibrium could also recommend itself on conceptual grounds, and on grounds of "consilience" with respectable ideas. Because it discounts anagenetic change, for example, it could commend itself as consonant with (though not necessitated by) the marked shift in recent decades toward a new kind of systematics, "phylogenetic systematics" or "cladism," which counts taxa only between branching points and is more or less indifferent to trends and grades. (See Chapter 10 for more on "cladism.") Then, too, punctuated equilibrium is also consonant with the stress on species as historical entities. (This in turn is related to the claim that species should be regarded, ontologically, as individuals – temporally finite, spatially continuous entities – not as classes [see Chapter 10 for more on the species = individuals controversy]). For another thing, Eldredge and Gould have argued that punctuated equilibrium is just what one should expect from Mayr's peripatric model of speciation: Most change is concentrated in the genetic reorganization that establishes a species (Eldredge and Gould 1972, p. 83). Then too, punctuated equilibrium was also viewed by its authors as supporting Lyell's, Simpson's, and Mayr's recognition that one should not *expect* to see much direct evidence for speciation events. These are unpredictable affairs involving at first only a few organisms. Punctuated equilibrium promotes itself as superior to the Synthesis in its way of dealing with the absence of evidence for intermediates. The evidence is missing, Eldredge and Gould argue, because the process of collinear change leading from one species to another does not actually exist (Eldredge and Gould 1972).

When it comes to conceptual considerations, however, perhaps the most important and controversial aspect of punctuated equilibrium is this: It has been taken by its sponsors to support the general idea, examined in the previous section, that selection operates at a variety of levels, some above the level of the individual organism. In particular, they take it to support the idea that selection operates at the level of species. Eldredge and Gould do not deny that evolutionary trends exist. But, as Gould later put the point,

> If species originate in geological instants and then do not alter in major ways then evolutionary trends cannot represent a simple extrapolation of allelic substitutions in populations. Trends must be the product of differential success among species.
>
> Gould 1980, p. 119

The notion of macroevolutionary entities being selected by a process that is not resolvable into standard microevolutionary forces operating on individuals within populations, or even into group selection, is a threat to Dobzhansky's, Mayr's, and Simpson's assurances to the contrary. However, just how big the threat is depends on whether and how far one is willing to "expand" the Synthesis to accommodate selection at a variety of levels, including species within clades, and on how one construes the selection process in question.

Trends might be the products of slight biases, such as ontogenetic constraints, that favor some lineages. However, if a "hierarchical expansion" of the Synthesis to accommodate selection at a variety of levels is to be genuinely Darwinian, and reasonably continuous in spirit with the Synthesis, species selection must be more than mere species sorting. To maintain the proper parallel behavior with organismic and group selection, the differential retention of species within the clades of which they are parts (or members) must be caused by some advantageous property of differentially retained species, *qua* species, and of their interactions with an environment of some sort. These properties must belong to species and not to any other level. In other words, there must be something like species-level adaptations.[17] However, just what such properties might be – and what sort of environment is envisioned in relation to which they are favored – is a matter of contention, and of some unclarity, even among those who agree that a combination of punctuation, species as individuals, and species-level selection forces a "decoupling" of macroevolution from microevolution.

Eldredge and Gould support this combination of ideas (as does the paleontologist Stephen Stanley, who was the first *contemporary* to use the phrase "species selection" [Stanley 1975][18]). Eldredge, for his part, has attempted to find the irreducibly species-level traits that are required for species selection by identifying properties that a species has in virtue of its interactions with other species in its ecological community (Eldredge 1985; Eldredge and Grene 1992).[19] Species-level properties are the result of what is going on in the "ecological theatre" – G. E. Hutchinson's phrase (Hutchinson 1965) – where the "evolutionary play" takes place. Eldredge postulates that the species-level traits in question are the often quite subtle set of adaptations that lead local representatives of a

---

[17] By his own account, it took Gould some time to realize this. He confesses that his initial formulations were more along the lines of species sorting than species selection (Gould 2002, p. 671).

[18] We say *contemporary* because the first person to use the phrase "species selection" was de Vries. He, too, meant selection in virtue of a species-level property, but he was a saltationist when it came to how the new species was formed. For de Vries species come first; those that cannot find a place in nature are eliminated by negative selection. The Synthesis, in any and all of its forms, asserts precisely the opposite (although, in historical context, de Vries's idea is not as odd as it might seem; as we noted in Chapter 8, in the later nineteenth and early twentieth centuries, natural selection, when it was given a role at all in the evolution of species, was generally viewed as eliminating the unfit, not bringing the fit into existence). See de Vries 1905, pp. 742–44; see also Gould 2002, pp. 446–451.

[19] Eldredge, like several other authors writing in the 1980s, develops the idea that the replicator-interactor distinction can be expanded into a genealogical and an ecological hierarchy. Eldredge defends a two-hierarchy view, which on the side of replicators – the genealogical side – includes genes, organisms considered as replicators, demes, species, and monophyletic taxa (taxa with a shared ancestor). On the other, "economic," side, interactors include cell lines, organisms considered as interactors, local representatives of species (avatars), local ecosystems, and regional ecosystems (Eldredge 1985; Eldredge and Grene 1992) For related views, with very different premises and implications, see Salthe 1985; Brandon 1990; Williams 1992.

species ("avatars") to recognize one another as good to mate with (and not good to eat). (These are the "specific mate recognition systems" of Paterson 1985; see Eldredge 1985; Eldredge 1989, Eldredge and Grene 1992.)

Gould's view is different. He argues that the properties in virtue of which species, *qua* species, are selected must be properties by which species compete, not with different species within an ecological community, but with very closely related species within their own clade. Species vary within a clade in terms of their ability to last longer in phylogenetic time (in virtue, perhaps, of a Dobzhansky-like ability to retain variation useful in changing environments); or to speciate more prolifically, or more often, thereby acquiring a better chance of driving species-level selection in a certain direction to form a trend; or perhaps in other ways. For Gould, such traits are components of fitness that are not resolvable into the properties of organisms or trait-groups. There is a difficulty in thinking of such species-level properties as acting in the way organisms, or even trait groups, do – by means of direct contact with competitors in a real environment. But in his last thoughts on the subject, Gould has asked us to free ourselves from the notion that species, to be individuals that can be affected by environments, must be like organisms (and perhaps from the related idea that group selection involves "superorganisms"). Somehow, there can be competition at a distance in environments that extend over phylogenetic and geological time scales (Gould 2002, p. 624; see also pp. 705–09). Ideas like this require further clarification.

The Developmentalist Challenge

While the Modern Synthesis was supposed by Huxley and Mayr to unify all biological disciplines around a population genetical core, it is generally conceded that the field that has variously called itself ontogeny, embryology, and finally "developmental biology" played no role in the formation of the Modern Synthesis, and was in fact never well integrated into it. Ever since T. H. Morgan, who was himself trained as an embryologist, there had been a tendency among geneticists to set aside embryological considerations "for the time being" in order to create space for an autonomous science of population genetics. The time being tended, however, to turn into forever as the conceptual and experimental orientations of the two communities continued to diverge. The result is that that "the major works that embody the Modern Synthesis . . . hardly mention embryonic development" (Hamburger 1980, p. 97).

Even more significant is the fact that several well-respected developmental biologists, some with Darwinian credentials, expressed skepticism from the outset about whether the Modern Synthesis would ever be put on solid foundations or would constitute a general theory of evolutionary change, until developmental genetics had matured to a point where it became central. "Changes in genotypes," wrote C. H. Waddington, "only have ostensible effects in evolution if they bring with them alterations in the epigenetic processes by which phenotypes come into being" (Waddington 1953a, p. 190). Variants of this view were expressed by Gavin de Beer, C. D. Darlington, Richard Goldschmidt, and I. I. Schmalhausen – all highly respected biologists with roots in, or a keen eye on, developmental processes. All of them were aware that the Synthesis, with its exclusive concentration on the phenotypes of adults in populations, was neglecting questions about how the remarkably stable processes of embryogenesis and life-cycles might have evolved, and how in consequence the Synthesis might be underestimating the possibility that significant evolutionary change, especially at the macroevolutionary level, is instigated by shifts in the timing of what came to be called "developmental programs."

Goldschmidt, Schmalhausen, and Waddington were aware that gene expression is buffered by the ability of embryos to recover their developmental trajectory when something goes wrong. In their view, then, a complete theory of evolution must be able to explain not only how "balancing selection" retains variation in population, but also how the mechanisms evolve by which organisms grow along a very reliable pathway by funneling the effect of genes into one or at most a few phenotypes. (Waddington called this "canalization" [Waddington 1953b]). These developmentalists knew, too, that, even if ontogeny is largely the "expression of genes," genes need to be triggered in the right order by chemical inducers in the cytoplasm, and that these chemical inducers themselves are induced by quite variable environmental factors (Gilbert 1991).[20] Waddington asserted that the causality of genes is negligible until it is acknowledged that interactions between the cytoplasm and the nucleus during different stages of development activate genes. Given this interactive approach to development, traits that are permitted by

---

[20] We qualify this sentence because many people, including Darlington, stressed that there is such a thing as cytoplasmic inheritance that does not reduce to genetic heritability.

underlying genes but require much reinforcement at the phenotypic level, are reliably heritable in part because what Schmalhausen called "stabilizing selection" narrows down the range of phenotypes that are possible for each genotype (Schmalhausen 1949, p. 90; Waddington 1953b).

For their part, the makers of the Modern Synthesis tended to minimize the importance of these claims and the facts they reported. While acknowledging interaction among genes and other cellular components, they tended to redescribe what was happening in embryogenesis in ways that showcased the action of genes in creating phenotypes and downplayed their interactive status (Keller 1995; Keller 2000). (Dawkins' ascription of importance to genes alone did not come out of thin air.) In other words, they pushed to the margins phenomena that could not be described in terms recognizable by the Synthesis – terms restricted to variation among genes and natural selection in accord with environments, with no independent or interacting role for developmental dynamics in between. Mayr and Simpson, for example, played off "genetic revolutions" and "quantum evolution" against Goldschmidt's "systemic mutations" and "hopeful monsters" in precisely this way. As a matter of principle deeply embedded in the Synthesis, mutations are only variation; by definition, it takes natural selection working in a trans-generational context to turn variation into adaptations – and into species. True, Schmalhausen's "norms of reaction" and "stabilizing selection" were subsequently taken up into the Synthesis as it reached the stage of textbook orthodoxy. But they were taken up in ways that bore no special relationship to developmental dynamics. Under the general term "normalizing" selection, for example, stabilizing selection became the mutation-eliminating counterpart of balancing selection (Dobzhansky, Ayala, Stebbins, and Valentine 1977, p. 117). It was all a matter of genes below and phenotypes above – with very little of evolutionary interest in between.

By the end of the twentieth century, this dismissive and cooptive attitude had changed markedly. In large measure, this is due to the cascade of knowledge that has been coming from the combined efforts of molecular geneticists and developmental biologists. This new knowledge has been affecting older attitudes of Synthesis-oriented evolutionists toward both molecularists and developmentalists, attitudes that have often included indifference, wariness, and hostility.

Watson and Crick's discovery of the structure and function of DNA in 1953 was generally greeted by the makers of the Modern Synthesis as a welcome reaffirmation of Mendel's laws, as well as of Morgan's hard-won belief that genes are real things located on the chromosomes in the nucleus of cells. Thus, in an address at the biological laboratory at Cold Spring Harbor on Long Island, New York, in 1959, Dobzhansky greeted "the brilliant hypothesis of Watson and Crick" to the effect that genes are DNA, and that copying errors in its base pairs are the ultimate source of genetic variation, as explaining the mechanisms underlying Mendel's laws without contradicting anything that the Synthesis had built on this basis by way of the Hardy–Weinberg formula (Dobzhansky 1959, p. 15). Moreover, Crick's "Central Dogma of Molecular Biology," which dictated that information runs from DNA to RNA to protein, and never the other way around, seemed to confirm Dobzhansky's own definitional presumption that "evolution *is* change in gene frequencies in populations," and is never a result of the inheritance of acquired characteristics, as well as the causal primacy that the Synthesis accorded to genes. There was a strong presumption latent in these assertions that molecular genetics would never turn up anything that would challenge the fundamentals of the Modern Synthesis. It was precisely this confidence that enabled evolutionists to resist the imperious expectation of molecular biologists such as Watson, who intimated, as the effort to determine the "genetic code" that linked sequences of base pairs in DNA to the production of amino acids and proteins came to fruition, that molecular biologists would soon be evicting ecologists, evolutionists, and natural historians from biology departments.[21] It was in this context that evolutionists committed to the Synthesis first made common cause with philosophers who were beginning to become disenchanted with the "received," logical empiricist philosophy of science; both communities were alert to the misuse of reductionist ideals implicit in the triumphalism of many molecularists.

Nonetheless, a sense that molecular genetics might eventually turn up facts that were challenging to the Synthesis eventually did emerge. Its first inklings can be felt, if only retrospectively, in François Jacob's and

---

[21] In his autobiography, E. O. Wilson recalls what happened when he recommended that his colleagues in Harvard's Department of Biology hire an ecologist: Watson said softly, "Are they out of their minds?" "What do you mean?" I was genuinely puzzled. "Anyone who would hire an ecologist is out of his mind," responded the avatar of molecular biology. Wilson goes on to comment: "The ranks of the molecular and cellular biologists swelled rapidly . . . No one knew how to stop them from dominating the Department of Biology to the eventual extinction of other disciplines." (Wilson 1994, pp. 220–2.)

Jacques Monod's discovery in 1961 of the *lac operon*, with its concomitant distinction between structural and regulatory genes. Regulatory sectors of the genome turn on and turn off the production of structural gene products – proteins – at various points in development. But the triggering of regulatory genes, and even the triggering of higher rates of mutation, of gene duplications, and of enhanced mobility of genetic elements within the genome itself, is highly dependent on what the embryologists of Waddington's day had called "inducers" and "enhancers." The days of "gene action," as opposed to "gene activation," were beginning to be numbered (Keller 1995; Keller 2000).

A key event was the discovery of the structure and function of Hox genes in the late 1970s. A single genetic sequence was shown to control the segmentation of the arthropod body plan. Indeed, these genes, which arose through gene duplication, are lined up on the chromosome in exactly the order of the body segments themselves (Lewis 1978). Whether each of the various arthropod species is to be more developed in its anterior or posterior segments, and whether it would sprout legs here, antennas there, and wings somewhere else – all these differences depend on the timing in which these genes turn on and off and on the strength of the chemical gradients that are controlled by these genes (Lewis 1978; Gilbert 1998; Gilbert, Opitz, and Raff 1996; Arthur 2002; Gould 2002, pp. 1095–1106). Even more interesting is the fact that against the background provided by the gene mapping and gene sequencing programs of the 1980s and 1990s – not just the Human Genome Project, but the less popularly known mapping and sequencing work on species of bacteria, yeast, flatworms, and fruit flies – Hox genes have been shown to be conserved not only among arthropods, but in vertebrates as well. Shades of Geoffroy's contention against Cuvier! The traits of quite widely separated taxa are undergirded by comparatively few, very old homologous genes and the inherited developmental pathways they stabilize. In a similar vein, it has turned out that, whereas evolutionists not long ago had postulated up to twenty-four separate origins for analogous traits called "eyes," they now believe that all of them depend on the same highly conserved, homologous PAX-6 gene.

News of this sort has been assimilated by the contemporary heirs of the mid-century Synthesis in strikingly different ways. Genic selectionists have greeted the modular nature of genetic units, with their "longevity, fidelity, and, fecundity," as a vindication of "selfish genes." Highly conserved genetic modules, they claim, are the materials out of which natural selection can build optimally adapted machines operating under the control of epigenetic programs that work like computer programs (Dennett 1995). However, those who take this view continue to push aside the way in which highly conserved genetic sequences bias and constrain the path of phylogeny, quite apart from any adaptive significance. They also underplay the interactive processes in ontogeny. Instead, they proclaim, in the idiom of "gene action," that development is "read out only" from "genetic programs" that are assumed to be products of adaptive natural selection. According to a wide variety of "new developmentalists," on the other hand, the lesson to be learned is the exact opposite. "Evo-devo," as the suite of discoveries coming from developmental genetics is informally called, portends an impending "New Evolutionary Synthesis" that will reduce the causal role of genes in ontogeny and assign a direct role to developmental changes in macroevolution. Such a synthesis would have developmental processes at its core, as Waddington and others had required (Gilbert, Opitz, and Raff 1996). Some evolutionists and philosophers of biology who take this view, calling themselves Developmental Systems Theorists, have even argued that the impression of the causal primacy of genes will never be alleviated until a variety of concepts long dear to the Synthesis have been abandoned: In an ontogenetic process that involves the presumptively equal participation of a whole raft of "developmental resources," received notions such as that genes "contain" information, that there are such things as "developmental programs," and that a fairly clear distinction can be drawn between replicators and interactors have come under fire (Griffiths and Gray 1995; Oyama, Griffiths, and Gray 2001).

Evo-devo reopens the idea that new species arise by changes, perhaps sudden, in regulatory genes (Gould 1977). Still, some biologists and philosophers of biology are beginning to doubt that the new developmentalism heralds anything as grand as the arrival of a "new and general theory of evolution" at all. That would require, among other things, far greater unanimity about the meaning of the concept "gene," and about answers to questions about the relationships between genes and other "developmental resources," than contemporary biological theory and practice can yield (Beurton, Falk, and Rheinberger 2000). What seems to be happening today is that the developmental gene concept, the gene concept of classical population genetics, and the molecular concept of the gene as a coding section of DNA, instead of drawing closer, as reductionists had expected, are actually pulling further apart (Gilbert

2000; Gilbert & Burian 2003; Keller 2000; Moss 2001). This splitting up of the concept of "gene" gives rise to ambiguities of the sort that philosophers of biology have helped clarify in the past. Still, even after ambiguities about the shifting meaning of the term "gene" have been sorted out, it may remain true that different gene concepts will figure in making true claims in different disciplinary and experimental contexts (Gilbert 2000, p. 180; Gilbert and Burian 2000).

The Modern Synthesis was formed on the basis of population, not molecular, genetics. Its makers knew nothing of neutral alleles, self-ish DNA, selfish genes, coding and non-coding sectors of the genome (exons and introns), structural versus regulatory genes, gene duplica-tion, or gene splitting. As these discoveries, along with the possibility of manipulating genes directly, have proliferated, the claim has always been that the Synthesis will remain consistent with the new knowl-edge. But mere consistency does not a fully adequate theory make. If a new general theory of evolution coalesces, it will almost certainly put developmental biology into a far more central position than it had un-der the mid-(twentieth) century Synthesis, and that it still lacks in most contemporary genic-selectionist thinking. But even as genetics goes from triumph to triumph, the ideal of a single background theory of evolution that commands the assent of all professional biologists as a framework within which these discoveries make sense – the ideal of the mid-century Synthesis – may prove elusive, or even illusory.[22]

A Note on Evolutionary Progress

It was an ambition of the researches that began with Mendelian ge-netics and culminated in the Modern Synthesis to deny that the course of evolution is inherently directional – "from monad to man," as the phrase has it – and especially to deny that it is directed toward an end. Views of this sort (which, as we saw in Chapter 8, went under a variety of names, such as "orthogenesis," "aristogenesis," and "nomo-genesis") were explicit in what, in retrospect at least, we might recog-nize as the "unmodern" evolutionary synthesis that has been displaced by the *modern* synthesis (Gilbert 1998, p. 170). On the "unmodern"

---

[22] We are alluding here to Dobzhansky's proclamation that "nothing in biology makes sense except in the light of evolution" (Dobzhansky 1973). True enough, but that doesn't mean that nothing in biology makes sense except in the light of the Modern Evolutionary Synthesis.

view, ontogeny was central; it provided the pattern of goal-directed, developmental change that could be seen in ecological succession and in phylogenesis. Conversely, phylogenetic order was seen as the cause of the order that was "recapitulated" in each developing organism. The Modern Synthesis exploded this confabulation into unrecognizable bits. Natural selection builds contraptions for dealing with particular environments out of whatever genetic materials are on hand. What gets built is always being broken up again, sometimes by entirely contingent events, such as a meteor striking the earth.

One might think that, in consequence, the Synthesis would have totally rejected the notion of evolutionary progress. It has certainly rejected the notion that evolution is directed toward an end, insisting that whatever direction can be discerned has "not been an inevitable trend, but rather an occasional by-product of certain kinds of adaptive radiation" (Stebbins 1969, p. 124). Nonetheless, within this constraint, the theme of evolutionary progress has been surprisingly tenacious in the thinking of the central figures in the Modern Synthesis (see Ruse 1996).

Julian Huxley is the most striking case. In *Evolution: The Modern Synthesis*, he asserted that "progress is an improvement in efficiency in living in general," rather than specialization; that this kind of progress can be recognized by the degree to which a species exhibits "control over nature" and "lives in greater independence" of its environment; and that by this standard, "man possesses control over nature and lives in greater independence than any monkey" (Huxley 1943, pp. 562–565). It is true that when, in November 1959, Huxley repeated this rather Promethean view at a convocation that the University of Chicago called to celebrate the 100th anniversary of the publication of the *Origin*, and inferred from it that henceforth religion would be displaced by humanism, his remarks fell flat. Many luminaries in an audience that included Ford, Dobzhansky, Mayr, Stebbins, and Wright wanted badly, it would seem, to portray themselves to the public and to other academics on this solemn occasion as practitioners of a purely professional, "value-neutral" discipline, who, at conferences and in peer-reviewed publications, were making piecemeal contributions to a resolutely empirical, and increasingly unitary, science. Huxley's ideologically charged remarks did not help the cause (Smokovitis 1999, pp. 302–305).

Still, Huxley's view of evolutionary progress was not all that different from Dobzhansky's sense that natural selection is progressive in favoring "mechanisms for evolutionary plasticity" (Dobzhansky 1941, pp. 338,

341);[23] or from Simpson's view that human beings are higher than other animals because they are aware of their environment and of their own individuality (Simpson 1949, p. 261); or from Mayr's later view that "an enlarged central nervous system" and the emergence of "parental care ... by internal fertilization, which provides the potential for transferring information non-genetically from one generation to the next," are marks of "objective progressiveness" (Mayr 1988, p. 252); or from Stebbins's prescient argument that the conserved complexity of organization is an objective mark of evolutionary progress. (In a remarkable anticipation of the new genetics, Stebbins wrote in 1969 that "whenever a complex, organized structure or a complex integrated biosynthetic pathway has become an essential adaptive unit ... the essential features of this unit are conserved in all of the evolutionary descendants of the group concerned" [Stebbins 1969, pp. 124–5]). To be sure, all these figures duly concede that their objective facts are visible only from a perspective, even a value-laden perspective, that has been adopted by the inquirer. Nonetheless, they did believe in general that there was something objective to be seen from these various angles.

Williams and Gould, among biologists, have resisted the idea of evolutionary progress of even this modest sort (Williams 1966; Gould 1989). For Gould, ascriptions of progress from any point of view are little more than ideology. This assertion is supported by Gould's demonstration that the progressivism of the "old" evolutionary synthesis dies hard, not just among the general public, but within expert communities themselves. For his part, Gould stressed the utter contingency in the overall history of life on earth. The pattern that has led "from monad to man" is not simply epistemologically suspect; even if it were real, so chancy is the history of phylogeny that it is very unlikely that "the tape of life," running from the beginning of life on earth until now, would ever be replayed the same way twice, no matter how many times it was rerun (Gould 1989).

In general, however, it has been the lot of professional philosophers of biology to throw the coldest water on the notion of evolutionary progress. Even aside from the hopeless ambiguities of the term "progress," David Hull argues that it is unpersuasive to say that what one sees from a value-laden angle can be anything other than a projection. For every progressive trend, measured by any of the criteria

[23] The first edition of Dobzhansky's *Genetics and the Origin of Species* does not contain a chapter on "evolutionary progress." The second, 1941, edition does.

mentioned here, there have been trends in exactly the opposite direction. What seems in retrospect to be evolutionary progress involves protracting adaptedness across different environments in ways that the Synthetic theory frowns on. But Hull's main reason for calling the very idea of evolutionary progress into question is that "the sequence of organisms from bacteria and paramecia to porpoises and human beings," on which every directional story is based, "are not grounded in the process of phylogeny, but in classification" – and classification grids by their very nature are overlaid on a pattern of evolutionary branching that contains no such information (Hull 1988b, pp. 32–33). The inference (shared with Williams) is that anything that can reasonably be called Darwinian, once it has freed itself from taxonomic thinking inherited from pre-evolutionary thought, cannot be called progressive, and that anything that can be called progressive cannot be called Darwinian.

✦

CHAPTER 10

Some Themes in Recent Philosophy of Biology

The Species Problem, Reducibility, Function, and Teleology

Introduction

In the main, it was only in the second half of the twentieth century that the philosophy of biology emerged as a distinctive sub-discipline in academic philosophy. Indeed, the philosophy of science as such – as distinct from "natural philosophy" or just philosophy – is a relatively recent phenomenon. At its start, however, with the rise of logical positivism, soon renamed logical empiricism, it was chiefly a philosophy focused on physics, or even, in its extreme reconstructionist forms, a philosophy based on a rather slanted view, even a caricature, of that "fundamental" science. The life sciences were usually ignored, or treated as an embarrassment to be explained away. True, J. H. Woodger produced in the nineteen-thirties what was supposed to be a statement of the principles of biology; but apart from a few followers in Great Britain, his effort had little influence (Woodger 1937).[1] In general, the hope of philosophers of science was that all the sciences would one day (perhaps even soon?) be unified in the terms of, and through the theories of, the most basic level of physics. For example, at one of C. H. Waddington's conferences on theoretical biology, there was one very vocal participant who kept deploring that the Volterra–Lotka equations, which express regularities in populations, could not (yet?) be reduced to terms of quantum mechanics. Nothing else was really science.

As we saw in the last chapter, this situation changed in part because of the interest of philosophers as well as biologists in conceptual problems associated with the flowering of the evolutionary synthesis in the middle third of the twentieth century. However, even if we admit

---

[1] But see Wiley and Mayden 2000a,b.

that biology now makes sense only in the light of evolution (Dobzhansky 1973), there have also been lively philosophical issues not directly connected with the structure of evolutionary theory. They are issues that we have met a number of times in our retrospective reflections. We will concentrate in this chapter on three of them, already familiar to our readers in one guise or another: the species problem (together with the question of the systematic foundations of taxonomy); the problem of the reduction of biology to physics and chemistry (with a bow to the small remnant of vitalism conspicuous in the early years of the century); and the problems connected with the concepts of function and teleology: their meaning and their role(s) in biological explanation.

The Species Problem

The question, "What are species?" has recurred a number of times in the previous chapters. Until the post-Darwinian acceptance of evolution, however, there was general consensus on their permanence (with the notable exceptions of Lamarck and Geoffroy). Not that it was necessarily easy to specify just what a species was. As we noted in our first chapter, Aristotle, in his biological writings, did not use the term *eidos* univocally as contrasted with, and less extensive than, *genos*. But he certainly did believe that there were permanent forms of living things (again, *eidos*, the elusive word that was rendered as both "form" and "species" in the Latin tradition). So, despite that ambiguity, it seems fair that Cuvier, in his insistence on the uniqueness and stability of each species (until extinction!), should consider himself a faithful follower of the Stagyrite. It is true, too, that Linnaeus focused rather on the genus than the species as the chief "natural" unit in nature. Yet despite the acknowledged artificiality of his actual divisions, Linnaeus was confident that there was a permanent, God-given hierarchy of living things that would always stay as it was in this well-ordered world. Buffon, distrustful as he was of Linnaean classification, looked at the species as a reproductive unit. Even in his case, however, the stress on genealogy did not entail, or even suggest to him, a transmutational reading. As we noticed earlier, naturalists spend their lives identifying this and that kind of plant, animal, or fungus, and, other things being equal, they are unlikely to think the objects of their study impermanent. It was probably Lamarck's love of grand theories, along with his dislike of the very thought of extinction, that moved him to devise his theory of life's advancement, and it was his work in embryology, along with his disagreement with Cuvier about the crocodiles of Caen, that finally moved Geoffroy to formulate a general transmutational view.

After Darwin, however, the situation has changed. Perhaps we should say it changed even after, or at the time of, the notorious *Vestiges*, whose many reprintings, as well as critiques, suggested a public ready for some transformist message, if a more cautious one than the Scottish journalist had offered. Whatever the correct reading of this story, and of Darwin's role in it, it is certainly the case that since 1859 there has been a rampant "species problem" – and it has not yet petered out. Indeed, a recent account finds twenty-two species concepts in the literature! (Mayden 1997). Darwin's own position, as we have noticed, was ambiguous: On the one hand, he was equating species and varieties, and so doing away with the species as special or unique; on the other hand, he was rereading the concept of species as one of a lineage rather than a special collection somehow to be classified as permanent. It seems he was both questioning the reality of species and affirming their reality in a more perspicuous sense (Beatty 1982; 1985).

When taxonomists, and philosophers interested in this question, look back at the pre-Darwinian approaches, they often refer to the earlier species concepts as "typological" or "essentialist." Every individual, as a member of a species, represented a type, and such a type appeared to have an essence that could be characterized, presumably by listing its necessary properties. Thus, as we noted in discussing Cuvier, Coleman, his most distinguished recent expositor, constantly speaks of "types" in discussing Cuvier's work, even though the Baron himself largely eschewed that term. Fair enough. We must be careful, however, not to caricature this view. Elliott Sober, for example, tells us in an otherwise very subtle textbook in philosophy of biology: "[Essentialism] ... holds that each natural kind can be defined in terms of properties that are possessed by all and only the members of that kind" (Sober 1993, p. 145; 2000, p. 148). It is to be questioned, however, whether either an Aristotelian least kind or a Cuvieran species can be so defined. For Cuvier, who is following what he conceives to be the spirit of Aristotelian comparative anatomy, it is the total "conditions of life," the unique style of this kind of organism's existence: its morphology, physiology, behavior, relationship to other organisms, and to its ecological niche; it is that inimitable whole that the naturalist is trying to understand. No little list of properties would suffice to capture it. If we want to call this a kind of essentialism, it is not the sort that can be itemized in lists of properties.[2]

Still, in the light of evolution, any concept of species that insists on the permanence of each kind does indeed seem antiquated. With a few exceptions, most biologists consider species in some sense historical entities.[3] They are not what they are forever; they come into being and pass away. So, if not typologically, how shall we define this minimal unit in our classification of living things? Without listing twenty-some conceptions, we will look at some of the main recent candidates.

Most conspicuous, and probably most authoritative, has been the biological species concept (BSC), developed and defended by Ernst Mayr, the spokesman for taxonomy among the architects of the synthesis. According to Mayr, biological species are "groups of interbreeding natural populations that are reproductively isolated from other such groups," where "interbreeding" indicates a propensity. (Clearly, individuals belonging to the same species but in widely separated populations cannot interbreed, but they could if the geographical barrier were removed [Mayr 2000, in Wheeler and Meier 2000, p. 17].[4]) Or, as Mayr also puts it in the same place, "a species is a reproductively cohesive assemblage of populations." The great merit of this conception, in its author's view, is that it substitutes populational for typological thinking. There were two obvious problems with the typological concept. First, many members of the same species are morphologically very different: males and females in many species of birds, for instance; and second, there are sibling species, which are morphologically indistinguishable though reproductively separate. The BSC overcomes these difficulties. Moreover, it has the merit, Mayr believes, of assigning a "why" to the species concept: It is isolating mechanisms that produce and maintain these segregated populations. No other species concept, in his view, is explanatory as well as merely descriptive. The BSC, it should also be noted, is clearly and solely a definition of the species category: It tells us what a species is, as a rank in the Linnaean hierarchy. The delimitation of particular species taxa – that is, of collections of actual entities each of which really is a species – is an entirely different and often more

---

[2] A recent book by Mark Ereshefsky, *The Poverty of the Linnaean Hierarchy* (Ereschefsky 2001), which deals with many of the views we are discussing here, is simplistic in its approach to the history involved. See the review by Peter J. Stevens (Stevens 2001).

[3] A striking exception is the work of Webster and Goodwin. See their 1996.

[4] Cf., e.g., Mayr 1970.

difficult matter, but the account of the category, in Mayr's view, is clear, adequate, and definitive.

For its critics, however – and they are legion – there are serious problems with the BSC. For one thing, it is, as Mayr puts it, "non-dimensional." By this he means that it refers to populations at particular points of space and time only, not in general, whether geographically or historically. But then the question about interbreeding as a propensity arises: If widely separated populations are said to belong to the same species, how can we know that? And if we cannot identify species over time, how does this concept fit in with our evolutionary view of their origins and histories? Further, the BSC applies only to sexually reproducing organisms. That's all very well for an ornithological systematist like Mayr, but biologists who work on organisms that sometimes or always reproduce asexually are likely to resent a conception that leaves the entities they study with no taxonomy at all!

In response to the temporal question, Simpson proposed an "evolutionary species concept." "An evolutionary species," he writes, "is a lineage (an ancestral-descendant sequence of populations) evolving separately from others and with its own unitary role and tendencies" (Simpson 1961, p. 153). This formulation, referring as it does to populations, was intended to include rather than to contradict the BSC. And it does seem reasonable to identify a species as a lineage. Still, the notion of a "unitary role and tendencies" appears too vague to be satisfactory.

A number of other authors who have tried – and are still trying – to identify species in historical, rather than "non-dimensional," terms claim to base their conceptions on the widely acclaimed "phylogenetic systematics" of Willi Hennig. Hennig, a German entomological systematist, published a work in 1950 entitled *Grundzüge einer Theorie der phylogenetischen Systematik* (Hennig 1950). Since it was published shortly after the Second World War, its author had been relatively out of touch with evolutionists outside his own country, but when a revised version was published in translation in 1966, it created a sensation (Hennig 1966). Hennig's aim was to establish a clear-cut scientific procedure for establishing phylogenetic relationships, resulting in branching diagrams. His terminology is formidable; we here introduce some of his basic concepts. First, what we are interested in is monophyletic groups – that is, groups comprising all, and only all, the descendants of a single original group. Consider an example, which Hennig diagrams in Figure 10.1. Hennig explains:

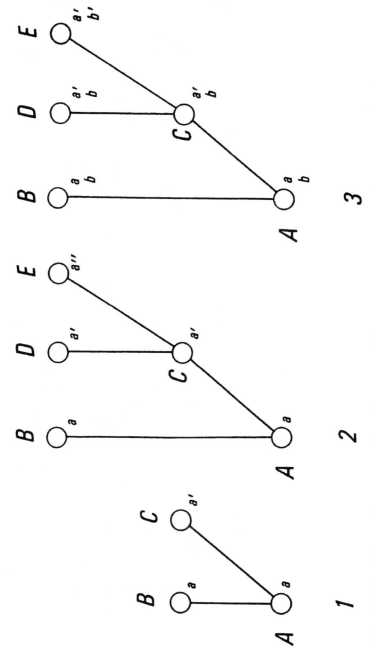

Figure 10.1. **Speciation and character transformation.** From *Phylogenetic Systematics*. Copyright 1966, 1979 by Board of Trustees of the University of Illinois. Used with permission of the University of Illinois Press.

We will call the characters or character conditions from which transformation started (a, b) in a monophyletic group *plesiomorphous*, and the derived condition (a', a", b', b") *apomorphous*. Simple reflection shows that these are relation concepts: the characters a' and a" are both apomorphous compared with character a, but a' is plesiomorphous compared with a".

We will call the presence of plesiomorphous characters in different species *symplesiomorphy*, the present of apomorphous characters synapomorphy, always with the assumption that the compared characters belong to one and the same transformation series. In our example, ... the species B and D are symplesiomorphous with respect to character b, the species D and E symplesiomorphous with respect to character a, ... or to the group of characters a' and a". ... It is evident that the presence of corresponding characters in two or more species is a basis for assuming that these species form a monophyletic group only if the characters are apomorphous, if their correspondence rests on synapomorphy.

Hennig 1979, pp. 89–90

Because of its insistence on speciation through splitting, Hennig's systematic method came to be known, after "clades" or branches, as "cladism" or "cladistics."[5] The enthusiasm – perhaps even the fanaticism – of its adherents, or at least of some of them, is evident in the foreword to the second, 1979 English-language edition. The editors write:

Hennig established a criterion of demarcation between science and metaphysics at a time when neo-Darwinism had attained a sort of metaphysical pinnacle by imposing a burden of subjectivity and tautology on nature's observable hierarchy. Encumbered with vague and slippery ideas about adaptation, fitness, biological species, and natural selection, neo-Darwinism (summed up in the "evolutionary" systematics of Mayr and Simpson) not only lacked a definable investigatory method, but came to depend, both for evolutionary interpretation and classification, on consensus or authority. Several authors besides Hennig recognized the shortcomings of neo-Darwinism ... but for us Hennig was the most successful because his method is simple, explicit, and tied to nature's hierarchy.

Rosen et al. in Hennig 1979, p. ix

Not every one who found merit in Hennig's approach was so vehemently anti-Darwinian. In a 1982 symposium, for example, A. J. Charig presented points for and against both cladism and what he called

---

[5] For a lively account of cladism and of the "cladist wars," see Hull 1988.

"Simpsonian classification" (Charig 1982, in Joysey and Friday 1982). There is no doubt, however, that the effect of the reception of Hennig's work was almost to produce a cult.

At the same time, as happens with many "schools" of thought, cladism has also produced disagreement, even among its ostensible adherents. Quentin Wheeler and Rudolf Meier conveniently present four of these positions in debate (Wheeler and Meier 2000), all of them in contrast with Ernst Mayr, with his biological species concept, who speaks first.[6] Let us look briefly at the alternatives offered by these alleged followers of Hennig's approach.

First, cladism in some quarters has been presented as a continuation or elaboration of Simpson's evolutionary species concept. Thus, Ed Wiley (Wiley 1981) and, more recently, Wiley and Mayden have attempted to add precision to Simpson's account, in part by adopting some of the lessons of the cladistic approach (Wiley 1981; Wiley and Mayden 2000a, b). Granting, however, that it is attractive from an evolutionary perspective to think of species as lineages, we find it difficult to specify the way in which this version has advanced over Simpson's original suggestion. Still, its authors consider it a Simpsonian–Hennigian theory.

Simpson himself, as we noticed, did not believe his evolutionary species concept to be in conflict with Mayr's account. And, in fact, some of Hennig's German disciples seem to find his theory not so far removed from the reigning BSC. Thus, Rudolf Meier and Rainer Willmann, in their presentation of what they call the "Hennigian Species Concept" (Meier and Willmann 2000, in Wheeler and Meier 2000, pp. 30–43; 167–68), stress reproductively isolated populations as central to their conception. Species are natural populations or groups of such populations that are potentially reproductively compatible, and they are ranked by the fact that they are absolutely reproductively isolated from other such groups. It does in fact appear to be the case that Hennig himself saw no particular conflict between his own phylogenetic theory and a more explicitly populational view of species; he does use expressions such as "reproductive community" (e.g., Hennig 1979, p. 46). Yet what interests him especially is rather the sequence of species, which he calls phylogeny, than the character of a given species,

---

[6] Although Mayr here appears willing to discuss the species problem, in a recent popular work he declares flatly that there are only two species concepts, essentialism and the BSC. Those who resist the latter, he declares, simply do not understand the difference between taxa and categories (Mayr 2001, p. 167).

whether as reproductively isolated or reproductively coherent. Thus the "Hennigian" concept does not seem to be especially Hennigian.

In addition to these two versions, at least two further species concepts claim to derive from Hennig, both of them, unfortunately, calling themselves "the phylogenetic species concept." In the Wheeler and Meier collection, they are distinguished as "The Phylogenetic Species Concept (*sensu* Mishler and Theriot)" and "The Phylogenetic Species Concept (*sensu* Wheeler and Platnick)" (Wheeler and Meier 2000, pp. 44–54, 119–32, 179–84, 55–69, 133–45, 185–97).

Brent Mishler and Edward Theriot are seeking to establish the phylogenetic trees that have in fact existed in the history of life – and it seems clear that that was Hennig's aim. They do this through seeking out synapomorphies – that is, characters marking sister groups as descendents (and as the only descendents) of an immediate ancestor. Thus they look at characters, whether of extinct or extant organisms, in search of clues to actual phylogenetic relationships. Moreover, in their search for the monophyletic groups characteristic of such phylogenies when they are rigorously analyzed, they claim no special status for species; it is monophyletic collections at whatever rank that interest them. Indeed, Mishler is among those who are now urging taxonomists to abandon species as a special rank, and to adopt a mononomial nomenclature rather than the traditional binomial system imposed on them by Linnaeus (Mishler 1999; Pennisi 2001). If we are going to get a natural system, it is monophyly that matters (see also Mishler and Donoghue 1982; Mishler and Brandon 1987).

Quentin Wheeler and Norman Platnick, on the other hand, represent a position that has also been called "transformed cladistics" (Wheeler and Platnick 2000). Their professed aim is to turn the attention of taxonomists away from what appear to them to be fuzzy questions about the *processes* of evolution to the *patterns* those processes happen to have produced. The vehement anti-Darwinism evident in the passage we quoted from the second Illinois edition of Hennig's text is characteristic of this group. The lesson seems to be: Evolutionary processes are complicated, unknown, perhaps even unknowable, so let's just look at what we have, which is a collection of characters at species level, rather than traits at the level of interacting and interbreeding individuals, which vary in and out confusingly, in ways irrelevant to the clear-cut differences of character evidenced when species split apart. One of the chief boasts of this program is that its method is maximally parsimonious – that is, it makes the fewest possible assumptions about what there is.

Indeed, its proponents have been given to arguing with one another as to who best exemplifies this unique scientific virtue. If this appears to be – as it does to us – a phylogenetic species concept deprived of phylogeny, it is certainly a position that has its loudly eloquent supporters.

As we indicated, there are still other species concepts in the field: an ecological species concept, a cohesion species concept, a mate recognition species concept, and so on and on (Van Valen 1976; Templeton 1989; Paterson 1985). In this situation, it is not surprising to find some biologists despairing of a definitive solution to the species problem, and proposing a purely "phenetic" solution (e.g., Sokal and Crovello 1970). That is to say: they allege, rather as John Locke did long ago, when he deplored our inability to achieve more than "nominal definitions" of kinds or classes of entities, that the best we can to is to pick some characters or others ("phenes" or "phenons") that are convenient for our classifying purposes, without making any claim to some foundation in reality. In effect, our classifications are purely conventional; so the species problem can wither away. In fact, even before the *floruit* of the BSC, the English taxonomist J. S. L. Gilmour had supported a purely conventionalist approach to the species question (Gilmour, in Huxley 1942, pp. 461–74).

The most conspicuous versions of the species concept have in turn served as the foundation for differing schools of taxonomy. Before we turn to these, however, we should pause to consider some philosophical issues raised by these disagreements.

First, are species real, or uniquely real? From a phenetic point of view, of course, species are purely inventions of the taxonomist. Others consider them real, and, indeed, as most of the participants in these debates aver, uniquely so. Species taxa are really out there in nature, although the designata of higher categories are more or less matters of convention. On this, Mayr and most of the Hennigians seem to agree – although the position of transformed cladists is not quite so clear. Since they focus on characters only, which are what we find to single out in the phenomena, their method may not seem to carry clear ontological implications. Yet what they single out is uniquely at the species level. For cladists of the Mishler–Theriot variety (perhaps we could call them "classical," as distinct from transformed, "cladists"), on the other hand, it is monophyletic groups at any level that count.

We may add in passing that one of the objections voiced against cladism has been that if it is synapomorphies that distinguish species, then an extinct "species" that never split would not be a species, but a phylum (Hull 1970a). Classical cladists could answer: For us, that's fine! If it's monophyletic groups as such that matter, we can overlook species as a rank and forget that seeming paradox. So, from their point of view, species are real, like all monophyletic groups, but have no special status.

A second question: Given, *contra* the phenetic stance, that species are real, what sort of entities are they? To the layman, it appears that they are classes, or somehow groups with members, but in the light of evolution, classes that come into being and pass away. However, according to a view first proposed by Michael Ghiselin, and seconded by David Hull, species are not classes with members, but individuals with parts (Ghiselin 1974; Hull 1976; 1978). Ghiselin's essay started from the premise that things are either individuals, with starting and ending points, or classes forever identified by the possession of certain essential properties, thus unable to come into being, to change throughout their history, and finally pass away. This is a strong disjunction; no compromise allowed. Much ink has been spilled over this proposal. Some celebrate the innovation, some lament it, but the debate goes on. Hull warns us that advancing science often forces us to abandon our "common sense" intuitions (Hull 1980). The sun doesn't really rise and set, does it? The earth turns. Similarly, species don't really have members; they have parts. You and I are parts of *Homo sapiens*, just as our hands or our noses are parts of our bodies. Each of us is an individual who was born and will die; the same for the species – we wanted to say, to which we belong, but, no – of which we are part. Otherwise, supporters of the species = individuals theory say, there could be no "origin" – or extinction – of species. But species do come and go; ergo, they are not classes, or natural kinds, but individual actors on the evolutionary stage.

Is that strong disjunction necessary? Why cannot there be, why are there not, classes whose existence is temporally restricted? There are indeed historical entities that are not classes. There are not only individual organisms (which could themselves be understood, if you like, as classes of cells or the like). There are phenomena such as American democracy, for example, which arose, has a history, and presumably will some day go extinct. But it isn't a class of anything; it is just itself. But apart from obviously individual historical phenomena, why must every class be non-historical? There is nothing in logic to suggest such a drastic conclusion. Ontologically, also, the species-individuals identity has some odd consequences. Individuals do not evolve, we are constantly told; only populations do. But species are commonly thought to be collections of populations. Then, in turn, such collections are themselves individuals who, therefore, do not evolve.

Further, one of the great virtues of the Ghiselin–Hull concept is supposed to be that it does away with the appeal to properties, which are typological, and so wicked. The only property allowed, it seems, is the ancestor-descendant relation. But isn't that a property? Species, like (other) individuals, are supposed to be baptized, not characterized. They just have the names they have, regardless of their characters, because they have been given them. Yet each one is characterized by the property of having a set place in a family tree. That one property is not only permitted; for the individualists, it is in fact the essential property. So what we have is an essentialism much narrower and more dogmatic than anything to be found in Aristotle or Cuvier.

Yet another difficulty: If life is monophyletic, there is only one family tree, and therefore only one case of a complete and exhaustive ancestor-descendant relation – life itself. All splitting into smaller units, including species, is arbitrary. Epistemologically, on the other hand, it may be argued that classes express contrary-to-fact, exceptionless natural laws. Things are this way and couldn't be otherwise. Species, however, exhibit variation; they don't follow laws as classes have to do. So they can't be classes. Does this sharp contrast hold? Paul Griffiths, defending the notion of classes with "historical essences," suggests that all we need to form a class or a natural kind is predictions better than chance: If we know what the characters of a group are likely to be, we can see it as a class. We don't need the contrary-to-fact necessity traditionally believed to belong to natural laws (Griffiths 1999, in Wilson 1999, pp. 187–208). Griffiths refers to Günther Wagner's recent work on evolutionary homology to show the kinds of considerations on which the assignment of historical natural kinds can be based (see, e.g., Wagner 1996). Robert Wilson also wants to reestablish a modified essentialism, according to which species are natural kinds defined by something called "homeostatic property cluster" concepts (Wilson 1999).[7] Their arguments may provide some comfort for those who resist the individualist stance.

But, again, we may ask, is essentialism versus individuality our only choice? John Dupré argues for a retention of natural kind concepts

---

[7] These ideas come from Richard Boyd. His argument is extremely inwrought and technical; a statement of it is included in Wilson 1999, pp. 141–185. Boyd lists there his previous writings developing the HPC idea, starting with its roots in Cornell moral realism. We would not venture to try to expound it.

with no trace of essentialism (Dupré 1993). Ecologists, he points out, need to refer to the species they study as classes, or kinds, whose members they identify. It is only strict evolutionary theorists, looking at a succession of species through time, who need to differentiate them as individuals. Now it does seem to be the case that biologists primarily concerned with evolutionary theory can be pretty imperious about the primacy of their way of thinking for biology as a whole. Ernst Mayr published a famous essay on "Cause and Effect in Biology" (Mayr 1961), in which he proclaimed that only evolutionary explanations are ultimate (Mayr 1961). And of course it is true, as Dobzhansky declared, that nothing in biology makes sense without evolution (Dobzhansky 1973). Evolutionary theory, and, by now, usually the theory of natural selection, is there in the background of any contemporary biologist's thought. But for an ecologist or a physiologist or a molecular biologist, it is not necessarily an evolutionary question that is being asked or an evolutionary answer that will "ultimately" provide an answer. One can ask how something works or what it's made of, as well as asking: How did it arise? If we look more generously at the scope of biological inquiry, we may find areas where genealogy is not the perspective of interest, and where species turn up as groups whose members interest us, rather than as lineages succeeding one another (assuming we have some "properties" by which to determine when one stops and another begins!).

In the presence of so much disagreement about the nature and status of species, a number of writers have abandoned the notion that there is one solution to these problems, and have advocated some kind of pluralism about the species concept. Dupré's argument about the point of view of the ecologist suggests such an attitude: Different disciplines may use the species concept in different ways, each legitimate. There are varieties or, better perhaps, degrees of such pluralism. One may consider the situation so confused that the hope for one single concept must be abandoned. The same group might have to be considered a species from one point of view, but not from another (Kitcher 1984). A more modest form of pluralism finds species in a variety of senses covered by its single, more inclusive concept. The classical phylogenetic concept is said to apply pluralistically in this way: all species are lineages, but there are lineages of different sorts: some characteristic of sexually reproducing organisms and some appropriate to organisms that multiply asexually (Mishler and Donoghue 1982; Mishler and Theriot 2000, 40–54, 179–84).

Kevin de Queiroz proposes a "population lineage concept" that is monistic in that it is single, definable concept, but pluralistic in that, in its author's view, it covers all the bases. Thus a BSC species is a very short such lineage (as good as "non-dimensional"), whereas a PSC or an evolutionary species stretches out over time – yet each one is plainly in some way a population lineage. Population lineages can even be read either as individuals or as classes. Thus, de Queiroz suggests, we may take *Homo sapiens*, as a lineage, to be an individual, whereas the sum total of human beings may be said to constitute a class (de Queiroz, in Wilson 1999, pp. 49–90).

Systematics

If there is a species problem, clearly there is also a problem about the foundations of taxonomy: The work of elaborating a system of classification will be affected by the approach we take to its basic unit. Accordingly, following some of the major lines of division in species concepts, there have been differing schools of taxonomy. Hull, in his classic account of contemporary systematic philosophies, distinguishes the pheneticists, who want to use observable characters, with (ideally) no theoretical weighting, from phyleticists, who base their classifications, in one way or another, on evolutionary considerations (Hull 1970a). Phyletic taxonomists, in turn, are either cladists, who follow, again, in one way or another, Hennig's method of analysis, or evolutionary taxonomists, who pay attention to what one might call evolutionary distance in addition to just the splitting apart of descendant from ancestral groups. The three approaches are both clearly distinguished and set in relation to one another, in Mayr's "Biological Classification: Toward a Synthesis of Opposing Methodologies (Mayr 1981). He calls the first school "phenetics." "Cladistics" we have already encountered. Mayr himself is the most distinguished exponent of the third approach, evolutionary classification.

Phenetics is sometimes called by its adherents "numerical taxonomy" (Sokal and Sneath 1963). The original idea was that characters should be taken as they come, on the basis of pure observation, without any weighting influenced by theory. When, with their reliance on listing properties, they were called typologists, they replied by asserting that if we are typologists, we are *empirical* typologists (Sneath and Sokal 1961). There are obvious difficulties with this approach. First, pure observation is hard to define and hard to locate. Some theoretical, or conceptual, considerations are always lurking in the background. And, besides that problem, numerical taxonomists boasted that the success of their method was evidenced by the fact that their results agreed in the main with classical taxonomy. In that case, why all the fuss about a new method? In fact, phenetics as a working school of taxonomy has almost completely faded away. Mayr points out, however, that its practice was indeed useful as a starting point in classification: Systematists do have to start out by observing perceptible differences among specimens before they can arrange their subjects in reasonable ways.

As to the second method, cladistics, despite the variety of more or less Hennigian species concepts, and the acrimony that sometimes reigns among their proponents, one can say that in general, followers of Hennig do practice the form of systematics that goes by that name. They try to cut phylogeny at the joints, looking for signs of recent splits in the many-branched tree that constitutes the history of life. And although Mayr has been, and still is, vehement in his opposition to the phylogenetic species concept, or to cladism as a founding doctrine for systematics, even he does acknowledge that, up to a point, cladistic analysis is a useful method in sorting out relations between descendant and ancestral species.

At the same time, evolutionary classification, while sharing with cladism a phyletic, rather than a purely phenetic, perspective, does differ sharply from cladistics in some of its inferences. Cladists want to admit as the ground of their classifications only recency of division, as evidenced by synapomorphies. According to Mayr and his disciples, however, differences in life style – evolutionary distances – must also be taken into account. We are very closely allied to chimpanzees, for example, sharing with them 97 percent of our genetic endowment. Yet our life style is so remote from theirs that we seem to constitute not only a different species among the *Hominidae*, but even a separate family. Or consider the status of birds. They are descended from crocodilians, which had split off from other "reptiles." So the crocodilians are the sister group of birds. Yet crocodiles and their kind are much more like other "reptiles" than they are like their feathered and flying descendants. The cladistic taxonomist, relying only on synapomorphies, has to ignore such striking innovations in life style; he can consider only the similarities between sister groups, not the evolutionary histories that may take one branch much farther from the origin than the other. Mayr's illustration of this possibility (Figure 10.2), in contrast to Hennig's synapomorphic diagram, shows plainly the difference between the two

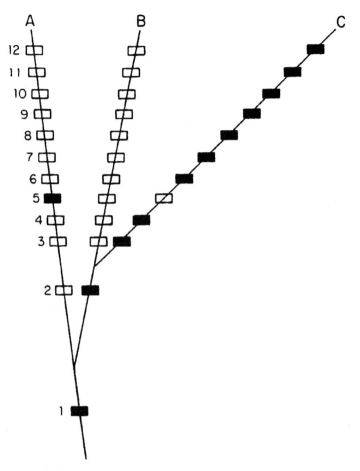

Figure 10.2. Cladogram of taxa A, B, and C. Cladists combine B and C into a single taxon because B and C are holophyletic. Evolutionary taxonomists separate C from A and B, which they combine, because C differs by nine autapomorph characters from A and B and shares only one synapomorph character (2) with B. Reprinted by permission of the publisher from TOWARD A NEW PHILOSOPHY OF BIOLOGY by Ernst Mayr, p. 279, Cambridge, Mass.: Harvard University Press, Copyright © 1988 by Ernst Mayr.

schools: On the face of it, the "evolutionary" approach seems more bi-ologically realistic than the rather abstract insistence of the cladists on counting sister groups only, without reference to evolutionary history. As Mayr points out, an evolutionary perspective can also catch conver-gences that cladism, looking only at characters rather than history, will sometimes miss.

Oddly enough, however, if Mayr is the leading exponent of evolutionary classification, it is also he who has developed and authoritatively expounded the BSC, which he describes as "non-dimensional," and so, one might think, not applicable, at least in any obvious or direct way, in a taxonomy that takes account of what in fact happened (so far as we can tell) through a long stretch of evolutionary time. How can a non-dimensional concept be used to give due weight, in taxonomic practice, to evolutionary divergences that have arisen in the course of a long and complex history? To cladists, indeed, such considerations appear unscientific; for evolutionary systematists, they are essential if, it may appear, in uneasy company with a purely contemporary rather than evolutionary or phylogenetic concept of species.

How have these three schools fared in recent history? Phenetics has more or less faded away; it has had to undergo such drastic modifications that, as a working school of taxonomy, we have to say it scarcely exists now. Evolutionary taxonomy, as we have just noted, seems to have common sense on its side – as well the ability to retrieve some information that cladism has to ignore. Yet in terms of public relations, at least, we have to admit that cladism seems to be winning. It is appealing for its neatness, perhaps. In any case, it is instructive to notice that for some time past, visitors to the Natural History Museum in South Kensington in London have been able to construct their own cladograms. And the American Museum of Natural History in New York is now so organized that in strolling through its series of rooms, one is in fact walking through a cladogram. Time will tell, but for the present, cladism is certainly conspicuous for its prestige and its successes.

Reducibility

A second major theme that has recurred in previous chapters is that of the uniqueness of the living, or, in opposition to that view, an insistence that the phenomena of life can be explained wholly in terms of the less than living. We found an interest in living things fundamental for Aristotle's philosophizing, with a turn to a more "mechanistic" thought style in his successors. In the Galenic tradition, opposition to mechanism introduced a more purposive kind of teleology into explanations of life. Then, in confrontation with Harvey's vitalism, we saw Descartes initiating the thoroughgoing mechanism that has recurrently marked modern biological thought. In Buffon, we noticed a tendency toward a more vitalistic approach. There was also an explicit revival of vitalism in the nineteenth century, unwittingly triggered by the work of Albrecht Haller, who had not himself intended such a philosophical inference from his introduction of "irritability" as a mark of the living (Roe 1981). Finally, it seems the vitalism–mechanism struggle reached its culmination, and its denouement, in the early twentieth century in the contrast between eminent supporters of each school: Hans Driesch and Jacques Loeb. Driesch had started out in the relatively mechanistic camp stemming from Wilhelm Roux's "*Entwicklungsmechanik* (mechanics of development)." In his own embryological work, however, he discovered that a whole sea urchin could develop from half an egg, and this led him to posit some kind of vital force supervising development, what he called an "entelechy," a special vital something unique to things that are alive (Driesch 1899; 1908). Loeb, in contrast, boasted of a heart that he had kept beating in solution for twenty-five years; so-called vital functions were thus plainly shown to be merely mechanical (Loeb 1912; Pauly 1987). By now, both these positions seem quaintly dated. Nobody would any longer proclaim the creed of vitalism, and Loeb's programmatic mechanism no longer appears necessary as a counterfoil to it.

Nevertheless, although no one would call himself a vitalist, something akin to the older controversy has continued in other terms. Nobody any longer wants to invoke some non-material something to explain biological phenomena. But despite that concession, many (both scientists and philosophers) have resisted the idea that biology, lacking some special vital force to sustain its subject matter, must be on principle reducible to the "more basic" sciences of physics and chemistry. Living systems are material, Simpson said, but they consist of matter organized differently; and studying the organization is not the same as simply studying the matter (Simpson 1968, p. 30).

Both biologists and philosophers entered into this debate. Leading evolutionists, finding themselves threatened first by the biochemical turn, and then by the molecular turn in biology, defended the irreducible nature of their subject matter. Thus, in an address to the American Society of Zoologists in 1964, Dobzhansky declared:

> Unlike the atoms of classical physics . . . the chromosome is not a heap of genes . . . A cell is not a conglomeration of chromosomes, but a supremely orderly contrivance . . . Cells are in turn complexly structured and well-organized entities . . . Individuals and species belong to ecological communities and ecosystems. The progress of biology would not be furthered by frenetic efforts to reduce organismic biology to physics. This is not

because there is anything in living things that is inherently irreducible. It is rather because a different research strategy is more expedient . . . These are levels of increasing complexity and they are hierarchically superimposed.

					Dobzhansky 1964, p. 446

Simpson, in the essay already referred to, first published in 1967 in *The American Scholar*, also appealed to the hierarchical nature of its subject matter to demonstrate the "compositionist" approach necessary to the study of biology. He considered first the phenomena of heredity, insisting on the informational character of DNA as distinct from the mere chemical sequence of bases. Then he turned to evolution, and pointed out the irreducible nature of the phenomenon of adaptation central to the theory of natural selection (reprinted in Simpson 1969, pp. 3–18).

A few years earlier, at the 1959 Cold Spring Harbor symposium, Ernst Mayr, too, had appealed, in an address entitled "Where Are We?" to the irreducibility of biology, in particular of evolutionary biology, to a merely molecular level. It takes the thinking of naturalists to face the problems we confront. In his peroration, Mayr exclaimed: "We live in an age that places great value on molecular biology. Let me emphasize the equal importance of evolutionary biology. The very survival of man on this globe may depend on a correct understanding of the evolutionary forces and their application to man" (Mayr 1959). Fighting words!

Mayr's argument for the primacy of evolutionary over molecular biology rests on the simple idea that we recounted in the last chapter – namely, that an organism must first succeed in dealing with its environment if it is to pass on its genes, an idea that resonates with Simpson's stress on the primacy of adaptation. Natural selection, Mayr argues, "sees" only the phenotype; organisms, not genes, are the "targets" of selection (Mayr 1982; 1988, pp. 120, 128). Thus, transgenerational selection of genes depends in the first instance on within-generation selection of phenotypes. It is a two-step process, with action at the phenotypic level taking the lead, even though the adaptive changes that result are grounded in genotypes. Accordingly, one must look at evolutionary change from a phenotypic perspective first, as naturalists do, and assume that genetic change follows as a trailing rather than a leading indicator. As William Wimsatt would put it, genes are the bookkeepers of evolution, not its agents (Wimsatt 1960). In later writings, Mayr reiterated this view by arguing that theoretical population geneticists – he mentions Fisher by name, although Dobzhansky held the same view, and he came close to holding it himself – had made a mistake when they defined evolution as "change of gene frequencies in populations" (Mayr 1982, p. 588).

To reinforce his point, Mayr and others argued that there is presumptively a many-many relation between genes and phenotypes. It takes many genes to make an adaptive trait. Conversely, many different traits can depend partly on the same genes. This means that you can know the entire molecular sequence of a genome without knowing how it will be manifested in the phenotype (see Lewontin 1974). The same point follows from the fact that the fitness of genes (alleles) is context-dependent. It depends, for example, on its frequency within a given population, as the example of mimicry shows. Mayr's stress on the relativity of fitness due to complexity and context-dependence is further intensified by his theory of speciation, according to which a new species originates through the isolation of a founder population from its conspecifics. Life at the edge of a species' biogeographical distribution requires a transition to a new environment, which in turn entails a radical reconstruction of the genome (Mayr 1988, p. 447).

Short of the defeat of many-leveled evolutionary thought by a triumphant molecular biology, however, there are also more or less reductionist positions that can be taken by evolutionists themselves. As we saw in Chapter 9, some writers, notably G. C. Williams in his influential *Adaptation and Natural Selection* (1966) and Richard Dawkins, in *The Selfish Gene* (1976; rev. ed. 1989), have insisted on the primacy of genic selection, resisting any move to something like group selection. The debate about units or levels of selection reflected a multiplicity of attitudes on this question (Brandon and Burian 1984; Sober 1984; Lloyd 1988). In a later work, Williams looks in detail at some of the difficulties faced by a maximally parsimonious selection theory. He concedes, at least, the reality of what he calls "clade selection" – that is, the selection of gene pools rather than single genes – but is doubtful of such phenomena as trait-group selection, let alone species selection, a concept, he argues, that depends on a gross misunderstanding of the term species (Williams 1992).

In the present context, however, the point is that even the most austere genic selectionist comes short of reducing his subject matter to physics and chemistry. Williams, for example, distinguishes between the "codical" and the "material" aspect of a DNA molecule. In the former role, as conveyor of information, it enters into a special historical and ontological context that the mere specification of its chemical nature, and even its position in the series of bases, would not suffice to specify (Williams 1992). Francis Crick, co-discoverer of the DNA code, on the other hand, proclaims in his lectures *Of Molecules and Men* a thoroughly one-level reductionist stance (Crick 1966). "The ultimate aim of the modern movement in biology," he writes, "is in fact to explain *all* biology in terms of physics and chemistry" (Crick 1966, p. 10; Crick's italics). Any resistance to this ultimate reduction, Crick believes, betrays remnants of vitalism – that is, of the notion "that there is some special force directing the growth and behavior of living systems which cannot be understood by our ordinary notions of physics and chemistry." The problem, he continues, "is in explaining what sort of thing this might be, and almost everyone who has written or spoken about it becomes very confused at this point" (Crick 1966, p. 16).

Crick proceeds to consider three areas in which such misguided notions arise: molecular biology, the origin of life, and the study of the central nervous system. In the first case, he thinks increasing knowledge of *E. coli* has taken us well along the way. The others are more difficult. The proper approach, he tells us, is to admit that, for the moment, we have to address these problems at a relatively high level of organization. However, unlike the evolutionists we quoted earlier, he sees no reason why, if we allow science and scientific education to advance as they ought, we should not ultimately reach that final reductionist goal. He is as keen as they are on the near omnipotence of natural selection, and thinks it should be instilled as a leading principle in every school curriculum, although he recognizes the difficulty of tracing out its actual course in any given case. Also, the complexity of the brain – a rather sloppy complexity, unlike the neat binary thinking of computers – makes the task of reduction still problematic there. But Crick is confident it will be done.

In the heyday of the philosophy of science as equivalent to the philosophy of physics, philosophers entered the reductionism debate under the influence of their dream of the unity of science: the notion that all scientific explanation would one day be unified within the scope of an all-inclusive, most basic science from which all more limited theories could be seen to flow. In fact, this is very like Crick's *desideratum*, which we have just been describing. Ernest Nagel's *Structure of Science* was authoritative for this perspective, although logical empiricists in general were certainly sympathetic to it (Nagel 1961). Such a fundamental science, of course, would be basic physics; so, for philosophers of biology, the question was: "Can biology be reduced to physics and chemistry?" Or, to put it positively, "Is biology an autonomous science?"

In fact, the question was put in three forms. The ultimate reduction would be that of one theory to others, or rather, of other theories to one definitive one. It has been, at least in part, a tendency to atomistic (or molecular) thought that has motivated this kind of argument. From Leucippus and Democritus on, there have recurrently been thinkers in our tradition for whom the only satisfactory explanation is one of wholes in terms of their (least) parts. The presumption is that theories of wider scope, embracing the "laws" of a more restricted science, will always range over smaller and smaller units. A few writers, notably Wimsatt and James Griesemer, appear to use the term "reduction" in either direction – that is, they would call the transformation of a lower level into a higher level theory also a reduction (Wimsatt 1980; Griesemer 2000). Generally, however, theory reduction has been thought to mean reduction downward – to lesser parts contained in, and, on this view, constituting, a larger whole. Such was the hope, for example, of the person we mentioned earlier who regretted that the Volterra–Lotka equation could not yet be stated in terms of quantum mechanics.

Short of the longed-for theory reduction, one could insist on two less extreme forms of reduction: ontological and methodological. If it means only the denial of vitalism – that is, the admission that living systems are material systems – ontological reduction seems to present no difficulty. On the other hand, if what it means is the actual reduction of all systems to their least parts – of ecosystems to molecules – this may be a more dubious stance. Dupré's argument, which we reviewed earlier, as well as those of Dobzhansky, Simpson, and Mayr, all indicate that something important about complex wholes is lost if they are conceived solely in terms of their least parts, even though it is those parts of which they are in fact composed. On principle, however, ontological reduction seems relatively harmless.

Methodological reduction, too, appears less drastic then theory reduction. Even Dobzhansky admitted, in the passage quoted earlier, that analytical methods could often be useful in studying complex systems. So, in some circumstances, non-reductivists may well advocate reductive methodologies on principle. In the other direction, it is also possible for investigations aimed at a reductive explanation to use non-reductive techniques. The late Lucille Hurley, a teratologist at the University of California, studied the swimming behavior of mice in order to learn the effects of the absence of trace minerals that she had removed from the diets of their pregnant mothers. A colleague reproached her for her non-reductive procedure, but her aim was reductive enough. It was to show that if there were no manganese in the mother's diet, unbalanced swimming in the offspring would follow. Even a believer in ultimate theory reduction could happily admit such techniques in the absence of the molecular explanations of all behavior that he would expect to eventuate as science advances.

As a general philosophical problem, the question of reduction has rather faded away, although it does still have some conspicuous true believers. Kenneth Schaffner, in his *Discovery and Explanation in Biology and Medicine*, makes much of the issue of reducibility and the unity of science (Schaffner 1993). He has dutifully followed Crick's suggestion by showing that an organism's genes completely specify its neurons, whereas its neurons completely specify its behavior (Schaffner 1998). A faithful follower of his master Nagel, Schaffner still hopes for an eventual utopia of a properly reduced, single basic science. On the other hand, Alexander Rosenberg, in his *Instrumental Biology or the Disunity of Science*, while expressing a firm belief in the reducibility of proper science, argues that biology, since it is not unified, reductive, and (he holds) "realistic," has to work merely instrumentally in a lot of diversified fields – hence the unfortunate "disunity of science" (Rosenberg 1994; contrast with Dupré's glorification of the "disorder of things" [Dupre 1993]). Both Schaffner and Rosenberg still believe in theoretical reduction and in the consequent unity of science. One still hopes for it, even in biology; the other despairs of its achievement in those stubbornly irreducible, and hence merely "instrumental," disciplines, which in his view include biology as well as the social sciences.[8]

In a limited area, the question of reduction was raised with respect to transmission, or classical Mendelian, versus molecular genetics. Hull made this issue central to his 1970 textbook, one of the first of many in the newly emerging field of philosophy of biology (Hull 1970b).[9] If science advances to "better" theories, and especially from the larger to the smaller, why hasn't molecular genetics simply replaced its more superficial predecessor? From the point of view of standard philosophy

---

[8] Rosenberg expresses the same view in Rosenberg 2000 as in his 1994, alongside confidence in "reductionism redux" for the case of molecular biology. Rosenberg 2000 makes it clear that when he refers to realism, it is the reality of least particles that he has in mind. When we talk about realism in the following chapter, this is not the sense of realism we have in mind.

[9] Philosophers of biology have used the textbook form as a means for conducting serious arguments with one another. Instances of the genre include Hull 1970b; Sober 1984; Rosenberg 1985; Sterelny and Griffiths 1999.

of science, this seemed quite puzzling. According to Sahotra Sarkar's analysis, however, each of those two disciplines has fostered its own variety of reductionism, molecular genetics more successfully than its Mendelian predecessor (Sarkar 1998). In terms of transmission genetics, recurrent attempts have been made to carry through what Sarkar christens "genetic reductionism," a position more familiar under the name of genetic determinism – that is, a program of deriving all adult human characteristics and behaviors from the individual's genetic endowment. Conspicuous here is the discourse surrounding the Human Genome Program, to which we shall return in the next chapter. Molecular genetics, on the other hand, carries out what Sarkar calls "physical reduction," investigating the physical structure of the genetic material. In terms of the older philosophical approach, this would be an ontological reduction. It poses fewer problems than genetic reductionism, which has strong theoretical, as well as practical, ambitions.[10]

Function and Teleology

If biology is not theoretically reducible to physics and chemistry, how is it different? There used to be a distinction between so-called "descriptive" and "exact" or experimental sciences, but given the recent history of biology, such a criterion no longer holds – if it ever did. Is there, then, some unique form of explanation characteristic of biology? If so, teleology seems the obvious candidate. Granted that Aristotelian final cause obtains everywhere in his universe, not only in living things; still, it is especially conspicuous in his biology, in the stress on development. Much later, that good Aristotelian, Georges Cuvier, identified his "conditions of existence" with final cause. And even in Darwin's case, his central use of the concept of adaptation seems to suggest a teleological component even in the theory of natural selection. In contrast, from Democritus on, opposition to the uniqueness of life often entails a denial of final cause. Everything happens by chance or necessity – to borrow Jacques Monod's famous title; there are no ends of nature. Following Descartes, the very notion of teleology appears "unscientific."

How has teleology fared in post-Darwinian discussions? For orthogeneticists, evolution has a direction. But although, as we noted earlier, such views were current early in the twentieth century, we also noted that by now they have disappeared. Does the theory of natural

---

[10] But see Wilson, E. 1998

selection contain a teleological component? As a theory of the origin of adaptations, it at least trades on a teleological concept: An adaptation must be to and for something. Darwin's position on this question was ambiguous; what about ours?

Robert Brandon answered this question definitively in an address to the Philosophy of Science Association in 1981, and again in his book in 1990 (Brandon 1981b; 1990). It is the *explanandum* of the theory of selection, he noted, that is teleological. We have here what Waddington called "the appearance of end" (Waddington 1957). The *explanans*, however, is straightforwardly causal. Although, as Brandon argues, the principle of natural selection has no special empirical content, it is a place holder for any number of particular when-then cause and effect connections that, in themselves, bear no trace of teleology. In his book, Brandon calls these "mechanistic explanations" of adaptive phenomena (Brandon 1990, pp. 184–189). Although the concept of "mechanism" may be ambiguous, in this context it seems plain enough: It is ordinary "billiard ball causality" that is involved – one thing after another, or many things after many others. The explanation includes no trace of teleology.[11]

The same point has been made by one of the present authors (Grene) in her "Explanation and Evolution" (in Grene 1974), where she contrasts various kinds of biological questions, in particular, "What is it for?" and "How did it arise?" (Grene 1974). The former question – the question of instrumental teleology – is teleological on the face of it, but its role is chiefly heuristic, and the question that follows, "How does it work?" demands answers in terms of operating principles rather than ends. In Aristotelian terms, it looks for formal rather than final causes. Grene concludes:

> The subject matter of evolution does not lend itself to teleological expla-
> nations. Even functional explanation . . . , though it responds to questions
> expressible in teleological language, is, as explanation, more formal than
> teleological in its import. It shows how a system, or subsystem, works,
> not how, as an end, it evokes its own means. But evolutionary explanation
> cannot even be functional, because there is no single organized system we
> are looking at, whose functions we could be examining. It is, in short,
> ordinary causal explanation of a well-developed and subtle sort, fitting

[11] Oddly, Brandon also published an article in 1981 that calls for teleological explanation in the very kind of evolutionary theory that he denies in his PSA address. This seems to be done by stipulatively defining teleology as function (Brandon 1981a).

smoothly into a naturalistic view of nature, and mediated in its details by the tools of mathematics.

Grene 1974, pp. 207, 227, 226

The problem of teleological explanation in biology has also been approached, or better, side-stepped, by speaking of "teleonomy" instead of teleology. The term was introduced by Colin Pittendrigh (Pittendrigh 1958). Presumably, we should be able to say, "The turtle came up on land to lay its eggs," without suggesting the unfortunate "purposive" implications of "teleology." Our statement is only "teleonomic." Mayr, in adopting this term, restricted its applications to systems governed by programs (Mayr 1976a). Such naming, though perhaps comforting, seems to have no effect on the kind of causality entailed in the theory of natural selection. As François Jacob has put it, evolution personified is more like a tinkerer than an architect (Jacob 1977). Within the limits permitted by history, circumstances arise that lead to the relative reproductive success of organisms carrying one trait rather than another. That's how it happens, with no goal and no program superintending the process. Some evolutionists have tried to make selection creative – like a poet or a sculptor. When they do, they are betraying their own best insights (Mayr 1963, p. 119; Dobzhansky 1970, pp. 430–431).

This all seems plain enough. The situation is complicated, however, by recent philosophical debates about function and its relation to teleology. There was lively debate on this question some thirty years ago. Two alleged solutions are still invoked, and their consequences are still subject to controversy. One answer, put forward by Larry Wright, proposes an "etiological" concept of function (Wright 1958; 1967). According to Wright, "the function of X is Z" means:

(i) Z is a consequence (result) of X's being there
(ii) X is there because it does (results in) Z.

Wright 1967, p. 81

That sounds sensible enough. The blood circulates because the heart beats. The heart exists because, by beating, it makes the blood circulate. But how do we cash this out? According to Wright, we look for a process of selection, conscious and intentional in the case of human behavior, natural in the sphere of biology. So any time we want to investigate the function of any biological organ or process, we are asking how natural selection produced it. As we have seen, however, that is not to give a teleological answer. On the contrary, it is to give a causal-statistically deterministic answer as to how something that looks telic, an adaptation, arose. The question may be metaphorically teleological: What was it designed for? But the answer will eventuate from a purely causal story of how one thing succeeded another. Yet Wright, and his later disciples, insist that their kind of explanation *is* teleological. Moreover, it gives us no clue as to the answer to the different, and more substantive, functional question, "How does it work?" We just assert, "X is there because it does Z," but this gives us no hint of how to pursue an inquiry into its function. So we have a conflation of function with teleology, as well as a suppression of the latter in the supposed teleological explanation itself.

The common alternative to "Wright functions" is a view put forward in 1975 by Robert Cummins (Cummins 1975). After a careful criticism of the standard logical empiricist attempt to explain, or explain away, functional analysis, Cummins concludes: "To ascribe a function to something is to ascribe a capacity to it which is singled out by its role in an analysis of some capacity of a containing system" (Cummins 1975, in Allen et al., eds., 1998, p. 193). This approach has the advantage of allowing us to ask of some process how it works. As Cummins points out, it also cleanly separates the question of function from that of teleology. In the case of artefacts, we can indeed ask why a feature is there, and can answer in terms of its artificer's reason for putting it there. In natural systems, however, Cummins points out, such a "design" answer is doomed to failure (Cummins 1975, in Allen et al., eds., 1998, pp. 175–76). We can only give a straight causal answer, and any "what for" consideration is irrelevant. Moreover, if we want to study such a feature, we need not look back at its prior causes, but can ask, straightforwardly, what role it plays in the system of which it forms part. Critics of Cummins's view object that in these terms it is impossible to separate function from accident or normal functioning from malfunction. But it seems to be precisely the role of a process or structure within a system that allows such evaluation.

The discussion of functional analysis in William Bechtel and Robert Richardson's *Discovering Complexity* (Bechtel and Richardson 1993, pp. 89–90) appears to follow the Cummins line. Moreover, there were other papers published in the 1970s that stressed, as Cummins did, the distinction between the study of function and a general teleological approach (not to mention the deteleologizing teleology of Wright functions). In a very subtle and detailed argument on "Function and Explanation," for example, David Hirschmann concluded, in seeming anticipation of Cummins, "A functional statement or analysis is explanatory because it shows the role that some item has in a system organised with respect to some characteristic" (Hirschmann 1973, p. 37). A few years later, in 1977, Peter Achinstein published an article called "Function Statements" in which he distinguished various senses in which statements of function may be asserted. In only some of these is the question "What is it for?" being addressed (Achinstein 1977). Achinstein's chief conclusion is that the goal question, "For what end does x exist?" if it is asked at all, is independent of the question, "How did it come to exist?" Arguments like these, together with Cummins's original essay, appear to answer – again, in anticipation – Kitcher's 1993 claim that the approach to function in both artefacts and natural objects can be unified by the appeal to design (Kitcher 1993).

For the most part, however, these earlier answers have been forgotten – except for Cummins, who is referred to chiefly in disagreement. Indeed, various forms of "Wright function" have been elaborated, to the extent that it now seems to dominate the philosophical scene, celebrated by its adherents in innumerable pages of argument.[12] Millikan has even introduced the term "proper functions" for the "selected effect" (SE) variety, which contrasts with the "causal role" (CR) variety (Millikan 1989). At the same time, however, there are other candidates in the field, some of them attempted modifications of the etiological view. To take care of the problem of changing selective values, for example, Peter Godfrey-Smith proposes what he calls a "modern history theory of functions," in which only *recent* selective processes are considered (Godfrey-Smith 1994, in Buller 1999, pp 199–220). Buller, in the concluding essay in his anthology, proposes a "weak etiological theory," which he believes will allow some CR analysis (Buller, ed., 1999, pp. 281–306). John Bigelow and Robert Pargetter articulate a forward-looking theory, in terms of present propensities, rather than past selective histories (Bigelow and Pargetter 1987, in Buller 1999, pp. 97–114). Denis M. Walsh has developed a "relational theory," which ties function to a given selection regime. This approach retains the evolutionary

---

[12] See, e.g., Millikan 1989; Neander 1991, both reprinted in Allen et al. 1998. More extensive bibliographies of these authors are given in the reference list of Buller 1999. See also Cummins et al. 2002. Peter McLaughlin, in *What Functions Explain* (2002), gives a detailed account of the two major views, deriving them from the original formulations of Carl Hempel and Ernest Nagel, respectively, and offering his own solution with reference to what he calls self-reproducing systems.

framework, while allowing present evaluation of a function in CR terms (Walsh 1966, p. 564).[13]

In the face of this multiplicity of function concepts, Walsh and André Ariew have presented a useful "Taxonomy of Functions," which analyzes the competing proposals and describes their relation to one another (Walsh and Ariew 1996). Amid these multifarious proposals, both Buller and Godfrey-Smith – along with the canonical etiological enthusiasts – celebrate the progress achieved in the light of Wright's 1973 formulation (Buller 1999, p. 20; Godfrey-Smith 1994, in Buller 1999, p. 185).[14] Compared with analyses such as those of Cummins, Hirschmann, or Achinstein in the 1970s, unfortunately, these developments strike us as rather an unhappy retrogression.

As some of the alternatives we have just mentioned indicate, however, not all participants in these debates have subscribed to the Wrightian concept. A paper by Ron Amundson and George Lauder clearly and persuasively demonstrates the use of the concept of causal role (CR) function in functional anatomy, a discipline fundamental to evolutionary biology (Amundson and Lauder 1994). As Amundson and Lauder point out, "Wright", or Selected Effect (SE), function permits the reading of present functions, or adaptations, only in the light of their selective history. But, as Darwin had already recognized, a given structure may have more than one function. For example, in a celebrated study, Kingsolver and Koehl showed that early insect wings probably served for thermoregulation rather than flight (Kingsolver and Koehl 1985). So there is no straightforward SE story to tell here. (Nor would it help to consider, as Godfrey-Smith suggests, only recent selective history; in the Kingsolver and Koehl case, we would miss the early history, and besides, some new trait might arise that no historical investigation would capture.) Further, it is difficult to tell what the original selective effect of a structure was; too much has happened over evolutionary time. It is also difficult to tell just what structure was selected for. In this situation, the functional anatomist can proceed in one of two ways: adopting an "equilibrium" approach, in which ecological factors are studied in order to assess present selective values (though not the history demanded

---

[13] Walsh's formula is as follows: "The/an evolutionary function of a token of type A with respect to selective regime R is to m if and only if X's doing m positively (and significantly) contributes to the average fitness of individuals possessing X in it" (Walsh 1996, p. 564).

[14] Godfrey-Smith declares: "I think of this trend as an example of real progress in philosophy" (Godfrey-Smith 1994, in Buller 1999, p. 185).

by the etiological school), or adopting a "transformational" approach, in which changes in form are analyzed to determine the effect of intrinsic constraints in design. Here, design is studied, not as the product of selection, but as a constraint on it.

Amundson and Lauder still give some place to SE function, while contrasting it with causal role analysis. Cummins himself, however, has recently come to the defense of his original position, defending the CR concept and arguing with devastating effect that the etiological view, or neo-teleology, as he calls it, is a non-starter. The neo-teleologist asks, for example, of a bird's wing, why is it there, and answers, because it enables flight – and therefore selection produced it. But the trait or traits that make flight possible have to come into existence before the function eventuates. So neo-teleology has its history backward. Admittedly, selection itself *produces* nothing; it helps spread whatever happens to be there. Yet neo-teleology fares no better, Cummins argues, in explaining why traits spread. It wasn't wings that were selected; it was just slightly more efficient wings as against their slightly less efficient competitors. Thus "[s]ubstantive neo-teleology misidentifies the targets of selection with the sort of complex generically defined traits – having eyes or wings – that have salient functional specifications" (Cummins 2002, p. 170). Nor will it do to try to save neo-teleology by identifying having a function with adaptiveness, since adaptiveness is a matter of degree, while having a function is not.[15] True, Cummins admits, "neo-teleology... dies hard." He continues:

> Its rejection sounds to many like rejection of evolution by natural selection. But it isn't. Darwin's brilliant achievement has no more need of neo-teleology than it has for its classical predecessor. What it needs is a conception of function that makes possession of a function logically independent of selection and adaptiveness. For it is only by articulating a reasonably illuminating functional analysis of a system that we can hope to understand what it is that evolution has created. If we want to

---

[15] One of us (Depew) has been sympathetic in the past to Wright's approach in view of the idea criticized here: that in Darwinian evolutionary biology, functions and adaptations can be equated. This no longer seems satisfactory both for the reason expressed here, and because it makes functions harder to know than they are; in many cases the relevant information about the etiology of the trait would, on the view criticized, be lost forever in the nooks and crevices of evolutionary history. This identification was also encouraged by the proposed stipulative reform of the term "teleology" to refer to traits that have come to be through concerted natural selection because they perform some function. See n. 11. It is true that adaptations are inherently etiological. But not functions.

understand how it was created as well, there is no avoiding the messy historical details by the cheap trick of assuming that all we have to do to understand trait proliferation and maintenance is to attribute a function. Neo-teleology thus amounts to a license to bypass the messy and difficult details.... The idea that evolution and development are goal oriented is precisely what makes classical teleology unacceptable. Neo-teleology creates the same impression while masquerading as good Darwinian science.

Cummins 2002, p. 170

It is clear that Cummins's analysis corresponds much more closely than the "etiological account" to the practice of scientists in many areas. Lauder's work in functional anatomy, for example, shows this clearly. So does Niko Tinbergen in his famous list of four questions asked by the ethologist. We can explain a given behavior, he says, (1) in terms of the physiological mechanism and the physical stimuli that lead to it (that is, by asking: "How does it work?"); (2) in terms of its current functions (or survival value) ("What is it for?"); (3) in terms of its evolutionary history; or (4) in terms of its development in the individual (Tinbergen 1963). What is important for us here is the distinction between (1) or (2) and (3). Only (3) is etiological.

Moreover, the distinction between (1), (2), and (3) is equally fundamental for paleontology, as Rudwick argues in an article on "The Inference of Function from Structure in Fossils" (Rudwick 1964). Suppose we are looking at a pterodactyl limb. It looks rather like a wing, something to fly with. But is it? We can answer that question only by thinking in engineering terms: "What kind of machine does this look like?" We cannot, of course, ask how it actually works – it is only the remnant of something that worked somehow long ago; but we can compare its structure with that of machines we know. In this case, we find our object rather like an aerofoil. The pterodactyl probably glided, but did not fly. So we try to answer the question, "What is it for?" in reliance on the prior question: "How could it have worked?" There is no question here, however, of the origin of the mechanism we are studying. As Rudwick puts it:

Functional inference involves an analysis of adaptation only as a static phenomenon. The perception of the machine-like character of the parts of the organism is logically independent of the origin of the structures concerned. Theories of their causal origin (e.g. by natural selection or by orthogenesis) and theories of their temporal origin (i.e. by a particular ontogenetic and phylogenetic sequence) are strictly irrelevant to the detection and confirmation of the adaptation itself. The functional reconstruction of fossils is thus logically unrelated to any and all evolutionary theories

Rudwick 1964, in Allen et al., eds. 1998, pp. 112–113

In this connection, Rudwick reminds us that Cuvier's brilliant paleontological inferences were in fact allied, in his thought, to an emphatic rejection of any form of transformism. And, although to us, a century and a half after Darwin, paleontology is one of the most obvious supports for the fact of evolution, for Darwin himself the fossil record counted first among the "difficulties on the theory." The point here is that, one way or another, etiology, so-called SE function, has simply no connection with strict paleontological functional inference.

The same holds of work in the growing field of functional genomics. Here, investigators may be asking, given a certain condition to which an organism needs to respond, what part of the genome is responding, and how. Thus, for example, in the Plant Functional Genomics Research Group in Japan, a group of workers studied cold-treated, drought-treated, or unstressed *Arabidopsis* plants. Their abstract states that "[f]ull-length c-DNAs are essential for functional analysis of plant genes." Clearly, they are establishing a function – cold or drought resistance – and asking, through delicate experimental techniques, what it is in the genome that carries out that function. The causal role is known; they are examining the means by which it is exerted. They are asking neither "What is this for?" nor "How did it get that way," but simply, "How does it do it?" Given these examples, it seems to us that CR functions are more conspicuous in much scientific research than are so-called SE functions. Confidence in natural selection as *a*, if not *the*, chief "agency" of evolution is there in the background, but the operation of natural selection is not what is being investigated (Sekia et al., 2001).

This has been a partial sketch (perhaps in both senses of that adjective) of a very extensive literature. The anthologies we have referred to can provide the curious reader with a much more thorough overview. We must admit, however, that we find Cummins's attack on the etiological view entirely convincing. His aim, he says, is to "nudge [neo-teleology] to a well-deserved extinction" (Cummins, 2002, p. 151–72). We can only hope that he succeeds.

CHAPTER 11

✦

Biology and Human Nature

Introduction

The evolution of human beings, and the meaning of "human nature" within a comparative, biologically grounded framework of inquiry, is a huge topic, extending well beyond the contours of this book. Nonetheless, much of the interest that human beings have in other living beings reflects our unquenchable interest in ourselves. In this chapter, accordingly, we will say a few things about several aspects of human evolution, if only by way of commending further inquiry to the reader. We will touch, first, on human origins – "the descent of man," in Darwin's phrase – with particular attention to the unity of the human species; second, on the vexed topic of nature and nurture; third, on the evolutionary mechanisms required to account for characteristics that human beings alone possess: large brains, language, and mind; and finally, we will touch on what implications, if any, can be drawn about the "future of man" from the Human Genome Project, and, more generally, from the fact that within the last several decades, we have begun to acquire the technical ability *directly* to manipulate genetic material. Much of what we will have to say involves revisiting some of the figures and theories we have encountered earlier in this book, with special attention this time to our own species.

The Descent of Man

Buffon is generally accorded the title of "father of anthropology." In his *Histoire naturelle de l'homme* (1749), Buffon resolved to study human beings in the same way he had been studying other animals. He set out to review their anatomical and physiological traits, when and why these traits appear in the life-cycle of this species, the biological functions of the human senses, and, finally, geographical variation. In doing this, Buffon became, according to Cuvier, "the first author to deal *ex professo* with the natural history of man" (Cuvier 1843, p. 173). As Topinard, the author of a late nineteenth-century work, *Elements d'anthropologie générale*, put it, Buffon

> founded what would soon be designated anthropology, whose main branches he sketched out: man in general, considered at all ages from the morphological and biological [= physiological] point of view; the description of the races, their origins and intermixing; and finally the comparison of man with the apes and other animals from the physiological point of view, from the study of his characteristics, from his place among other beings, and from his origin. These amount to the three branches of anthropology made distinct by [Paul] Broca: general, special, and zoological.
>
> Topinard 1885, p. 48

To accept this account it is necessary to circumscribe it a little. First, as the author implies, Buffon did not call his natural history of man anthropology. The term "anthropology" goes back to the seventeenth century, when it referred to anatomical studies of human beings of the sort conducted by Vesalius (Blanckaert 1989; Sloan 1995). Although Buffon incorporated human anatomy into his work, what he studied was human beings from the perspective of natural history. This allowed his successors (including Kant, whose *Anthropology from a Pragmatic Point of View* can hardly be said to be about anatomy) to transfer the term anthropology from the anatomist's functional to Buffon's historical biology. Second, Buffon was confident that his study of human beings as natural entities – geographically dispersed, and open in their differences to the influence of climate and other aspects of their environments – would be protected from theological and philosophical objections because he carefully sequestered man's "moral" characteristics – the "metaphysical" attributes of reason, free will, and so forth – from his natural history of the species. It is true that the masters of the Sorbonne lodged their usual complaints, delaying the publication of the *Histoire naturelle* until Buffon judiciously affirmed his belief in "all that is told [in the Scriptures]" (Roger 1989/1997, pp. 187–89). Nonetheless, the enduring influence of the Cartesian separation of mind from matter now made it possible, ironically enough, to study human beings in everything *but* their rational life, to study them, that is to say, as animals among other animals, and thereby to pose a question

that is still with us: whether man's "moral" characteristics can be reduced to, or shown to emerge from, his biological nature. (Rousseau
thought the latter, having discounted the rational side of man.)

Substantively, Buffon's greatest contribution to anthropology in the
sense just explicated was his categorical rejection of polygenism – that
is, of the idea that different races of human beings have the status of different species or sub-species. Polygenism, which had been defended by
Paracelsus in the sixteenth century, always had about it a whiff of het-
erodoxy. Among Buffon's reasons for attacking it was the defense of it
by Voltaire and Rousseau, less as science, to be sure, than as a deliciously
scandalous way of rejecting the Bible. Happily, the overwhelming consensus among biologists, and philosophers of biology, has always been
with Buffon; evidence in favor of monogenism, most recently genetic,
has steadily accumulated.

Buffon's defense of monogenism was based on his conception of a
species as the "constant succession and uninterrupted renewal of the
individuals that comprise it" (Buffon 1753). Human beings of every sort
can produce fertile offspring, and so, by this criterion, fall into a single
species. In making this point, Buffon was animated by a desire to deflate
Linnaeus's "artificial" system of classification, which, as we have already
noted in Chapter 4, was giving Linnaeus unwarranted conceptual cover
for a classificatory view of various human races, and, potentially at least,
for polygenism (see also Sloan 1995). We also discussed in Chapter 4
how Blumenbach and Kant, writing independently on the subject in
1775, sided with Buffon, and developed his thought in ways that could
accommodate the notion of various human races without compromising
monogenism. The main difference between Kant and Blumenbach, as
we have also mentioned, was that Kant retained Linnaean classifications
as natural descriptions, but not natural history, whereas Blumenbach,
in rejecting that distinction, also rejected underlying assumptions about
the Great Chain of Being. "I am very much opposed," he wrote, "to the
opinions of those who, especially of late, have amused their ingenuity so
much with what they call the continuity or gradation of nature ... There
are large gaps where natural kingdoms are separated from one another"
(Blumenbach 1775, 3rd ed, 1795, in Blumenbach 1865, pp. 150–51).
In this chapter, we will see how these differing, but related, approaches
were applied, not so much to the problem of human races, which we
have already addressed in Chapter 4, but to the relationship between
the single human species and other closely related species, an issue that
Buffon's naturalism had made central to anthropology.

Linneaus, for his part, extended John Ray's category of *anthropomorpha* – changed, in 1758, to primates – to include the genus *Homo*, as well as *Simia* (apes and monkeys), and *Bradypus* (sloths, including lemurs). In doing so, Linneaus was claiming that man is *naturally* a quadruped. "He has a mouth like other quadrupeds," Linnaeus wrote, "as well as four feet, on two of which he locomotes and the other two of which he uses for prehensile purposes" (Linnaeus 1754, from English trans. 1786, pp. 12–13). At first, Linneaus placed only one species in the genus *Homo* – our own. Later, however, he added a *"Homo troglodytes"* or *"Homo sylvestris,"* which he derived, none too accurately, from anatomical and ethological descriptions of the East Asian orangutan. For Blumenbach, this tendency to lump our species together with great apes (which in his opinion sprang from the continuationist biases of the *scala naturae*) underestimated the vast differences between our species and apes. Human beings, says Blumenbach, are *naturally* upright and bipedal, *pace* Linnaeus. That is clear to anyone who looks at a large number of traits and their mutual relations, as the Göttingen biologists proposed to do, instead of a few arbitrary traits that are useful, at best, for identifying species. Alone among species, Blumenbach argues, human beings exhibit a complex, interlocking set of traits that support their upright posture and bipedality: a bowl-shaped pelvis; hands that are not, like those of apes, merely pressed into service as grasping devices, but are anatomically structured to do so; close-set, regular teeth, instead of the menacing canines and incisors of apes; sexual availability and eagerness at every season, and relatively shortened gestation times, which, together with non-seasonal sexuality, produce a constant supply of infants who must be enculturated if they are to survive and flourish at all (Blumenbach 1795). In consequence of all these mutually reinforcing traits – a fairly good list even to this day, as lists go, if we add an enlarged brain – Blumenbach concluded that the human species is naturally in possession of its upright posture, as well as of the dominion that it exercises over other living things.

Darwin's thesis of descent with modification from a common ancestor affected the state of this question greatly. Once Darwin's claim about descent was generally accepted (even if his idea of natural selection were discounted – see Chapter 8), it was no longer possible to regard apes as degenerate human beings, or to see races as so many departures from a civilized, rational, white, European norm, as Buffon and Linnaeus had done. On the contrary, it now appeared that human beings must have come from apes – or rather, from the common ancestor of both apes and human beings – and that civilized races must have come from the most primitive. Darwin, as we mentioned, was amazed with what the unity of the human species actually entails when he cast his first glance at the naked savages of Tierra del Fuego (see Chapter 7). Oddly enough, however, the acceptance of Darwinism, or at least of Darwinism as people construed it in the second half of the nineteenth century, had the effect of belatedly reconciling Linnaean classification with natural history. Although Darwin rejected the Great Chain of Being, he thought of orders, families, genera, species, sub-species, and varieties as points at which lineages split, forming thereby a "natural" rather than an artificial system. In consequence, Darwinism tacitly restored the anti-saltationist ideas that Blumenbach and other early professional biologists had fought to discount. When Huxley protested against this implication – Darwin had saddled himself, he famously wrote, with an "unnecessary burden" in clinging to the maxim "*natura non facit saltum*" (Huxley to Darwin, November 23, 1859, in Darwin 1985-, Vol. 7, 1991, p. 391) – he may well have had in mind this apparent regression to an older conception of biological order.

One result of the Darwinian revolution of the second half of the nineteenth century was that African great apes tended to displace the orangutan as the closest living human relative. Darwin wrote of the gorilla and the chimpanzee: "These two species are now man's nearest allies" (Darwin 1871 [1981], p. 199). Darwin reasoned that only strong, continuous selection pressure (whether natural or sexual, on the second of which he put such stress in *The Descent of Man* [see Chapter 7]), could drive a slowly diverging speciation process, and, accordingly, that the evolution of man must have occurred in a very challenging environment. At one point, Darwin speculated that Africa might be just the place (Darwin 1871, p. 199). The result was that the earlier enthusiasm for the shy orangutan, who dwells sleepily in the deep, rich forests of Southeast Asia, was displaced by enthusiasm for African species. Huxley, for his part, favored the gorilla. He turned out to be wrong. Darwin did not make a choice. So he turned out to be half-right; we now know that chimps share more of their genes with human beings – at least 97 percent – than other extant apes.

There was another alleged reason behind the preference of Darwinians for an "African genesis." Nineteenth-century Darwinism did not do away with the tendency to distinguish and rank-order human races. The prejudicial notion that the black race is the oldest (because least civilized) variety of man was combined with the Darwin–Wallace presumption that new species are generally to be found in the same geographical areas as their closest relatives to support an inference that Africans are closer than other races to the common ancestor of *Homo sapiens* and to the extant great apes. Since few extinct hominids or other primates had been discovered during the nineteenth century, it was also tacitly presumed that the number of intermediates between great apes and human beings would turn out to be relatively small.

This foreshortening of the presumed distance between apes and human beings began to give way in the twentieth century with the discovery of australopithecenes, on the one hand, and, on the other, a wider number of species of *Homo* (moving more or less backward in time – *H. neanderthalensis, H. heidelbergensis, H. antecessor, H. erectus, H. ergastor,* and *H. habilis*). In *Mankind Evolving,* Dobzhansky wrote, "We recognize two genera [of hominids] – *Homo* and *Australopithecus*" (Dobzhansky 1962, p. 186). Since then, most anthropologists have recognized *Paranthropus* (*P. robustus, bosei,* and *aethiopicus*) as a distinct genus lying between the australopithecenes and *Homo*. Nonetheless, anthropology must continually struggle with the fact that whereas the nearest relative of most species is usually another living species, the nearest *living* relative of *H. sapiens,* the chimpanzee, is phylogenetically very distant from us. (The separation from a common ancestor occurred between 5 and 7 million years ago.) In consequence, there has been a tendency, which only a fuller fossil record could diminish, to underestimate the complexity of what happened since the lineage of great apes diverged from the lineages that resulted in ourselves (Schwartz 1999; Tattersall and Schwartz 2000).

The distance has begun to be made up by the discovery of the fossil remains of *A. aferensis* ("Lucy"), who lived some 3.5 to 4 million years ago, and of *A. africanus.*[1] *Paranthropus* aside, it is possible that the genus *Homo* split off from *A. africanus,* or something like it, in the form of *H. habilis,* notable for its more concerted and skilled use of tools; that *H. habilis* gave way in turn to *H. erectus;* and, finally, that *H. erectus* gave rise to *H. sapiens sapiens,* as well as to *H. [sapiens] neanderthalensis,* which Dobzhansky treated as a distinct race of *H. sapiens* (Dobzhansky 1962, p. 191). An impression of linear progress of this sort can easily be given by the fact that, throughout this history, there has

---

[1] Recently, the world of anthropology was excited by the discovery in Chad of a 7 million-year-old skull (*Sahelanthropus tchadensis*) that seems to lie close to the point when chimpanzees and hominids diverged.

been an increase in both the absolute and relative size of the brain. *A. africanus* had a brain of only 441 cc in a body weighing about 50 pounds. *H. erectus*, though heavier, had a brain of about 950 cc. The upper end of Neanderthal brains is about 1,450 cc. The brain of *H. sapiens sapiens* averages a massive 1,500 cc.

Although it discredited racial rank-ordering, the Modern Evolutionary Synthesis did surprisingly little to dampen the presumed progressiveness in this picture. The concepts of grades and trends were dear to its heart. These notions tended to put a directional spin on what might well have been a very bushy process. Thus, Mayr, having discounted *Paranthropus* as not very different from *Australopithicus*, wrote in *Animal Species and Evolution*, "The tremendous evolution of the brain since *Australopithecus* permitted [hominids] to enter so completely a different ecological niche that generic separation [from australopithecenes]) is definitely justified" (Mayr 1963, p. 631). It is admitted on all sides that many species and sub-species of *Homo* may have come and gone since the period that began some 250,000 years ago, when the climate began markedly to oscillate, bringing with it concerted selection pressures of a kind that rewarded cooperation, communication, and cleverness. Still, the impression is often given that, once up and running, modern human beings were soon dominating the scene. By 30,000 years ago, from an original population of some 10,000 individuals, populations of modern human beings were to be found nearly everywhere in the Old World, but remained in sufficient contact that they constituted, as they still do, a single species.

This linear picture may have to be refined by data coming from the comparative analysis of mitochondrial DNA and other methods. These suggest that there may have been species of bipedal apes prior to *A. aferensis*; that there were outmigrations of *H. erectus* from Africa long before the evolution of modern human beings; and that when *H. sapiens sapiens* began its own adaptive radiation out of Africa a mere 90,000 years ago, it may have killed off Neanderthals (and perhaps others) in the process, with or without some gene flow (Templeton 2002). In their attack on trends and grades, some contemporary anthropologists who are also cladists have called on genetic data to discredit the tendency of the Modern Synthesis to lump together separate species of *Homo* (including Neanderthals) and to see the descent of man as more tree-like and less bushy (Schwartz 1999; Tattersall and Schwartz 2000). Nonetheless, it must be admitted that the orthodox, mid-century view has the advantage that it can readily explain the very fact that makes establishing the descent of man so difficult. If we are separated by a great gulf from our living relatives, unlike other lineages, the reason might well be that the powerful evolutionary innovations in our generalizing, polytypic, culturally based species had sufficient power to eliminate competitors, leaving, as Blumenbach had long ago suggested, an enormous hole in the phylogenetic continuum.

Nature and Nurture

Some prominent versions of Darwinism in the late nineteenth century, notably the biometric research program that in England was to mark out the path that eventually led toward the genetic Darwinism of the twentieth century, were deeply implicated in eugenic thinking – that is, in schemes for "improving the race" – which generally meant the white, European race – by preventing the "unfit" from having children ("negative eugenics") and by encouraging, at the other end of the statistical curve, the presumably most fit to marry one another and to produce large, fecund families ("positive eugenics"). English eugenics, under the influence of Charles Darwin's first cousin William Galton, was dominantly positive. The science of biometry at the University of London, which eventually brought forth the world's first department of statistics, was founded with the purpose of identifying, tracking, and marrying off to one another (before it was too late) members of families in which "hereditary genius" was presumed to run (see Chapter 8 for more on Galton). The tendency in the United States, on the other hand, was toward negative eugenics, backed by state laws preventing the "shiftless," "degenerate," "feeble-minded," and so forth from having offspring. Here, the originating influence was Charles Davenport, whose Eugenics Record Office was the seed from which sprang, ironically enough, the great biological laboratory at Cold Spring Harbor on Long Island (Kevles 1985).

R. A. Fisher (who, with Sewall Wright, was a pioneer of population genetics – see Chapter 9) became Galton's heir when he was appointed to the Chair of Eugenics at the University of London. The second half of Fisher's the *Genetical Theory of Natural Selection* (1930) is replete with assertions that human beings were appointed by their own evolutionary success to take over the evolutionary process (presumably from God) in order to beat back the melancholy effects of the Second Law of Thermodynamics. Natural selection, followed by the discovery of ever more powerful means by which human beings themselves can intervene in and direct biological processes, would sustain the notion of progress that had been undermined by the realization in the last of third of the nineteenth century that the Second Law dictated the ultimate "*heat death*" of the universe. On this view, artificial selection – the breeding of human beings – would be carrying on the work of God (Norton 1983; Hodge 1992).

Whereas the resistance of Wright and Dobzhansky to Fisher that we discussed in Chapter 9 was grounded in purely scientific objections, neither was unaware of Fisher's ideological baggage, of which they were critical. To be sure, the heyday of eugenics in the United States had for many reasons waned by the late 1930s, when the Modern Synthesis began to take shape. Still, the kind of Darwinism that the founders of the Synthesis envisioned, institutionalized, and defended set out to secure its scientific credentials partly by dissociating Darwin's name from the eugenic enthusiasms and distortions that had previously marred it. This effort, as well as a more diffuse shift from "nature" to "nurture," had been made all the more salient after the revelation of eugenics' ultimate fruit, the Holocaust.

Perhaps more importantly, Dobzhansky's theory of balancing selection seemed to provide good scientific reasons for thinking that eugenics is both impossible and unwise (see Beatty 1994). For Dobzhansky, it is impossible because genetic variation is plentiful in natural populations, and is not bunched up at one end of the curve or the other. So the very possibility of identifying "hereditary geniuses" and the "feeble-minded" makes no sense. (The opposite, so-called "classical" view of population structure, according to which one can identify good and bad outliers in populations whose genes are generally presumed to be fairly homogeneous, was presupposed by eugenicists. Although not every one who held this view was a eugenicist, the last significant defender of the classical view was the left-eugenicist Hermann Muller, who for a time abandoned the United States for the Soviet Union, where he naively assumed that socialism would include an effort to select for the traits of Beethoven, Goethe and, alas, Lenin, rather than the likes of Jack Dempsey and Babe Ruth (Muller 1935; on Muller's persistent defense of the classical view of population structure, see Chapter 9; Muller 1948).

But even if it were a genuine possibility, Dobzhansky suspected that *any* eugenic program would be biologically unwise. Balancing selection stresses the idea that certain genetic combinations that are fit in one environment may not be fit in another. Having one allele for sickle cell anemia, for example, confers some protection against malaria in Africa, where slash-and-burn agriculture had produced swamps infested with malaria-carrying mosquitoes. The single-dose pattern thus spread through the population under the control of natural selection, even at the cost of killing off a predictable number of offspring with two sickling alleles. In malarial environments, sickle cell anemia is an adaptation. But in environments in which malaria is not a problem, it is far from adaptive. Combine this fact with the equally relevant fact that environments often change, especially through the powerful agency of human beings themselves, and you might well infer that it would be counterproductive to try to second-guess nature by declaring what traits are and are not fit, as eugenicists did. Nature itself creates a buffer against the effects of environmental change by preserving genetic diversity in natural populations and by using that variation to produce, when it can, adaptations that enable a species to deal with changing environments by its behavioral, and in our case cognitive, flexibility. "Populations that live in different territories, allopatrically," wrote Dobzhansky, "face different environments. Different genotypes may, accordingly, be favored by natural selection in different territories, and a living species may respond by becoming polytypic" (Dobzhansky 1962, p. 222). Thus it would seem that nature itself teaches us the best eugenic – or rather anti-eugenic – lesson. A diverse, panmictic population, and the democratic beliefs necessary to sustain it, produce the most adapted, and adaptable, populations (Beatty 1994).

At the 1959 centennial celebration of the publication of Darwin's *Origin of Species* at the University of Chicago, many speakers stressed that the Modern Synthesis, which was on this occasion presenting itself in public as the scientific fruit of Darwinism, had shown that nature prizes diversity, and that Darwin's vision of man did not pose any real threat to liberal, and even religious, values. Eugenics was off the table. A prominent theme at the Centennial celebration was that culture itself is an adaptation – maybe the ultimate adaptation – for dealing with changing, unpredictable environments and, more generally, for avoiding adaptive dead ends. Far from cosseting the unfit, then, as both Social Darwinians and eugenics enthusiasts had argued earlier in Darwin's name, culture, with all the different forms of nurture that it signifies, is, in a real sense, the highest product of natural selection. At the Chicago Centennial, speaker after speaker (except for Julian Huxley – see Chapter 9, on evolutionary progress) repeated the point that the event's principal organizer, the anthropologist Sol Tax, had hoped they would state. Waddington put that point as follows in introducing a plenary discussion on human evolution:

> Conceptual thought and language constitute, in effect, a new way of trans-mitting information from one generation to the next. This cultural inheri-tance does the same thing for man that in the subhuman world is done by the genetic system.... This means that, besides his biological system, man has a completely new 'genetic' system dependent on cultural transmission.
>
> Waddington 1960, pp. 148–149

Several years later, Dobzhansky was arguing that "culture arose and developed in the human evolutionary sequence hand in hand with the genetic basis which made it possible" (Dobzhansky 1962, p. 75). One way in which genes and culture are implicated with each other can be seen in the selection pressure for shorter gestation times that character-ize human beings, which is evidenced in the immature, paidomorphic features of human neonates (hairlessness, for example) and which makes maturation radically dependent on child care and a host of other cultural practices. Many mutually interacting causes are at work here. Among the most salient are the fact that the birth canal was narrowed with the evolution of a upright posture and, concomitantly, of a bowl-shaped pelvis, thereby selecting for earlier, less painful, less fatal deliveries, as well as the fact that early delivery requires massive care, and so involves the development of bonds among parents. However it may have hap-pened, it would seem that in *H. sapiens*, and perhaps our closest, extinct relatives, nature had brought forth a "promising primate," to borrow Peter Wilson's title. As Wilson puts it, in contrast to species that are adapted only to a single or narrow range of environments, *H. sapiens* is

> most generalized not only in its morphology, but also in its total inven-tory of dispositions and capacities. It is both uncertainty and promise. Whereas in other species we may find, for example, that the relation be-tween the sexes for the sake of reproduction is specified and particularly adaptive, in the case of humans we should find no determined and species-specific mode of relationship, but rather generalized features from which it is necessary to define specific modes ... [This means that] an individual has little advanced information that will help him coexist with others on a predictable basis ... If the human individual is to coexist with other such individuals, he must arrive at some ground for expectation and reciproca-tion. He must work out some common form of agreement about actions and reactions.
>
> P. Wilson 1980, p. 43

The "common forms" for sharing information of this sort are what we call cultures. Hence, on the conception of man as a polytypic, generalized, behaviorally plastic, enculturated sort of being, members of our species are animals who realize their biological nature in and through culture (Grene 1995, p. 107; this is a more accurate way to put the point than to speak, with Marshall Sahlins, of man as a biologically "unfinished" animal [Sahlins 1976]). Given this general point of view, the modern discipline of anthropology contained within itself two mutually supporting branches: physical anthropology, which relies on natural selection to deliver man into culture, and cultural anthropology, which appeals to natural selection to free the notion of culture from the discredited crypto-Lamarckian idea that culture itself is a site and form of biological, evolutionary progress. It is true, writes Dobzhansky, that "the genetic basis of culture is uniform everywhere; cultural evolution has long since taken over" (Dobzhansky 1962, p. 320). But this does not mean that the genetically controlled aspects of our behavior are fixed and deterministic, or that the cultural aspect is free and variable. "This is the hoary misconception that 'heritable' traits are not modifiable by the environment," Dobzhansky says, "and that traits so modifiable are independent of the genotype" (Dobzhansky 1962, p. 74). Nor does it imply that natural selection ceased once human beings had developed culture. "It is a fallacy to think that specific or ordinal traits do not vary or are not subject to genetic modification; phenotypic plasticity does not preclude genetic variety" (Dobzhansky 1962, p. 320; see also p. 306). The spread of sickle cell trait in Africa after the introduction of agriculture, mentioned earlier, is a case in point.

Without an appreciation of the mid-century consensus about how cultural and biological factors interact, it is hard to appreciate the furor that greeted the publication in 1975 of E. O. Wilson's *Sociobiology* (Wilson 1975). Wilson's application of inclusive fitness, kin selection, and game-theoretical models for calculating the "advantage" of genes to the study of ants – about which he knew more than almost anyone – as well as to other social species would not have raised such opposition if Wilson had not framed his argument, in both the early and closing pages of his book, by making a number of provocative remarks about human evolution. In these remarks, Wilson proposed, in effect, to shift the boundary between the human, cultural, or social sciences and biology. There is more in human behavior that is under genetic control than had been appreciated, he argued. And although some of the impulses that undergird "division of labor between the sexes, bonding between parents and children, heightened altruism [only] toward closest kin, incest avoidance, other forms of ethical behavior, suspicion of strangers, tribalism, dominance orders within groups, male dominance above all, and territorial aggression over limiting resources" can be resisted, the genes do indeed "have us on a leash," even if it is a long one (Wilson 1994, p. 332; for "on a leash," see Wilson 1978, pp. 167, 207).

What lies behind Wilson's enthusiasm for these views is Hamilton's idea that genes will code for cooperative traits if this "strategy" enhances their own replication rate (see Chapter 9). This theory did much to liberate the Darwinian tradition from objections to the effect that, because it portrays human beings (and perhaps other animals) as more competitive than we observe them to be, Darwinism must be false. We are, sociobiologists say, programmed to be cooperative, at least with our genetic relatives. Predictably, philosophers' debates soon broke out about whether human morality, if it is an adaptation with a genetic underpinning, remains normatively binding, or whether it is merely a trick our genes play on us to get us to cooperate for their sake. Wilson exposed himself to some obloquy when, like the philosopher Michael Ruse, he took the second view, as well as when he and others used his theory to license adaptationist stories – "just so stories," Lewontin and Gould called them – about all manner of highly variable human practices (Ruse and Wilson 1985; on "just so stories" see Gould and Lewontin 1979). Still, by his own account, Wilson was taken aback completely when a more incendiary objection was raised. Some highly respected evolutionary biologists, including heirs of the version of the Synthesis celebrated in Chicago in 1959, read Wilson's proposal for human sociobiology as a shift back toward genetic determinism, and to the spirit, if not the letter, of eugenics (Lewontin, R., S. Rose, and L. Kamin 1984; for Wilson's reaction, see Wilson 1994, pp. 337–341).

We cannot go into the ins and outs, or the rights and wrongs, of the "sociobiology controversy."[2] Wilson was more orthodox than he was made to appear, sometimes even by himself. He did not believe that we are adapted only to the environments in which our hominid ancestors arose. On the contrary, he argued (or at least was soon to argue, in a book written with Charles Lumsden) that gene-culture interaction means that cooperative genes keep competitive culture on a

---

[2] We do not think that Ullica Segerstrale, who has studied the sociobiology controversy extensively, has been quite judicious enough in assigning praise and blame. Segerstrale 2000 is mostly a defense of Wilson against enemies on the right and the left.

leash! (Lumsden and Wilson 1981). The real issue is not the importance of culture, but what kind of scene culture presents. For Wilson, it is a scene of potential fanaticism. When he had asserted that "the genes have us on a leash," he claims to have meant that genes (à la inclusive fitness) perform the admirable service of damping down the runaway effects of culture, such as ritual cannibalism (Lumsden and Wilson 1981, p. 13).

Nonetheless, it is fair to say at the very least that Wilson was somewhat unfortunate in his allies. Sociobiological arguments, though not necessarily Wilsonian sociobiological arguments, were used, or abused, by Richard Herrnstein and Charles Murray to suggest that legislation aimed at improving the lot of minorities, such as the Head Start Program, would yield only marginal improvement, so that the game might not be worth the candle (Herrnstein and Murray 1994). And in a context in which the mid-century stress on nurture was rapidly shifting back toward nature, the anthropologist-turned-sociobiologist Napoleon Chagnon argued that among the Yanomami, a very primitive and often violent tribe who live in the upper reaches of the Orinoco and Amazon, the fitness of dominant males – their enhanced reproductive output – was made possible by their ability brutally to commandeer women, kill rivals, and as a result pass on their high-grade genes to a disproportionately large number of offspring, thereby enhancing the fitness of the population as a whole (Chagnon 1988). Chagnon imagined that among the Yanomami, natural selection is free to do its brutal work unimpeded by cultural constraints that had lowered the fitness of civilized societies and undermined social practices that are consistent with human nature. Old notions die hard, especially if a novel, ultra-Darwinian conceptual framework seems to breathe some new life into them.

In recent decades, the sociobiology controversy has intensified the hitherto manageable tension between anthropology's two sides. On the one hand, sociobiology has been trying to pull cultural anthropology into its orbit, as in the case of Chagnon. On the other, strong versions of cultural relativism have been used to discredit the very possibility of value-neutral, scientific cultural anthropology and, in its post-modern version, to dismiss the very idea of human nature as ideologically contaminated. For our part, we have no trouble acknowledging the existence of a human nature, characterized by a species-specific array of highly plastic and variable traits, which, just because they *are* plastic, forbid easy normative conclusions about what behaviors, practices, institutions, laws, moral codes, and so forth are "natural." We do not see how the fact that our species is individuated by its position between two events of phylogenetic branching – the "species as individuals" argument that we discussed in Chapter 10 – undermines our ability to identify our species by its traits (Grene 1990). While we believe, too, that Darwinism suggests a naturalistic world view, and so tends to pull the rug out from under religious dogmatism and "enthusiasm" in the old sense, we fail to see how appealing to Enlightenment commonplaces, ultra-Darwinian adaptationist stories, implicitly fixist views about the narrow range of long-gone environments to which alone we are supposedly adapted, and suspicions about abilities, including moral and cognitive abilities, that we clearly possess – does much to advance either anthropology or the explanatory scope of Darwinian thinking. Strenuous proclamations of materialist metaphysics are not Darwinism. Rather, they are attempts to use philosophical concepts to advance *versions* of Darwinism that demand a skeptical, truncated, even eliminativist view of human capacities for caring, reflecting, and thinking. Rather than telling us that we don't really have our faculties after all, an adequate Darwinian anthropology will provide us with well-grounded accounts of the faculties we do have. These are, for the most part, the faculties we think we have, informed by philosophical reflection on scientific discoveries, including evolutionary discoveries (Grene 1995, pp. 109–112).

Brains, Language, Mind

Human consciousness is unique. Clearly, our form of consciousness involves what Daniel Dennett calls "the intentional stance" (Dennett 1987). Mind is minding, attending to, and acting in a specific environment in accord with what steps might fulfill, or help fulfill, our hopes, dreams, desires, plans. First, our consciousness is also social. Second, our symbol systems, especially language, allow us to pursue the curious mix of cooperation and competition that is our species' "form of life." Third, our consciousness is reflective. Members of our species are able not only to plan what they want to do in specific circumstances, by cooperating with others or scheming against them, but have the ability to think about their own thoughts in reflective privacy.

The reflective cast of our minds may well be a necessity if we are to think in a practical way about the future; good plans fold back on themselves as a way of anticipating how things might go, monitoring their progress, and taking corrective measures after missteps. Language is clearly both a presupposition of this kind of reflection and a means by which we hold our ideas "in mind." But the reflective nature of our consciousness is so different from what, to the best of our knowledge, we observe in other species, and so ethereal in its qualities, that it has suggested to many philosophers the notion that mind is independent of nature and that it has a contingent relationship to the brain. This independence has been conceived either functionally, as Aristotle had it, or, more radically, as substantial, as Descartes notoriously thought. Either way, dualism is the result. Those who take a non-natural view of mind generally assume, too, that practical reason – the thinking in which we engage in order to affect our environment, including, prominently, the other human beings with whom our fates are so closely entangled – is a derivative use of what is essentially a contemplative, reflective sort of consciousness.

There can be little doubt that the diffusion of evolutionary thinking since the middle of the nineteenth century has initiated a shift away from anti-naturalism about mind. Nobody (religious folk aside) wants to be a straightforward dualist today. As we saw in Chapter 7, the idea that mind is a "secretion" of the brain occurred to Darwin when he was quite young. Fearing scandal, he tucked the thought away in the privacy of his reflective mind. Since then, philosophers who favor a naturalistic stance with respect to mind have come up with a large array of ways in which mind might be non-contingently related to brain. We do not need to pursue these hypotheses here. The point we wish to make is that until recently, surprisingly few of these philosophical hypotheses have been informed by detailed biological theory, especially evolutionary theory. It is this connection – the connection between evolution and mind – that will be the focus of the few remarks (most of a general, and sometimes a cautionary, nature) we will make here about a much larger topic.

One obvious way to get evolutionary insight into the human mind is to try to imagine what biological function or functions its various aspects serve. In precisely this spirit, William James, who was among the first to think of himself as a *Darwinian* psychologist, argued that "unless consciousness served some useful purpose, it would not have been superadded to life" (James 1875, p. 201; see Richards 1987, pp. 427–33). This question places the focus back onto the primacy of practical (and technical) reason, and on the embeddedness of thinking animals like ourselves in environments, both natural and social, in which we must figure out what to do. That is all to the good (although pragmatists who were influenced by James, notably Dewey, perhaps went too far in insisting that the practical nature of our intelligence means that the possibility of objective knowledge for its own sake, the greatest triumph of reflective reasoning, is merely an ideology common among those who live idly from the labor of others [Dewey 1925]).

From this useful starting point, it is easy to conclude that "the intentional stance" is adaptive. It is tempting, in fact, to think of it as an adaptation – as a way of representing the world to ourselves that has been produced by natural selection as a fitness-enhancing trait that enables us to make our way around in a complex, often unpredictable environment. Dennett, for one, asserts this (Dennett 1987). In its first stirrings, this trait would have enabled our hominid ancestors to deal with the fluctuating environments in which the genus *Homo* evolved. Concerted selection (to the point of fixation in all normal members of the species) for a planning-oriented form of consciousness, combined with room for private reflection, would have resulted from the fact that hominid species, especially our own, had to respond to the planning and cunning of their fellows, who form the most prominent part of the hominid's environment and who can be intentionally deceptive. A plausible story. But we must be careful. It is also possible that the ability to represent past, present, and future states of affairs to ourselves in the context of satisfying desires is a by-product of other adaptations, rather than an adaptation in and of itself.

Certainly, the purely reflective side of our intentional consciousness – the side on which Descartes concentrated, and that has traditionally been viewed as necessary for scientific knowledge – seems to go beyond its practical utility. That explains why naturalistic philosophers of mind who think that our form of consciousness *is* an adaptation have tended to discount, or even eliminate, the higher dimensions of reflection. Dennett makes this case by combining his arguments about the evolution of the intentional stance as a fitness-enhancing trait with genic selectionism of a Dawkinsean cast (Dennett 1995).[3] The genes "use" intentional consciousness as an interactor in order to get themselves multiplied more successfully. Indeed, actions mediated by intentions are, by their very nature, a ruse played on us by our genes, like the urge to be moral that we discussed above (Dennett 1987; 1991; 1995).

---

[3] Among contemporary naturalistic philosophers of mind, Paul Churchland is more eliminative than Dennett; once mental phenomena are explained, most of the object to which they seemingly refer disappears. Jerry Fodor is less adaptationist than Dennett. Neither, however, appeals as extensively as Dennett to evolutionary arguments. Hence our concentration on Dennett's views.

Dennett supports this position by appealing to the contemporary penchant for modeling consciousness on computers. (The penchant is an old one; Descartes and Locke talked about the mind as a spring mechanism.) On this view, consciousness is like the monitor of a computer. It displays representations before our inspecting mind in a sort of private theater, à la Locke. This being so, we might well be, as Dennett suspects we are, little more than "dumb robots," machines that are very good at passing Turing tests (Dennett 1987; 1991).

For us, the cost of understanding the intentional stance on these terms is high. Our minds are certainly adapted to deal with our environments by way of ideas. But our environments are largely cultural, and the orienting role of means-end reasoning, and of consciousness and self-consciousness more generally, evolved because the tie that binds us to the cultural world as agents, caregivers, competitors, speakers, and thinkers affords us direct (rather than representational) access to the environments in which we act responsively and, ultimately, responsibly (see Chapter 12 for a discussion of the relevant epistemological points).

If we suppose that our minds are adaptations, we must also ask what environments they are adapted to. There has been a marked tendency among reductionistic (and especially eliminativist) naturalists to think that our minds are adapted to the Pleistocene environment and social structure of our hominid ancestors. The reason alleged is that too little time has passed since the dawn of civilized life for adaptations to our present form of life to take hold. "A few thousand years since the scattered appearance of agriculture," write Leda Cosmides, John Tooby, and Jerome Barkow, "...is...less than 1% of the two million years [since "Lucy" that] our ancestors spent as Pleistocene hunter-gatherers" (Cosmides et al., in Barkow et al. 1992, p. 5). The first generation of sociobiologists thought that the adaptations in questions were specific behaviors to which we are prone, leaving the impression that we are full of natural urges (to male promiscuity, female coyness, and the like) that must be suppressed if we are to live in contemporary societies. Richard Alexander modified this behaviorist penchant somewhat by arguing that our adaptations are situation-specific rules of action whose underlying imperative is: "maximize inclusive fitness" by whatever means possible (Alexander 1979). More recently, a successor program to sociobiology known as Evolutionary Psychology (EP), has amended psychological adapationism still further.

Advocates of EP do indeed agree with their colleagues in assuming that "the evolved structure of the human mind is adapted to the way of life of Pleistocene hunter-gatherers, and not necessarily to our human circumstances" (Cosmides et al., in Barkow et al. 1992, p. 5). (Implicit in this argument is a markedly gradualist vision of the evolution of species-specific traits, which may entail underestimating the rapid evolution that can be brought about by changes in developmental timing and differential expression of the same genes.) But what EP's supporters believe to have emerged are a set of *capacities* that, because they are not reducible to specific hard-wired behaviors, or even to rules of action triggered by specific situations, involve the mediating role of cognition. This shift from behaviorism to cognitivism is a feature of contemporary psychology. But the cognition in question is viewed by EP's advocates as adaptationist because, together with many contemporary neurobiologists, they regard the brain as a collection of dedicated modular structures each of which is adapted to deal with a particular set of problems (Tooby and Cosmides 1992, in Barkow et al., p. 97). There are, we are told, distinct modules for color vision, locomotion, language-acquisition, motor control, emotional recognition, and so forth. Each such module is asserted to have been brought into existence by natural selection (usually interpreted in a gene-selectionist way). Although each mental function is supposed to have been optimally adapted to a hunter-gatherer life style rather than to the kinds of environments that have come onto the scene since the emergence of agriculturally based civilization, they still function well enough in our world for us to get by (though at the cost of our making fallacious inferences on a fairly regular basis).

The evolution of the capacity for language illustrates the style of this sort of inquiry, as well as its difficulties. Judging by the air of conviction with which various people put forward their views about the origins of the ability of neonates to acquire language, one might infer that confidence reigns in this area. However, the exact opposite is the case. Language acquisition by *H. sapiens* remains a difficult, largely speculative subject. Most contemporary controversies about it take place against the background of the revolution in linguistics initiated by Noam Chomsky. Language, according to Chomsky, is a rule-governed activity. It involves a series of syntactic transformations of simple noun-verb phrases by something like a computer program whose most basic form, its "machine language," is presumably inscribed into the neurological architecture of our brains (but not into the brains of chimpanzees, which, even though they share some of our emotional and communicative life [De Waal 1996], lack the capacity for syntactical transformations). For his part, Chomsky assumes that this capacity evolved somehow, and lets the subject go at that. He thinks "it is perfectly safe to attribute this development to 'natural selection,' so long as we realize that there is no substance to this assertion, that it amounts to nothing more than a belief that there is some naturalistic explanation for these phenomena" (Chomsky 1972, p. 97). Gould says something only slightly more substantive when he asserts that language capability (*sensu* Chomsky) is a side product of the expansion and connectivity of the brain, what he calls a "spandrel" (Gould 1987). It is precisely this claim that has provoked advocates of EP to argue that linguistic ability must be, straightforwardly, an adaptation.

Steven Pinker is a well-known advocate of the adaptationist view of language ability. He holds that the brain is modular (rather than a "general purpose" computational device of the sort that was presupposed by most mid-twentieth-century thinkers, including the architects of the Modern Synthesis), that its various modules must each perform a distinct function, and that each is a biological adaptation. From these premises, Pinker concludes that there *must* be an evolved module for language competence – a "language instinct" – that grounds what happens in the brain in what happens in the genes (presumably a large number of them) (Pinker 1994). "There *must have been* a series of steps leading from no language at all to language as we now find it," write Pinker and Paul Bloom, "each step small enough to have been produced by a random mutation or recombination, and each intermediate grammar useful to its possessor" (Pinker and Bloom 1990, reprinted 1992, p. 475, our italics).

This is a rather a priori argument. For Pinker, indirect, but potentially empirical, support for it comes from the presumed regularity of syntactical combination and permutation itself, as well as the universality with which it is acquired by human beings (Pinker and Bloom 1990, reprint 1992, p. 463). According to Pinker, these features suggest design, and (assuming naturalism) suggest, too, that natural selection is the designer (Pinker and Bloom 1990, reprint 1992, p. 463). Yet from a view of natural selection in which natural selection is not a designer, but a process that takes the place of design, this is the very problem. According to the Modern Synthesis, traits that are selected typically vary a great deal from environment to environment (Piattelli-Palmarini 1989). Language competence does not. It is an all-or-nothing affair, occasional defects notwithstanding. It is possible, of course, to argue that language acquisition is an evolved adaptation even if it shows no variation. One might hypothesize, for example, that the emergence of a cultural environment creates enormous selection pressures in favor of neurological capacities for participating in that culture, including symbolic and ultimately linguistic ability; those lacking this capacity would be so disadvantaged that they would not have survived at all (or, if one prefers a group-selectionist view, sub-populations that did not do business in this way would have been crushed by those that did). This hypothesis has encouraged some to think of language acquisition as an example of the so-called "Baldwin effect," which treats culture as an environment that creates selection pressures that push whatever genetic variation happens to crop up in the direction first pointed by culture. Dennett argues for this view (Dennett 1995). So does Terrence Deacon (Deacon 1997). In Deacon's version, however, there are no genes directly for language acquisition, *pace* Pinker. Instead, changes in gene frequencies favor traits that indirectly support language (Deacon 1997). So one might accept an evolutionary account of language acquisition, but reject some aspects of EP (see Weber and Depew 2003).

A second difficulty is closely related. The necessity of telling an adaptationist story about the acquisition of language competence described as "designed" will tend to single out particular causes – "mutations," for example, that change the position of the larynx, making vocalization (as well as choking) easier. The difficulty is that these scenarios, even if they are combined and given a sequential order, probably underestimate the complex interaction and mutual feedback among a whole variety of factors in the relatively sudden emergence of language. It is possible, for example, that if we project at least minimal symbolic capacity back to some of our closest hominid ancestors, the enlargement of the brain and the acquisition of intelligence (means-end reasoning, mostly) may be as much effects of language-acquisition as its causes (Deacon 1997). EP discounts this possibility by tacitly supposing the existence of a prior urge to communicate on the part of an already big-brained, intelligent hominid – an urge that until the larynx descended or some other triggering event occurred, had been "dammed" up or otherwise held back. A similar picture is implicit in the presumption that at the exact point where rules, neurons, and genes are hooked together – EP's "pineal gland" – there exists no particular language, such as English or Swahili, but the universal (and hypothetical) language that Chomsky's MIT colleague, the philosopher Jerry Fodor, calls "mentalese" (Fodor 1975). But the very notion of "mentalese" reveals the persistence of an old, not-very-Darwinian view of the mind as a sort of theater in which pre-linguistic representations of the external environment are played before an audience of one. It is this picture that gives rise to the presumption that non-speaking hominids are just bursting to say something, but, lacking a requisite morphological feature, cannot. This is not a very coherent idea; it is, in fact, simply a recycling of the old "idea idea" of Locke.

The Human Prospect: The Human Genome Project and Genetic Biotechnology

In 1987, Robert Sinsheimer, who had once worked on determining how the genetic code specifies proteins, Renato Dulbecco, president of the Salk Institute, and Charles De Lisi, an administrator at the Department of Energy (DOE) and a former administrator at Los Alamos National Laboratory (where a good deal of genetic data had been stored, originally from post-war surveys of Hiroshima and Nagasaki), proposed that every single base pair in the human genome be sequenced. Their proposal initially struck some evolutionary biologists and medical researchers as grandiose, utopian, and misguided in ways that, from these opposing perspectives at least, were thought to be typical of physicists and their fellow-travelling molecular biologists (Lewontin 1991b; Hubbard and Wald 1993). At the time, very little sequencing had actually been done, and what had been done was painstakingly slow. Why consume massive amounts of scientific manpower in a long effort to decode the entire genome when only a small percentage of it (the part that codes for proteins) makes sense, and when, even in the part that makes sense, it might be more useful to concentrate on areas of the chromosomes where there is some evidence of genetic diseases that might, conceivably, be treated by somatic gene therapy?

Nonetheless, legislation authorizing the Human Genome Project (HGP) was approved by Congress in 1987. It assigned the information-processing aspects of the program to the DOE, which would develop automated gene-sequencing and computational programs for analyzing what eventually would turn up, and gave the initial mapping, sequencing, medical, and policy-making roles to the National Institutes of Health. James D. Watson, of DNA fame, would head the project. Along the way, scientists who found short cuts defected from the "government" program, and went into business on their own. Largely as a result of rapid technological development, as well as of a race that set in between the private and public versions of the project, it was announced in the summer of 2001 (by no less a pair of personages than the President of the United States and the Prime Minister of Great Britain) that a "rough draft" of the entire human genome was about to be published (simultaneously in *Science* and *Nature*; International Human Genome Sequencing Consortium 2001; Venter et al. 2001).

The popular press dispersed the news in pretty much the same terms that its early champions had used when they were seeking funding. "The Book of Life," – the title of a 1967 book by Sinsheimer (Sinsheimer 1967) – "is now opening," it was said. Human identity was at last to be revealed. We would learn not only what makes human beings different from other species, but, in the opinion of the pioneering gene sequencer Walter Gilbert, how each of us differs from others. One will be able, Gilbert wrote, "to pull a CD out of one's pocket and say, 'Here's a human being; it's me'. . . . Over the next ten years we will understand how we are assembled [sic] in accord with dictation by our genetic information" (Gilbert 1992, p. 97).[4] In the same spirit, a new era in medicine was proclaimed: an era of genetic medicine that would prove even more significant than the nineteenth-century germ theory of disease, which had dominated twentieth-century medicine. Genetic diseases caused by point mutations that interfere with the cell's ability to make a particular protein would be cured either by therapy aimed at repairing the somatic cells of individuals, or, eventually, by elimination of the defect from a whole population through the manipulation of germline cells, egg, and sperm. Beyond that, enthusiasts expected that multiple-locus traits would in time be identified on the chromosomes, and, where defective, changed, since complex traits were assumed simply to be compounds

[4] Gilbert might have been on slightly more solid ground about human diversity – about how we are each ourselves because we each have a unique genome – if the HGP had been complemented by the Human Genome Diversity Program (HDGP) that has been proposed by the population geneticist Luigi Cavalli-Sforza. The genomes that were actually sequenced by HGP came from very small populations of Europeans. HDGP would regain the proper perspective by demonstrating the width of genetic polymorphism in and across human populations that had been predicted by Dobzhansky and others. Of greatest interest would be small isolated populations, which are presumably the most different. However, HDGP was never approved at the international level, in part because it was felt by officials and field anthropologists that identifying the genetic traits of particular populations would in practice call forth the very racism and prejudice that the project was designed to refute in theory. Unfortunately, this might be true. Particular genetic markers often do correlate with particular traits of perceived races. We are grateful to Dr. Jeffrey Murray of the University of Iowa for helpful discussion of this difficult issue.

of basic gene-protein units. These traits would include behavioral tendencies such as alcoholism and homelessness (Koshland 1989). (In the notion that there might be genes for homelessness we cannot help hearing echoes of the "shiftlessness" that early twentieth-century eugenicists such as Charles Davenport thought was hereditary in some families.)

Behind this sort of talk about genetic medicine lies the fact that since the 1970s human beings have begun to acquire skills in manipulating the genetic material directly, rather than by using selective breeding, the relatively slow mimicry of natural selection that our species has deployed ever since the beginning of agriculture. (The development of cloning goes hand in hand with genetic engineering; one of its main jobs is to provide the reliably identical platforms that are required for experimenting on "designed" organisms.) Since the development of restriction enzymes – agents that can be used to snip portions of DNA out of genomes and to reinsert them into other portions of the genome, or into the genomes of other organisms – it has become possible to think that biology can, for the first time, join physics and chemistry as a "technoscience." This has in turn made it possible to imagine life as something other than a lottery (with all the problems about privacy and informed consent that uncovering what was hitherto hidden and chancy brings with it). Life, it would appear, is something that can now be engineered in an industrial sense, even to the point where the evolution of our own and other species might be directed. Perhaps this is what Fisher had in mind when he thought of eugenics as a way of taking over God's work.

We suspect that the presumed link between the information that is being "revealed" by the HGP and the promises of genetic medicine could not have been forged as quickly or confidently as it has been if the lived body had not already come to be viewed by many as a "printout" from a "genetic program." The mere fact that moving genes from one organism to another, sometimes of a different species ("transgenics"), is popularly imaged as a matter of uploading and downloading information testifies to the truth of this claim. To a large extent, this picture is the result of the triumph of molecular genetics. The conception of the gene presumed by the original advocates of the HGP is the molecular gene. And the notion of the organism as a "printout" from a genetic program represents a convergence of cybernetic, and ultimately computational, talk with the Central Dogma of Molecular Biology, which was originally little more than a working maxim to the effect that information flows from genes to proteins and from proteins to organisms and their traits, and not the other way around.

At the same time, it should also be noted that ultra-Darwinian versions of natural selection, which ascribe a good deal of causal agency to "selfish" genes, which construe adaptations as discrete modules of antecedently-identified mental, behavioral, or physical functioning, and which minimize or deny the difference between design and evolution, have done little to disturb, and indeed much to encourage, what is at root a technological vision of the living world. Natural selection is seen as mixing and matching genes in the spirit of a genetic engineer who uses a computer to model what Dennett calls "searches through design space" (Dennett 1995). Dobzhansky's famous remark that "nothing in biology makes sense except in the light of evolution" is well taken. But in recent decades, Dobzhansky's maxim has been given a twist. Rather than appealing to the contingencies of the evolutionary process as a constraint on Promethean ambitions – eugenics in Dobzhansky's day, "designer babies" in ours – evolution is now being asked by those who have construed natural selection as a designing biotechnician to bless the transition to the coming age of biotechnology. In the resulting biotechnological vision, the inherently complex relationships among genes, traits, fitness, and diverse environments is displaced by the homogenizing, standardizing, "quality control" tendencies on which genetic engineering depends, and which it is explicitly aimed at producing and reproducing. It is no doubt true that we are entering into an era in which genetic medicine can be expected to do a great deal of good. But genetic medicine will not, we suspect, be able to achieve its promise so long as it is seen as licensed by a simplistic, utopian (or, depending on your point of view, dystopian) view of organisms as technological objects – a view that contrasts with the main line of thinking about living things that we have followed in this book.

Luckily, the preliminary draft of the sequenced human genome contained surprises that might well disconcert those holding any such technological view of life. Originally, it had been estimated that there might be as many as 100,000 human genes (Gilbert 1992, p. 83). It has turned out (in part because of definitional shifts in the meaning of the term "gene") that there are only about 30,000. To make matters even more interesting, it has also turned out, on the basis of sequencing the genomes of other species – efforts that had been in part funded along with the HGP for comparison purposes – that most of the 30,000 genes we do have are also possessed by most other species: not simply by chimpanzees, but by fruit flies and flatworms as well. This turn of events has provoked many of those who had convinced themselves and others that all specific and individual differences must be encoded in the genes – that the HGP would "tell us who we are" – to lay plans for an even bigger project in "proteomics." In comparison to the relative simplicity of the DNA molecule, perhaps protein sequences contain enough information to capture what makes each of us be ourselves. A more reasonable possibility is that the HGP has actually been bringing into view not the simple molecular gene of the 1950s and '60s, but, perhaps in spite of itself, the developmental genes that constitute the fundamental architecture of life (Keller 2000; Gilbert 2000; Moss 2002). As we noted in Chapter 9, changes in how these highly conserved gene sequences are expressed – in their timing, in increased or decreased quantities of the protein products they specify, in chemical marking of DNA itself, and in a myriad of other subtle processes – cause evolutionary change, and mark this species off from that. Allelic variations in genes (as well as in the much larger part of the genome that codes for no proteins at all) do characterize each individual. Population genetics tracks changes in these variations. Still, there is little or no reason to think of these bits of DNA as containing a coded and programmed identity for each of us, or to think that slow, point-for-point changes are the proximate causes of large-scale evolutionary change, or to think of genes as selfish "more makers" rather than as "developmental resources" that interact with cellular processes and environmental signals (not least, social cues and individual reflections) in the process by which we come to be ourselves.

Most professionals know better. Still, from now on, philosophers of biology and philosophically alert biologists must think about how evolutionary and developmental biology can inform, and if necessary constrain, an overly technoscientific approach to living things that is widely disseminated in contemporary societies.

CHAPTER 12

✦

The Philosophy of Biology and the Philosophy of Science

In concluding our retrospective of the relations between philosophy and biology past and present, we may ask what the emergence of a philosophy of biology can contribute to the philosophy of science in general. What can the study of biology teach us if we take it either as our model field or as a model for our field?

There have so far been two major movements in recent philosophy of science. First, there was the so-called received view, initially logical positivism, rechristened logical empiricism. Taking fundamental physics, or a caricature of it, as its model, it separated the process of discovery (which it ignored) from the context of justification. Within the latter context, it aimed at a logical reconstruction of science, a science that rigorously followed a single hypothetico-deductive method, and that was to issue in the utopian structure of a unified science. In reaction, sociologists, and even some philosophers of science, have practiced a sociological deconstruction of science, which has left that family of disciplines with no claim whatsoever to epistemic justification. For the first school, science, with its sacrosanct method, stands serenely outside society, or else deigns to direct it by applying its superior procedure. For the second, science is reduced to politics: In effect, there is only society, no science.

What if we come to the philosophy of science through reflections on biology rather than physics, or some abstract dream of physics, as the received view used to do, or in preference to taking as our model for philosophy a rather naive sociology? From the present writers' perspective, we have a chance, or so we hope, of developing a more fruitful approach to the philosophy of science in general, what we might call an "ecological-historical" view.

348

Admittedly, we are not the first to ask what biology can teach the philosophy of science. As we noted in Chapter 10, some latter-day defenders of the once received view have stated contrasting positions in this respect. Believing that a unity of science must be possible, Kenneth Schaffner hopes that biology, for all its present resistance, will finally enter the utopian land he envisages (Schaffner 1993). Alexander Rosenberg, sharing the same ideal, but entertaining no such hope, finds biology a set of merely practical endeavors, which exhibit the unfortunate disunity of science (Rosenburg 1994).

There have also been seemingly more positive applications of biological lessons to the philosophy of science. There is evolutionary epistemology, for example, which purports to naturalize the theory of knowledge by explaining the growth of scientific knowledge in terms of the theory of natural selection. This sounds tempting. Life evolves, science evolves, so why not apply the principles of biological evolution to the development of science? Yet the situation, as we see it, is not so simple. While, as we shall emphasize, we believe that investigations in the philosophy of science need to be grounded in its history, our questions do not concern simply a matter of what succeeded what; they have an epistemological bent. We want to understand what scientific knowledge claims amount to in the context of this or that discipline, or in the context of a developing discipline, rather than simply chronicling the relative frequencies of these or those slightly differing assertions, succeeding one another by a kind of unnatural selection (see Hahlweg and Hooker 1989).

There is indeed a weak form of so-called evolutionary epistemology, which simply points out that we have come into existence as animals finding our way around the world, and taking scientific knowledge to be a subset of such varieties of way-finding. The slogan one of us is fond of repeating, and which we will repeat again shortly – "all knowledge is orientation" – echoes this sentiment. But the stronger form of evolutionary epistemology, which attempts to identify survival by natural selection with the history of science, seems to us to ignore entirely the epistemological aspect of our subject.

David Hull, although he rejects the label "epistemologist," takes a position allied to that of the strict evolutionary epistemologists. He has developed a general theory of selection, which, he argues, applies to such diverse areas as biological evolution, immunization, operant learning, and the history of science (Hull 1988a; Hull 2001, especially Chapters 3 and 4). Together with two colleagues, an immunologist and a behavioristic psychologist, he offers the following definition:

> We define selection as repeated cycles of replication, variation, and environmental interaction so structured that environmental interaction causes replication to be differential. The net effect is the evolution of the lineages produced by the process.
>
> Hull, Langman, and Glenn in Hull 2001, p. 53

That's all very well as far as it goes, but what does it tell us about science in particular? Yes, there is some kind of selection process there too, but how does that help us understand the odd, complicated history that started somewhere in Western Europe in the sixteenth and seventeenth centuries, or maybe in the fourteenth, depending on your point of view?

In an essay entitled "The Trials and Tribulations of Selectionist Explanation," Ronald Amundson asks, not just when there is selection, but when selection has explanatory force. Taking Darwinian selection as his model, he enumerates three "central conditions" that need to be fulfilled if selection is not only to exist, but to explain some biological, psychological, or social process. There must be (1) richness of variation, (2) non-directedness of variation, and (3) a non-purposive sorting mechanism that results in the persistence of those variations better suited to the needs of the organism or species in question in its particular environment. Only where these conditions are met, Amundson argues, is the invocation of selection explanatory rather than merely metaphorical. Darwin, he reminds us, "did not simply say that nature works as if there were an intelligent breeder selecting between variants. He said that there was no such breeder – that the forces of nature produced their riches without prior resources of direction and foresight" (Amundson 1989, p. 430). Selectionist explanations may hold, Amundson admits, for the immune system or for operant conditioning – two of the examples that would also be put forward by Hull et al. But what about science? Here, surely, conditions (2) and (3) fail: The variations in that case as well as the sorting process do surely appear to involve some conceptual, and hence purposive – intentional as well as intensional – components. To be sure, Hull has tried to give an account of scientific development that seems to eliminate such factors. He takes "curiosity" for granted as an element in our makeup, and then analyzes the activity of scientists, or scientific communities, in terms of the two further factors of "credit" and "checking." Are these agencies undirected? In any case, with no account of conceptual input, let alone of the nature and power of experiment, or of the relation of conceptual to experimental moves in science, this story wears very thin. In one essay, Hull does invoke, though somewhat lamely, considerations of "truth" (Hull 2001, Chapter 8). But this, again, is rather vague and general. Amundson is surely correct in maintaining that calling the history of science a selective process *explains* very little. One way to put it is to point out that selective explanations are purely causal, whereas science involves reasons as well as causes (see Donohue 1990).

Given, then, that we find a selectionist approach to science less than satisfactory, what alternative insights can we suggest if we start our reflections from biology rather than the "exact" sciences?

First, if we take the biological sciences as our model for the philosophy of science, we have a better chance of accepting a realist point of view as fundamental for the philosophy of science. For the present writers, realism is a principle. It is not something to be argued for, but where we start. The objects of study of physicists, theoretical or experimental, may lie far from ordinary experience. A physicist such as Mach in Vienna or Wigner in Princeton may proclaim that he is only making mathematical constructions on the basis of his sensations. In fact, the major tradition of the past century in philosophy of science was founded on this Viennese theme. Even when it was discovered that there are no pure observations – that all observation is "theory-laden," as they put it – the theoretician was still floating on a surface of little points of atomic or isolated sensations. In this way, the problem of "scientific realism" was reduced to the question of the reality of theoretical entities, atoms, electrons, quarks, and so on. The world as environment – even the impoverished world of Newton, according to which God probably formed matter out of hard, solid, impenetrable particles – no longer existed. In contrast, it is difficult for biologists to deny the reality of living things. Given this insight, moreover, they can more easily recognize that they are themselves living things among living things. It is true that molecular biologists, too, work, like physicists or chemists, far enough from ordinary life. But even they have to cultivate the organisms on which they do their research: from *Arabidopsis* or *Dictostelium* to pine trees or pigs, and so on and so on. Without adopting the science of Aristotle, we are returning here in some sense to the Aristotelian starting point of science. We find ourselves as living things in an environment that (up to now, at any rate) has permitted life. We find ourselves, too, among beings that are born, mature, grow old, and die. In short, we find ourselves from the beginning in a real world. Although at least one great biologist, Sewall Wright, adopted an extreme idealism as his own philosophy, biologists in general do not habitually deny the existence of their object of study, nor, by implication, of themselves as students of them. Science is – or, better, sciences (in the plural) are – communally organized efforts of real people to find their way in some section of the real world. Of course they don't always succeed, and they never come to an endpoint beyond which there is no further inquiry, but that doesn't undercut the reality both of their objects and of their efforts to understand them.

In short, what *we* are trying to understand, as philosophers, is the *life* of science: how scientific practices originate and continue as epistemic enterprises. In this context, perhaps there is some biological discipline we can take as our model. Ethology seems the obvious choice. True, philosophy is not an experimental science. Indeed, in terms of the English sense of "science," no branch of philosophy, including the philosophy of science, is itself a science. Philosophy consists in reflections, more or less systematic, on the structure or functions of certain disciplines or certain human interests. And the philosophy of science in particular tries to reflect on the aims, the successes, the failures of the practitioners of the sciences. That is, of course, what the ethologists do with their animals.

Granted, in our case – in the case of the philosophy of science – it is profoundly encultured animals that we are observing: ourselves, or rather, a small group among ourselves. It will be asked: Why not anthropology as our model? There are two reasons to prefer ethology. On the one hand, the sort of anthropology that has been used in the study of science is too externalizing an anthropology to take account of the sciences as scientific. True, that is not the only style that exists in anthropology, but it is the style that has been dominant in social constructivism.

On the other hand – and this is a more substantive reason – we must insist again that it is the *life* of the sciences we are trying to understand. The logical skeleton that was the ideal of the old orthodoxy in the philosophy of science had no connection with that reality. In our view, social constructivism offers no better choice. It is true that the sciences, like all human vocations, are social enterprises. But the Hobbesian vision that characterizes one sort at least of social constructionism is too far removed from a reasonable conception of the sciences as special segments of human life, segments in which the enterprise of learning and knowing for their own sake is central. And in general the emphasis on "construction," with its stress on the artificiality of language as the carrier of our practices, not only distances the scientists from their objects, with catastrophic results, but, unless qualified in ways we will mention later, prejudices questions about the very nature and scope of learning and knowing. For these reasons, it is in the efforts of a living being to understand the activities of other living beings that we want to look for a model for the practice of the philosophy of science.

Consider a particular example. Deborah Gordon, who studies the behavior of harvester ants (*Pogonomyrmex barbatus*), has described the way she carries on her research in the field (Gordon 1992). When she started, Gordon tells us, she saw only little bodies moving pell-mell on the ground. Little by little she succeeded in recognizing certain distinct patterns of behavior among the ants. She observed patrollers, who look for sources of nourishment; foragers, who bring food; guardians of the ant-hill; and trash collectors. Similarly, the philosopher of science is trying to understand the formations of research workers in a given discipline, the task they undertake, the goals that define those undertakings. And, continuing our analogy, we should note that Gordon not only studies the behavior of those particular ants; she observes at the same time the history of the colony, which does not correspond exactly to the behavior of individuals. In the philosophy of science, too, it is not only the history of the individual as research worker that we want to understand; it is the history and structure of the discipline itself: what has been called in the Canguilhem school, "l'institution de la science," the establishment of (a) science.

However, if we find in this analogy a useful lesson for the philosophy of science, we certainly do not want to deny the great differences that exist between the two practices. Trying to understand the life of a population of ants is far from being the same thing as trying to understand the activities of a population of human beings. In the latter case, we have to do with our peers, our kind, who are enmeshed, as we are, in language, in culture, in history. And up to a certain point, we can cultivate our imagination with the purpose of entering, by a kind of Humean sympathy, into a tradition that is not entirely our own. We are trying to understand what Ludwig Fleck called a different thought style, to practice what he called comparative epistemology. In this sense, it is true that the philosopher of science is more like an anthropologist than an ethologist. Nevertheless, we prefer the ethological analogy. For, as we have already said, the anthropology that was used in at least one famous case of the sociology of science appears to suffer from a barbarous reductivism. (The case we have in mind is Latour and Woolgar's *Laboratory Life*; a more recent instance of the same genre is Steven Shapin's *Social History of Truth* [Latour and Woolgar 1979; Shapin 1994]).

Further, as we have also said already, but it bears repeating: We must insist on the fact that when, as philosophers of science, we study a scientific discipline, or an episode in the history of science, or a particular variety of scientific knowledge, it is the ongoing practice, the life, of science that we want to describe, analyze, and understand. It is neither an abstract logical formulation that we are looking for, nor a caricature of scientific practice as a pure Hobbesian war of all against all. What we are aiming at is a multidimensional analysis that displays the complex and subtle elements that constitute science, or rather, a science. Again, science is a family (in the Wittgensteinian sense) of occupations of certain people and certain groups that have the common aim of seeking the truth, but of seeking it in a particular domain and by specific methods that we recognize in some sense or other as "scientific." Knowledge is a form of orientation, finding one's way in an environment. For men and women of science, that means orientation in a discipline, a language, a type of laboratory, a style of experimentation, and so on. There is no single, all-inclusive formula for such activities. It is a question of immersing oneself in the detailed history of some particular scientific enterprise, and, it is to be hoped, gaining philosophical insight from that study.

Richard Burian's study of the work of Jean Brachet may serve as an example of this kind of work (Burian 1997). Burian examines in some detail the exploratory work of Brachet and his colleagues on the localization of nucleic acids and part of the pathway that led him and his coworkers to consider the problem of protein synthesis. Burian writes:

> The tools he devised, appropriated, and adapted, were put to different uses in the service of a variety of problems, explored in parallel. In all of these studies, he sought to employ a wide range of organisms and the widest possible range of techniques in order to provide a basis for reconciling the findings obtained and for achieving broad understanding of the process in question.
>
> Burian 1997, p. 32

This wide-ranging style of experimentation, based in biochemistry and biochemical embryology, and using techniques from a variety of sources on a variety of organisms, contrasts strikingly with those of the more celebrated pioneers of molecular biology, who "typically employed tools imported from physics in combination with the new methods of genetic analysis applied to microorganisms" (Burian 1997, p. 39). Yet, by difficult and devious paths, these different styles of investigation converged on the same entities. Brachet's methods themselves shed important light on one variety of scientific practice, in which problems and techniques from a number of disciplines are brought to bear on a developing field of investigation. But this story also suggests a broader lesson. Comparing Brachet's research with that of other workers engaged in the same search, Burian observes:

> The very fact that work in three such different styles (and with aims as diverse) as Brachet's, Crick's, and Zamecnik's could be brought into concordance in about one decade serves as an important marker of the mobilization of the entities and phenomena in the enormous domain covered by these three distinctive research programs. One of the most important tasks in the philosophy of biology – indeed, the philosophy of science generally – is to understand how we achieve concordance in the interpretations of the findings of workers who, in [Hans Jörg] Rheinberger's terminology, work with such different 'epistemic objects' as those that preoccupied these three key figures. It is in handling such topics as this that the philosophy of experiment will find its liberation from the excessive theory-centrism, now waning, of recent philosophy of science.
>
> Burian 1997, pp. 40–41

Similar case-based philosophizing can also be found elsewhere. Almost classically by now, there is what Fleck did with the Wassermann test, deriving from that history his concepts of thought style and thought collective. Hans Jörg Rheinberger bases his philosophy of experimental systems and epistemic things on a detailed study of protein synthesis in Paul Zamecnik's laboratory. On a more general scale, but still through the study of particular cases, Jean Gayon also furnishes philosophical insights through his account of the destiny of Darwin's hypothesis of natural selection (Gayon 1998). In such studies, history is fundamental. Just what were the disciplinary backgrounds, interests, techniques involved in each case must be carefully described and analyzed. Even when the activity under study is contemporary, it is still history, like Thucydides' *Peloponnesian War* – or better perhaps, especially for readers of this book, like "natural history" from Aristotle to

Cuvier – that is, the study of life styles in their concrete distinctness from one another.[1]

So far, so good. But to say that our work is like that of an ethologist is not enough. Ethologists have different ways of approaching their subject matter. Some have been radically reductive in their methods, like Fraenkel and Gunn (Frankel and Gunn 1940). It is simply motions they claim to be studying. As we have indicated, it is more than such purely externalist description we are after. On the other hand, there are ethologists, such as Cheney and Seyfarth, who want to penetrate if they can "inside the mind of another species" (Cheney and Seyfarth 1990). Following Thomas Nagel's famous inquiry, "What is it like to be a bat?" they want if they can to penetrate the subjectivity of their object of study, in this case vervet monkeys (Cheney and Seyfarth 1990; see Nagel 1974). A similar intention underlies the work of Donald Griffin, whom Cheney and Seyfarth cite approvingly (Griffin 1984; 1992). Moving from a vain hope for total objectivity, such writers turn to a search for subjectivity, for a "secret inner something," to set against purely external, or purely material, appearances. However, that seems to us an equally vain hope – Cheney and Seyfarth certainly admit its difficulty – and, when carried to extremes, a foolish one. Thus Griffin, for example, wants bees to be conscious of their directed food-searching, as we are when we go to the supermarket. Surely the evolution of the central nervous system has made some difference in the nature of animals' experience.

In short, if we are not looking for pure locomotor descriptions, neither is it inner feels we are after in our reflections on the sciences. However, there is, we believe, a third, more promising, way to go. Another name for ethology is behavioral ecology (Cheney and Seyfarth 1990, p. 10). We are trying to understand certain ways of coping in, and with, certain environments. If we read it in terms of the principles of ecological psychology, Cheney and Seyfarth's subtitle, "How Monkeys See the World," may suggest a direction for such investigations. Traditionally, "seeing," and other perceptual systems, have been described in terms of isolated, private sensations from which we somehow infer hypotheses about what is out there beyond us. The relatively new discipline of ecological psychology takes a different approach.[2]

---

[1] Clearly there is no reason why careful case studies should be limited to biology. For example, Friedrich Steinle considers Charles Dufay's work in electricity as a case of exploratory experimentation similar to the Brachet case treated by Burian (Steinle 2001).

[2] The basic text is J. J. Gibson 1979.

Whatever any animal perceives entails three equally essential aspects. First, there are always things or events occurring in its environment. Again, we are considering real animals in real situations. No science fiction or possible world nonsense about it. Second, there is always information in that environment that the animal has the capacity to pick up. This takes the form of invariants: constancies within change which the animal's perceptual systems have evolved to be able to pick up. On the perceiver's side, such information pick-up consists in a process of differentiation within a structured context. It is important to stress this feature. On the older view of perception, it was a question of somehow associating together small, meaningless sensations. In terms of the ecological approach, on the contrary, perceivers, from early infancy on, are understood to be discriminating between, or differentiating, distinguishable features within their environment.

Third, there are *affordances*, opportunities the environment affords the perceiver, or dangers it presents. It may offer the chance of a mate, the threat of a predator, and the like. "Affordance" is a coinage of J. J. Gibson's, which, he says, he substituted for the term "value," with its subjective connotations (J. J. Gibson 1966, p. 285). It is not a question, as Köhler put it, of "the place of value in a world of facts." The world *itself* exhibits values, or meanings: relations between perceivers and features of their environments that offer them goals to seek or avoid. An animal's world is, from the beginning, a world full of meanings, and evolution has endowed it with the potentialities to respond to such a world.

In other words, there is a real world in which the animal finds itself, and to the constitution of which it in turn contributes through its activity. This world is structured: In the flux of events there are constants, invariants, stable proportions that characterize for the terrestrial animal, for example, the ground, the horizon, and so on. Again, it is these invariants that constitute the information that the animal picks up in its environment. And it is these same invariants that allow the animal in its turn to perceive affordances – that is, the advantages and the dangers that its environment presents to it.[3]

This situation exists for all animals. However, there are also peculiarities, so far as we can tell, in the human situation. Although perception as such is direct, there are also three kinds of indirect perception mediated

---

[3] We are here following an account by Eleanor J. Gibson (E. J. Gibson 1983). See also Gibson and Pick 2000.

by our cultural inventions. (They are nevertheless still to be found in the real world, in nature; there is only one world, within which culture arises.) These three inventions are *tools*, *language*, and *the use of pictures*. From birth, the perceptions of the infant, then of the child and of the adult are saturated by these human and cultural ingredients. But the fundamental structure of perception remains the foundation of these accomplishments – and, as we shall argue, if not the foundation, at least the analogue, of all knowledge (see J. J. Gibson 1979).

What lessons does this new theory of perception offer to the philosophy of science? We believe there are three.

First, the ecological approach assists us in maintaining our realist position. We can definitively finish with the phenomenalism that has haunted the philosophy of science since its inception. We can finally forget the picture of Mach counting his sensations, and try to understand the situation of scientific workers as engaged, each in his or her discipline, in an ongoing dialogue with reality.[4] We need to see ourselves, and the scientists among us, each in his or her own discipline or emerging discipline, as live, if enculturated animals in a complex, ongoing environment, full of meanings, whether soothing or startling, beckoning or alarming. The traditional theory of perception set the observer, let alone the thinker, apart from his milieu, isolated with a congeries of meaningless sensations that had somehow to be correlated with expectations in a constructed, perhaps fictional, "out-there." This has surely contributed a great deal to the difficulty of understanding scientific practice and formulating a more concrete and more adequate picture of such practice.

Second, if we adopt the ecological point of view, we find that perception plays an important part even in our more cerebral, or more enculturated, knowledge. Consider the three categories of indirect perception that characterize human knowledge: tools, language, and the use of pictures. Up to the beginning of the computer age, tools served chiefly to improve perception. Hooke looked at his cells, and Leeuwenhoeck at his little animals. Physicists still have to look at the traces in a Wilson cloud chamber. We never wholly escape the basic need

---

[4] We owe this expression to Dr. Frank Quinn of the Virginia Tech mathematics department. Michael Polanyi spoke of the scientist's confidence of being "in contact with reality" (Polanyi 1958). The concept of a dialogue further implies something like what Polanyi called "indwelling." The language in such a dialogue will itself be partly established, partly as yet developing.

of picking up information that allows us to perceive the affordances of our environment, be it in a laboratory or a field. Perception is always the fundamental knowledge on which other knowledge rests. Depiction clearly also involves perception, if, as J. J. Gibson, describes it, a kind of double vision: We are always aware both of picture and frame (J. J. Gibson 1979).

The case of language seems at first sight more difficult, since there is a long tradition that separates language from reality. Yet language does not separate us radically from the essential process of perception. Rather it enriches it; naming draws our attention to objects. Moreover, language itself has to be perceived: heard, seen, or in the case of the blind, touched. It is true that when the child begins to speak, it enters into a world not previously known, and from that moment its perceptions are caught up in its linguistic life. This is an infinitely subtle relation, and difficult to analyze. But it does not remove the foundation of all human activity in the perceptual processes by which we conduct ourselves in the world, at once cultural and natural, that surrounds us.

Finally – and this is the most significant implication of the ecological approach for the philosophy of science – the three components of the perceptual situation hold as well, analogically, for more cerebral kinds of knowledge. The term "perception" is used metaphorically for forms of insight other than strict sense perception, and with good reason. In his induction into any discipline, the student finds himself in a new world, surrounded by events and objects formerly unfamiliar. What he learns, in this new ambiance, is to pick up information in the form of invariants, and with their help to perceive the affordances – the meanings – available in this new environment. A beginning medical student, observing an X-ray, sees only lines; he *learns* to read these lines *as* an infected lung or a fractured limb or whatever. This is still a question of sense perception; but a similar process of increasing awareness marks the initiation into any new discipline, or sub-discipline. There are objects and events from whose constancies over change we pick up information that allows us to grasp formerly hidden meanings within the new world we have now come to inhabit.

This is very like a child's perceptual learning – mediated by language as well as tools and depictions – except that with scientific exploration, the process remains open-ended. While some features within the discipline become routine, it is the still not-so-clear features the scientist is always groping for. As François Jacob put it, "Unpredictability is in the nature of the scientific enterprise. If what is to be found is really new, then it is by definition unknown in advance. There is no way of telling where a particular line of research will lead" (Jacob 1982, p. 67). There is no single, over-all algorithm for such a process; styles of investigation will differ with the context.

In every discipline, however, or, as Rheinberger puts it, within every experimental system, there will be some procedures, some entities, that have become sufficiently established to serve as technical tools in the next stage of the investigation, while other entities or relations are still unclear. Rheinberger calls the upshot of this process "the production of epistemic things" (Rheinberger 1997). That is a puzzling phrase, since we usually think of scientists as discovering things, not making them. But what they make, presumably, are objects of knowledge: When we seize on the meaning of a phenomenon previously unknown, we are making into an object (for us) what was previously at most a shadow, foreshadowed in our search, but not "objectified." In a way, this is a Kantian insight, stressing the role of the knower in knowledge, but certainly without the sweep of the Kantian principles. Again, there is no overall rationale to be found here; we are restricted in every case to a given historical context, in a way that goes far beyond the dreams – or better, the nightmares – of the sage of Königsberg. Still, through careful case studies, we can gain insights into the ongoing practices of the sciences in ways that are philosophically revealing. Abandoning the goal of a grand overall synthetic view of science, or of "the scientific method," we can strive, like scientists though in a different style, to find ourselves, as we hope, engaged in a dialogue with reality.

Moreover, from careful case studies and carefully limited generalizations gained from comparing them, we can gain insights into the ongoing practices of the sciences that are philosophically revealing. To take an example from an earlier chapter, we found Cuvier insisting that only the careful study of each animal for its own sake was worth pursuing, whereas his colleague Geoffroy thought the search for large underlying generalities would set comparative anatomy on a new and more fruitful path. Yet by the middle of the nineteenth century, their opposing theses were taken as complementary rather than contradictory models for biologists' work. Contemporary investigations such as those of Burian or Rheinberger, which we have mentioned, also illustrate the variety of differing experimental systems, as well the possible interactions among them. Such interactions can be fruitfully studied without the old insistence on an ultimate unity of science of a monolithic and reductive sort. The growth of new disciplines also illustrates such interactions, as, for example, in the current development of bio-informatics, which combines the insights and methods of cell and molecular biology, mathematics and computer science, and statistics to achieve an understanding of gene structure and function that biologists would not have attained without the help of their neighbors – who in turn must learn something of the material their expertise is called in to interpret.

References

Achinstein, P. (1977). "Function statements." *Philosophy of Science*. 44: 341–67.

Adams, M. (1980). "Sergei Chetverikof, the Kol'tsov Institute, and the evolutionary synthesis." In Mayr and Provine 1980, pp. 242–78.

—— ed. (1994). *The Evolution of Theodosius Dobzhansky*. Princeton: Princeton University Press.

Adanson, M. (1763). *Famille des Plantes*. Paris: Vincent.

Adickes, E. (1924–5). *Kant als Naturforscher*. Berlin: DeGruyter.

Alberti, S. (2001). "Amateurs and professionals in one county: Biology and natural history in late Victorian Yorkshire." *Journal of the History of Biology*. 54: 115–47.

Alexander, R. (1979). *Darwinism and Human Affairs*. Seattle: University of Washington Press.

Allen, C., M. Bekoff, and G. Lauder, eds. (1998). *Nature's Purposes: Analyses of Function and Design in Biology*. Cambridge, MA: MIT Press.

Amundson, R. (1989). "The trials and tribulations of selectionist explanation." In Hahlweg and Hooker 1989, pp. 412–32.

Amundson, R. and G. Lauder. (1994). "Function without purpose: The uses of causal role function in evolutionary biology." *Biology and Philosophy*. 9: 443–69 (reprinted in Allen et al. 1998, pp. 335–69).

Appel, T. (1987). *The Cuvier–Geoffroy Debate*. New York: Oxford University Press.

Aquinas, T. (1964–76). *Summa Theologiae*. Oxford: Blackfriars.

Ariew, A., R. Cummings, and M. Perlman. (2002). *Functions: New Essays in the Philosophy of Psychology and Biology*. Oxford: Oxford University Press.

Ariew, R. (1998). "Condemnations of Cartesianism." In Ariew, R., J. Cottingham, and T. Sorell. (1988), *Descartes' Meditations: Background Source Materials*. Cambridge: Cambridge University Press, pp. 252–60.

—— (1999). *Descartes and the Last Scholastics*. Ithaca, NY: Cornell University Press.

Ariew, R., and M. Grene. (1997). "The Cartesian destiny of form and matter." In *Early Science and Medicine*. II: 300–25.

Aristotle. *Aristotelis Opera*. Ex recognitione I. Bekkeris edit Academic Regia Borussica. Berlin: G. Reimer, 1831. 2 vols. (References to Aristotle's texts are to this Greek edition. Unless otherwise specified, English translations are from Barnes 1984.)

Arthur, W. (2002). "The emerging conceptual framework of evolutionary developmental biology." *Nature*. 415: 757–64.

Asquith, R. and R. Giere, eds. (1981). *PSA 1980, Volume Two*. East Lansing, MI: Philosophy of Science Association.

Balme, D. (1962). "*Genos* and *eidos* in Aristotle's biology." *Classical Quarterly*. 12: 81–98.

———— (1987a). "Aristotle's biology was not essentialist." In Gotthelf and Lennox 1987, pp. 291–312.

———— (1987b). "Aristotle's use of division and differentiae." In Gotthelf and Lennox 1987, pp. 69–89.

Barkow, J., L. Cosmides, and J. Tooby, eds. (1992). *The Adapted Mind: Evolutionary Psychology and the Generation of Culture*. New York and Oxford: Oxford University Press.

Barlow, N., ed. (1963). "Darwin's ornithological notes." *Bulletin of the British Museum [Natural History], Historical Series*. 2: pp. 203–78.

Barnes, J., ed. (1984). *The Complete Works of Aristotle: The Revised Oxford Translation*. Princeton: Princeton University Press. 2 vols.

Barnes, J., J. M. Schofield, and R. Sorabji, eds. (1975). *Aristotle I: Science*. London: Duckworth.

Bartholomew, M. (1973). "Lyell and evolution: An account of Lyell's response to the prospect of an evolutionary ancestry for man." *British Journal for the History of Science*. 9: 261–303.

Barsani, G. (1992). "Buffon et l'image de la nature: de l'échelle des êtres à la carte géographique et à l'arbre généalogique." In Gayon 1992a, pp. 255–96.

Bates, W. H. (1862). "Contributions to an insect fauna of the Amazon valley." *Transactions of the Linnean Society of London*. 23: 495–556.

Bateson, W. (1894). *Materials for the Study of Variation, Treated with Special Regard to Discontinuity in the Origin of Species*. London: Macmillan.

———— (1907). "The progress of genetic research: An inaugural address to the Third Conference on Hybridisation and Plant Breeding." *Report of the Third Conference on Genetics, Hybridisation (The Cross-Breeding of Genera or Species), Cross-Breeding of Varieties, and General Plant Breeding*. London: Spottiswoode [for the Royal Horticultural Society], pp. 90–7.

Beatty, J. (1981). "What's wrong with the received view of evolutionary theory?" In Asquith and Giere 1981, pp. 341–55.

———— (1982). "What's in a word? Coming to terms in the Darwinian revolution." *Journal of the History of Biology*. 15: 215–239.

———— (1985). "Speaking of species: Darwin's strategy." In Kohn 1985, pp. 265–80.

———— (1994). "Dobzhansky and the biology of democracy: The moral and political significance of genetic variation." In Adams 1994, pp. 195–218.

———— (1995). "The evolutionary contingency thesis." In Wolters, G., and J. Lennox, eds. (1995). *Concepts, Theories, and Rationality in the Biological Sciences*. Pittsburgh: University of Pittsburgh Press, pp. 45–81.

———— (1998). "Ernst Mayr and the proximate-ultimate distinction." *Biology and Philosophy*. 9: 333–56.

Bechtel, W. and R. Richardson. (1993). *Discovering Complexity*. Princeton: Princeton University Press.

Beck, L. (1969). *Early German Philosophy*. Cambridge, MA: Harvard University Press.

Bellon, R. (2001). "Joseph Dalton Hooker's ideals for a professional man of science." *Journal of the History of Biology*. 54: 51–82.

Berg, L. (1926). *Nomogenesis*. London: Constable.

Bernasconi, R. (2001). "Who invented the concept of race? Kant's role in the Enlightenment construction of race." In Bernasconi, R., ed. (2001), *Race*, Oxford: Blackwell, pp. 10–36.

Bernasconi, R. and T. Lott, eds. (2001). *The Idea of Race*. Indianapolis: Hackett.

Berry, R. J. (1986). "What to believe about miracles." *Nature*. 322: 321–22.

Beurton, P., R. Falk, and H. J. Rheinberger, eds. (2000). *The Concept of the Gene in Development and Evolution: Historical and Epistemological Perspectives*. Cambridge: Cambridge University Press.

Bigelow, J., and R. Pargetter (1987). "Functions." *Journal of Philosophy*. 84: 181–96 (reprinted in Allen et al. 1998, pp. 241–59; Buller 1999, pp. 97–114).

Bitbol-Hespériès, A. (1998). "Descartes, Harvey, et la tradition médicale." In A. Bitbol-Hespériès 1998, *Descartes et ses oeuvres aujourd'hui*, Paris: Madrega, pp. 29–45.

—— (1990). *Le principe de vie chez Descartes*. Paris: Vrin.

Blanckaert, C. (1993). "Buffon and the natural history of man: Writing history and the 'foundational' myth of anthropology." *History of the Human Sciences*. 6: 13–50.

Blumenbach, J. F. (1775). *De Generis Humani Varietate Nativa*. Göttingen: Rosenbusch. 2nd ed. 1781; 3rd rev. ed. 1795. All passages are from the English translation of the 1st and 3rd eds in Blumenbach 1865, pp. 5–276.

—— (1789). *Über den Bildungstrieb und das Zuegungsgeschäfte*. 2nd ed. Göttingen: Heinrich Dieterich.

—— (1795). *Beiträge zur Naturgeschichte*. Göttingen: Heinrich Dieterich. (A second edition was published in 1806 and partially translated into English under the title *Contributions to Natural History*, in Blumenbach, 1865, pp. 281–340.)

—— (1802). *Handbuch der Naturgeschichte*, 6th ed. Göttingen: Rosenbusch.

—— (1865). *The Anthropological Treatises of Johann Friedrich Blumenbach*. Edited and translated by Thomas Bendyshe. London: Longmans, Green (reprint Bergman Publishers, New York, 1969).

Bolton, R. (1987). Definition and Scientific Method in Aristotle's *Posterior Analytics* and *Generation of Animals*. In Gotthelf and Lennox, eds., 1987, pp. 120–66.

Bourguet, L. (1729). *Lettres philosophiques sur la formation des sels et des crystaux et sur la génération et la mécanique organique des plantes et des animaux*. Amsterdam: F. L'Honore.

Bowler, P. (1983). *The Eclipse of Darwinism*. Baltimore: Johns Hopkins University Press.

—— (1988). *The Non-Darwinian Revolution*. Baltimore: Johns Hopkins University Press.

Boyd, R. (1999). "Homeostasis, species, and higher taxa." In R. Wilson 1999, pp. 141–85.

Brandon, R. (1978). "Adaptation and evolutionary theory." *Studies in the History and Philosophy of Biology.* 9: 181–206 (reprinted in Brandon 1996, pp. 3–29).

—— (1981a). "Biological teleology: questions and explanations." *Studies in the History and Philosophy of Science.* 12: 91–105 (reprinted in Brandon 1996, pp. 30–45).

—— (1981b). "A structural description of evolutionary theory." In Asquith and Giere 1981, pp. 427–39 (reprinted in Brandon 1996, pp. 46–57).

—— (1985). "Adaptive explanations: Are adaptations for the good of replicators or interactors?" In Depew and Weber 1985, pp. 81–96.

—— (1988). "Levels of selection: A hierarchy of interactors." In Plotkin, H., ed. (1988), *The Role of Behavior in Evolution.* Cambridge, MA: MIT Press, pp. 51–71 (reprinted in Brandon 1996, pp. 46–57).

—— (1990). *Adaptation and Environment.* Princeton: Princeton University Press.

—— (1996). *Concepts and Methods in Evolutionary Biology.* Cambridge: Cambridge University Press.

Brandon, R. and R. Burian, eds. (1984). *Genes, Organisms, Populations: Controversies Over the Units of Selection.* Cambridge, MA: MIT Press.

Brooke, J. (1979). "The *Natural Theology* of the geologists: Some theological strata." In Jordanova, L. and R. Porter, eds. (1979), *Images of the Earth: Essays in the History of the Environmental Sciences*, London: British Society for the History of Science, pp. 39–64.

Browne, J. (1996). *Charles Darwin: Voyaging: A Biography.* Princeton: Princeton University Press.

Buckland, W. (1836). *Geology and Mineralogy with Reference to Natural Theology.* London: Pickering. 2 vols. (Bridgewater Treatise).

Büchner, L. (1855). *Kraft und Stoff.* Frankfurt-am-Main.

Buffon, G-L, comte de (1735). "Preface du traducteur" [of Hales, S., 1727/1969] (reprinted in OP, pp. 5–6).

—— (1749a). *Preuves de la Théorie de la Terre* (Proofs of the Theory of the Earth), in Buffon 1749–1767, Vol. 1 and in OP, pp. 65–105. An English translation appeared in 1792 under the title *Buffon's Natural History*, London: J. S. Barr. This translation is reprinted in Lyon and Sloan, 1981, pp. 151–64.

—— (1749b). "Premier discours." In Buffon 1749–1767, Vol. 1. In OP, pp. 6–16. Translated as "Initial Discourse" in Lyon and Sloan, 1981, pp. 89–121.

—— (1749c). "Second discours." In Buffon 1749–1767, Vol. 1. In OP, pp. 45–64. An English translation appeared in 1792 under the title *Buffon's Natural History*, London, J. S. Barr. This translation is reprinted in Lyon and Sloan, 1981, pp. 134–50.

—— (1749d). "Histoire naturelle de l'homme." In Buffon 1749–1767, Vols. 2–3. Republished in Buffon, 1971, *De l'homme*, edited with an introduction by M. Duchet. Paris: Maspero.

—— (1749–1767). Histoire naturelle. Paris: Imprimerie royale. 15 vols.

—— (1753). "L'asne." *Histoire naturelle.* Vol. 4. In OP, pp. 353–58).

—— (1764). "Le mouflon et les autres brébis." *Histoire naturelle.* Vol. 11.

—— (1765). "Second vue." OP 35–41.

___ (1770). *Histoire naturelles des oiseaux*. Paris: Imprimerie Royale.

___ (1777). *Essai d'arithmétique morale*. (Translated by J. Lyon. In Lyon and Sloan 1981, pp. 56–7.)

___ (1778/1962). *Les époques de la nature*. Paris: Editions du Muséum (re-edited 1968).

___ (1884). *Oeuvres complètes de Buffon*. Paris: Pilon.

___ (1954). *Oeuvres philosophiques*, ed. Jean Piveteau. Paris: Presses Universitaires de France. Hereafter OP.

Buffon, G.-L. and L. J. M. Daubenton (1798). *Particulariers de Buffon et Daubenton*. 2 vols. Paris: Charles Pougens (reprinted by Slatkine Reprints, Geneva, 1971). English translation under the title "Observations on Buffon's *Natural History*." In Lyon and Sloan 1981, pp. 329–45.

Buller, D., ed. (1999). *Function, Selection and Design*. Albany: State University of New York Press.

Burian, R. (1983). "Adaptation." In M. Grene, ed., 1983, pp. 287–314.

___ (1992a). "Adaptation: Historical perspectives." In Keller and Lloyd 1992, pp. 7–12.

___ (1992b). "How the choice of experimental organism matters." *Synthèse*. 92: 151–66.

___ (1994). "Dobzhansky on evolutionary dynamics: Some questions about his Russian background." In Adams 1994, pp. 129–40.

___ (1997). "Exploratory experimentation and the role of histochemical techniques in the work of Jean Brachet." *History and Philosophy of the Life Sciences*. 19: 27–45.

___ (2000). "On the internal dynamics of Mendelian genetics." *Comptes rendus de l'Académie des Sciences, Serie III: Sciences de la vie*. 323: 1127–37.

Burkhardt, R. W. Jr. (1977). *The Spirit of System: Lamarck and Evolutionary Biology*. Cambridge, MA: Harvard University Press.

___ (1997). *The Spirit of System*, 2nd ed. Cambridge, MA: Harvard University Press.

___ (1997). "Lamarck in 1995." In Burkhardt 1997, pp. 11–39.

Burnet, T. (1691). *The Sacred Theory of the Earth*. London: Norton.

Byrne, P. (1989). *Natural Religion and the Nature of Religion*. London and New York: Routledge.

Cain, A. J. (1954). "Natural selection in *Cepaea*." *Genetics*. 39: 89–116.

Cain, A. J. and P. M. Sheppard (1950). "Selection in the Polymorphica land snail *Cepaea nemoralis*." *Heredity*. 4: 275–94.

Cain, J. (2002). "Co-opting colleagues: Appropriating Dobzhansky's 1936 lectures at Columbia." *Journal of the History of Biology* 35: 207–19.

Cannon, W. F. (1961). "The impact of uniformitarianism: Two letters from John Herschel to Charles Lyell, 1836–1837." *Proceedings of the American Philosophical Society*. 105, No. 301.

Caron, J. (1988). "'Biology' in the life sciences: A historiographical contribution." *History of Science* 26: 223–68.

Cassirer, H. W. (1938). *A Commentary on the Critique of Judgment*. London: Methuen.

Chagnon, N. (1988). "Life histories, blood revenge, and warfare in a tribal population." *Science*. 239: 985–92.

Chambers, R. (1844/1994). *Vestiges of the Natural History of Creation and Other Evolutionary Writings*. J. Secord, ed., Chicago: University of Chicago Press.

Charig, A. (1982). "Systematics in biology: A fundamental comparison of some major schools of thought". In Joysey and Friday 1982, pp. 363–40.

Charles, D. (1991). "Aristotle on substance, essence, and biological kinds". In J. Cleary, ed., 1991, *Proceedings of the Boston Area Colloquium in Ancient Philosophy VII*, pp. 227–62, Lanham, MD: University Press of America.

_____ (1997). "Aristotle on the unity and essence of natural kinds." In Kullmann and Föllinger 1997, pp. 28–42.

Charlton, W (1992). *Aristotle, Physics I and II*. Oxford: Clarendon Press (1st ed. 1970).

Cheney, D. L. and R. M. Seyfarth (1990). *How Monkeys See the World: Inside the Mind of Another Species*. Chicago: University of Chicago Press.

Chetverikoff, S. (1926). "On certain aspects of the evolutionary process from the standpoint of genetics." *Zhurnal Experimental'noi Biologii* 1: 3–54. English translation in *Proceedings of the American Philosophical Society* (1959). 105: 167–95.

Chomsky, N. (1972). *Language and Mind*. New York: Harcourt Brace and Jovanovich.

Church, R. W. (1847). "Review of *Explanations* [of *Vestiges*]." *The Guardian*. 18 March 1847, pp. 141–2.

Coleman, W. (1964). *Georges Cuvier: Zoologist*. Cambridge, MA: Harvard University Press.

Conimbricenses (1592). *Commentarii Collegii Conimbricensis Societatis Jesu in octo libros physicorum Aristotelis Stagyritæ*. Coimbra, Portugal.

Cooper, J. (1982). "Aristotle on natural teleology." In M. Nussbaum and M. Schofield, eds. (1982). *Logos and Language*. Cambridge: Cambridge University Press, pp. 197–222.

Correns, C. (1900). "G. Mendels Regel über das Verhalten der Nachkommenschaft der Rassenbastarde". *Berichte der deutschen botanischen Gesellschaft*. 18: 158–68.

Corsi, P. (1978). "The importance of the French transformist ideas for the second volume of Lyell's *Principles of Geology*." *British Journal for the History of Science*. 11: 221–44.

_____ (1988). *The Age of Lamarck*. Berkeley and Los Angeles: University of California Press.

Cosmides, L. J. Tooby, and J. Barkow. "Introduction: Evolutionary psychology and conceptual integration." In Barkow et al. 1992, pp. 3–15.

Creath, R. and J. Maienschein (2000). *Biology and Epistemology*. Cambridge: Cambridge University Press.

Crick, F. (1966). *Of Molecules and Men*. Seattle: University of Washington Press.

Cummins, R. (1975). "Functional analysis." *Journal of Philosophy*. 72: 741–65. Reprinted in Allen et al. 1998, pp. 169–196, and in Buller 1999, pp. 57–83.

Cummins, R. (2002). "Neo-teleology." In Ariew, Cummins, and Perlman 2002, pp. 151–72.

Cuvier, Georges. (1800–1805). *Leçons d'anatomie comparée.* Paris: Baudoin. 5 vols.

―― (1812a). *Recherches sur les ossemens fossiles.* Paris: Déterville. 4 vols.

―― (1812b). "Sur un nouveau rapprochement à établir entre les classes qui composent le règne animal." *Annales du Muséum.* 19: 73–84.

―― (1817a). *Essays on the Theory of the Earth,* 3rd ed. Translated by Robert Jameson. Edinburgh: Blackwood. Reprinted by Arno Press, New York, 1978.

―― (1817b). *Le règne animal distribué d'après son organisation pour servir de base à l'histoire naturelle des animaux et d'introduction à l'anatomie comparée.* Paris: Déterville. 4 vols. (vol. 3 edited by P. A. Latreille).

―― (1825). "Nature." *Dictionnaire des sciences naturelles.* XXXIV: 261–68.

―― (1843). *Histoire des sciences naturelles depuis leur origine jusqu'à nos jours.* Paris: Fortin, Masson. 4 vols.

Cuvier, G. and A. Valenciennes (1828–1833). *Histoire naturelle des poissons.* Paris: Baudouin. 8 vols.

Daniel, G. (1694). *Novae Difficultates.* Amsterdam: Wolfgang. Bound with Daniel, *Iter Per Mundum Cartesii.*

Darden, L. (1991). *Theory Change in Science: Strategies from Mendelian Genetics.* New York: Oxford University Press.

Darwin, C. (1859). *On the Origin of Species by Natural Means of Selection.* London: John Murray. Facsimile reprint, Cambridge, MA: Harvard University Press, 1964. Hereafter *Origin.*

―― (1868). *The Variation of Animals and Plants under Domestication.* London: John Murray. 2 vols.

―― (1871). *The Descent of Man, and Selection in Relation to Sex.* London: John Murray. (revised edition 1874).

―― (1872). *The Expression of the Emotions in Man and Animals.* London: John Murray.

―― (1873). "On the males and complemental males of certain *Cirripedes,* and on rudimentary structures." *Nature.* 8: 431–2. Reprinted in Darwin 1977, pp. 177–82.

―― (1881). *The Formation of Vegetable Mould Through the Action of Worms, and Observations on Their Habits.* London: John Murray.

―― (1958). *The Autobiography of Charles Darwin.* Ed. Nora Barlow. London: Collins.

―― (1973). *Charles Darwin's Natural Selection: Being the Second Part of His Big Species Book Written from 1856 to 1858.* Robert C. Stauffer, ed., Cambridge: Cambridge University Press.

―― (1977). *The Collected Papers of Charles Darwin.* Ed. Paul Barrett. Chicago: University of Chicago Press. 2 vols.

―― (1985–). *The Correspondence of Charles Darwin.* Eds. F. Burkhardt, S. Smith, et al. Cambridge: Cambridge University Press.

―― (1987). *Charles Darwin's Notebooks.* Eds. Paul H. Barrett, Peter J. Gautrey, Sandra Herbert, David Kohn, and Sydney Smith. Ithaca, N. Y.: Cornell University Press.

_____ (1990). *Charles Darwin's Marginalia*. Ed. Mario A. Di Gregorio, with the assistance of N. W. Gill. New York: Garland Publishing.

Darwin Research Archives. Cambridge University Library. Hereafter DAR.

Darwin, C and A. R. Wallace (1958). *Evolution by Natural Selection*. Gavin De Beer, ed., Cambridge: Cambridge University Press.

Darwin, F., ed. (1887). *Life and Letters of Charles Darwin*. London: John Murray. 3 vols. American Edition, New York: Appleton. 2 vols. Page numbers in the text are from the American edition.

Daston, L. (1998). *Classical Probability in the Enlightenment*. Princeton: Princeton University Press.

Daston, L. and K. Park (1998). Wonders and the Order of Nature, 1150–1750. New York: Zone.

Daudin, H. (1926). *De Linné à Jussieu*. Paris: Alcan.

Dawkins, R. (1989). *The Selfish Gene*. 2nd ed. Oxford: Oxford University Press.

Deacon, T. (1997). *The Symbolic Species*. New York: W. W. Norton.

Dear, P. (1995). *Discipline and Experience: The Mathematical Way in the Scientific Revolution*. Chicago: University of Chicago Press.

Delbrück, M. (1971). "Aristotle-totle-totle." In J. Monod and Borek, eds., 1971, *Of Microbes and Life*, New York: Columbia University Press, pp. 50–5.

Dennett, D. (1987). *The Intentional Stance*. Cambridge, MA: MIT Press.

_____ (1991). *Consciousness Explained*. Boston: Little, Brown.

_____ (1995). *Darwin's Dangerous Idea*. New York: Simon & Schuster.

Depew, D. and B. Weber, eds. (1985). *Evolution at a Crossroads*. Cambridge, MA: MIT Press.

De Queiroz, K. (1999). "The general lineage concept of species and the defining properties of the species category." In R. Wilson 1999, pp. 49–90.

De Vries, H. (1899). *Intracelluläre Pangenesis*. Jena: Fischer.

_____ (1900a). "Das Spaltungsgesetz der Bastarde (Vorläufige Mitteilung)." *Berichte der Deutschen Botanischen Gesellschaft*. 18: 83–90.

_____ (1900b). "Sur la loi de disjonction des hybrides." *Comptes Rendus Hebdomadaires des Séances de l'Académie des Sciences*. 130: 845–847.

_____ (1901–03). *Die Mutationstheorie: Versuche und Beobachtungen über die Entstehung der Arten im Pflanzenreich*. Leipzig: Veit. 2 vols.

_____ (1905). *Species and Varieties: Their Origin by Mutation*. Chicago: Open Court.

_____ (1910). *Intracellular Pangenesis*. Chicago: Open Court.

Descartes, R. (1985–1991). *The Philosophical Writings of Descartes*. Translated by J. Cottingham, R. Stoothoff, D. Murdoch and A. Kenny. Cambridge: Cambridge University Press. 3 vols. Hereafter CSMK.

_____ (1964–1982). *Oeuvres*. Edited by C. Adam and P. Tannery. Paris: Vrin. 12 vols.

Des Chène, D. (1996). *Physiologia: Natural Philosophy in Late Aristotelian and Cartesian Thought*. Ithaca: Cornell University Press.

Desmond, A. (1989). *The Politics of Evolution: Morphology, Medicine and Reform in Radical London*. Chicago: University of Chicago Press.

_____ (1994). *Huxley: The Devil's Disciple*. London: Michael Joseph.

_____ (1997a). *Huxley: Evolution's High Priest*. London: Michael Joseph.

—— (1997b). *Huxley: From Devil's Disciple to Evolution's High Priest.* Reading, MA: Addison-Wesley (single volume edition of Desmond 1994/1997a, b).

—— (2001). "'Professionals,' 'amateurs' and the making of mid-Victorian biology – a progress report." *Journal of the History of Biology.* 54: 3–50.

Detel, W. (1997). "Why all animals have a stomach: Demonstration and axiomatization in Aristotle's *Parts of Animals.*" In Kullmann and Föllinger 1997, pp. 63–84.

De Waal, F. (1996). *Good Natured.* Cambridge, MA: Harvard University Press.

Dewey, J. (1925). *Experience and Nature.* Chicago: Open Court.

Diderot, D., ed. (1751–1757). *Encyclopédie.* Paris: Neuchâtel.

Dobzhansky, T. (1937). *Genetics and the Origin of Species*, 1st ed. New York: Columbia University Press.

—— (1941). *Genetics and the Origin of Species*, 2nd ed. revised. New York: Columbia University Press.

—— (1959). "Evolution of genes and genes in evolution." *Cold Spring Harbor Symposium on Quantitative Biology.* XXIV: 15–30.

—— (1962). *Mankind Evolving.* New Haven: Yale University Press.

—— (1962–63). "Reminiscences of Theodosius Dobzhansky." Typed transcript, two parts. Oral History Research Office, Columbia University, New York.

—— (1964). "Biology, molecular and organismic." *American Zoologist.* 4: 443–52.

—— (1970). *Genetics of the Evolutionary Process.* New York: Columbia University Press.

—— (1973). "Nothing in biology makes sense except in the light of evolution." *The American Biology Teacher.* 35: 125–9.

Dobzhansky, T., F. Ayala, G. Stebbins, and J. Valentine (1977). *Evolution.* San Francisco: Freeman.

Donohue, M. J. (1990). "Sociology, selection, and success: A critique of David Hull's analysis of science and systematics." *Biology and Philosophy.* 5: 459–72.

Driesch, H. (1899). *Philosophie des Organischen.* Leipzig: Quelle und Meyer.

—— (1908). "Zur Analysis der Potenzen embryonaler Organzellen." *Archiv für Entwickungsmechanik.* 2 (195): 169–201. London: Ada and Charles Black.

Ducheneau, F. (1973). *L'empirisme de Locke.* The Hague: Nijhoff.

—— (1998). *Les modèles du vivant de Descartes à Leibniz.* Paris: Vrin.

du Lauren, A. (1610). *L'histoire anatomique.* Translated by F. Sizé. Paris: J. Bertault.

Dupleix, S. (1992). *La métaphysique.* Paris: Fayard.

Dupré, J. (1993). *The Disorder of Things.* Cambridge, MA: Harvard University Press.

Eimer, G. H. (1897). *Die Entstehung der Arten: Zweiter Teil: Orthogenesis der Schmetterlinge. Ein Beweis bestimmt gerichteter Entwicklung und Ohnmacht der natürlichen Auslese bei der Artbildung.* Leipzig: Engelmann.

Eldredge, N. (1985). *Unfinished Synthesis: Biological Hierarchies and Modern Evolutionary Thought.* Oxford: Oxford University Press.

—— (1989). *Macroevolutionary Dynamics: Species, Niches, and Adaptive Peaks.* New York: McGraw Hill.

Eldredge, N. and S. J. Gould (1972). "Punctuated equilibria: An alternative to phyletic gradualism." In T. J. M. Schopf, ed., 1972, *Models in Paleobiology*, San Francisco: Freeman, Cooper, pp. 82–115.

Eldredge, N., and M. Grene (1992). *Interactions: The Biological Context of Social Systems*. New York: Columbia University Press.

Ent, G. (1641). *Apologia Pro Circulatione Sanguinis*. London.

Ereshefsky, M. (2001). *The Poverty of the Linnaean Hierarchy*. Cambridge: Cambridge University Press.

Eustacius a Sancto Paolo. (1629). *Summa Philosophica Quadripartita*. Paris.

Ferejohn, M. (1991). *The Origins of Aristotelian Science*. New Haven: Yale University Press.

Fisch, M. (1991). *William Whewell: Philosopher of Science*. Oxford: Clarendon.

Fisher, R. A. (1918). "The correlation between relations on the supposition of Mendelian inheritance." *Transactions of the Royal Society of Edinburgh*. 52: 399–433.

——— (1930). *The Genetical Theory of Natural Selection*. Oxford: Oxford University Press; reprinted 1958, New York: Dover).

Fisher, R. and E. Ford. (1947). "The spread of genes in natural conditions in a colony of the moth *Panaxia dominula*." *Heredity*. 1: 143–74.

Fodor, J. (1975). *The Language of Thought*. New York: Thomas Crowell.

Fontenelle, B. de (1683). *Lettres diverses. . . .* Paris.

Forster, G. (1786). "Noch etwas über die Menschenrassen." *Teutsche Merkur*. 56: 73–80.

Fox, C., R. Porter, and R. Wokler, eds. (1995). *Inventing Human Science: Eighteenth Century Domains*. Berkeley and Los Angeles: University of California Press.

Frank, R. G., Jr. (1980). *Harvey and the Oxford Physiologists*. Berkeley: University of California Press.

Frankel, G. and D. Gunn (1940). *The Orientation of Animals*. Oxford: Clarendon.

Frede, M. (1985). "Substance in Aristotle's metaphysics." In A. Gotthelf 1985, pp. 17–26.

Fuchs, T. (1992/2001). *Die Mechanisierung des Herzens*. Frankfurt-am-Main. Suhrkamp (English translation, M. Grene, *The Mechanization of the Heart*, 2001, Rochester, NY: University of Rochester Press).

Furth, M. (1988). *Substance, Form, and Psyche: An Aristotelian Metaphysics*. Cambridge: Cambridge University Press.

Galton, F. (1865). "Hereditary talent and character." *Macmillan's Magazine*. 12: 157–166, 318–327.

——— (1875). "Statistics by intercomparison, with remarks on the law of error." *Philosophical Magazine*, 4th series. 49: 33–46.

——— (1875–1876). "Typical laws of heredity." *Journal of the Royal Institution*. 8: 282–300.

——— (1894). "Discontinuity in evolution." *Mind*, new series. 3: 362–72.

——— (1889). *Natural Inheritance*. London: Macmillan.

——— (1892a). *Finger Prints*. London: Macmillan.

——— (1892b). *Hereditary Genius*, 2nd ed. London: Macmillan.

Garber, D. (1992). *Descartes' Metaphysical Physics*. Chicago: University of Chicago Press.

Garden, G. (1691). "A discourse concerning the modern theory of generation." *Philosophical Transactions of the Royal Society of London*. 17: 474–83.

Garland, M. (1980). *Cambridge Before Darwin: The Ideal of a Liberal Education 1800–1860*. Cambridge: Cambridge University Press.

Gayon, J., ed. (1992a). *Buffon 88: Actes du colloque international Paris-Dijon-Montbard*. Paris: Vrin.

____ (1992b). "L'individualité de l'espèce: Une thèse transformiste?" In Gayon 1992a, pp. 475–489.

____ (1992c). "L'hypothétisme de Buffon." In Gayon 1992a, pp. 207–22.

____ (1998). *Darwinism's Struggle for Survival: Heredity and the Hypothesis of Natural Selection*. Cambridge: Cambridge University Press. Translation by Mathew Cobb of *Darwin et l'après Darwin: Une histoire de l'hypothèse de selection naturelle*, 1992. Paris: Editions Kiné.

Geoffroy, Saint-Hilaire, E. (1796). "Mémoire sur les rapports naturels des *Makis Lémur L*, et description d'une espèce nouvelle de mammifère." *Magazin encyclopédique, ou Journal des sciences, des lettres et des arts*. II: 20–50.

____ (1807a). "Premier mémoire sur les poissons, où l'on compare les pièces osseuses de leurs nageoires pectorals avec les os de l'extrémité antérieur des autres animaux à vertèbres." *Annales du Muséum d'histoire naturelle*. 9: 357–72.

____ (1807b). "Second mémoire sur les poissons: Considérations sur l'os furculaire, une des pièces de la nageoire pectorale." *Annales du Muséum d'histoire naturelle*. 9: 413–27.

____ (1807c). "Troisième mémoire sur les poissons: ou l'on traite de leur sternum sous le point de vue de sa détermination et de ses formes générales." *Annales du Muséum d'histoire naturelle*. 10: 249–64.

____ (1817). "Du squellette des poissons ramené dans toutes ses parties à la chapente osseusse des autres animaux vertèbres, et prèmièrement de l'opercule des poissons." *Bulletin de la societé philomatique*: 125–7 (reprinted in Geoffroy, Saint-Hilaire 1822).

____ (1818). *Philosophie anatomique: Des organes respiratoires sous le rapport de la détermination et de l'identité de leurs pièces osseuses*. Paris: Mequignon-Marvis.

____ (1820). *Mémoire sur l'organisation des insectes*. Read at the Royal Academy of Sciences, Paris.

____ (1822). *Philosophie anatomique: Des monstrosités humaines*. Paris: Mequignon-Marvis.

____ (1823). "Marsupiaux." *Dictionnaire des sciences nouvelles:* no publisher identified.

____ (1830). *Principes de philosophie zoologique discutés en Mars 1830 au sein de l'académie royale des sciences*. Paris: Pichon et Didier (reprinted in Le Guyader 1998/2001, pp. 129–237).

Ghiselin, M. (1974). "A radical solution to the species problem." *Systematic Zoology*. 23: 536–44.

Gibson, E. J. (1983). "Perceptual development from the ecological approach." *Advances in Developmental Psychology.* 3: 243–86.

Gibson, E. J. and A. Pick (2000). *An Ecological Approach to Perceptual Learning and Development.* Oxford: Oxford University Press.

Gibson, J. J. (1966). *The Senses Considered as Perceptual Systems.* Boston: Houghton Mifflin.

——— (1979). *The Ecological Approach to Visual Perception.* Boston: Houghton Mifflin.

Giere, R. (1979). *Understanding Scientific Reasoning,* 1st ed. New York: Holt, Reinhart and Winston.

Gilbert, S. (1991). "Induction and the origins of developmental genetics." In Gilbert, S., ed., 1991, *A Conceptual History of Modern Embryology,* Baltimore: Johns Hopkins University Press. pp. 181–203.

——— (1998). "Bearing crosses: A historiography of genetics and embryology." *American Journal of Medical Genetics.* 76: 168–82.

——— (2000). "Genes, classical and developmental." In Beurton, Falk, and Rheinburger 2000, pp. 178–92.

Gilbert, S., and R. Burian (2003). "Development, evolution, and evolutionary developmental biology." In Hall, B. and W. Olson, eds., *Keywords and Concepts in Evolutionary Developmental Biology* (Cambridge, MA: Harvard University Press), pp. 68–74.

Gilbert, S., J. Opitz, and R. Raff. (1996). "Resynthesizing evolutionary and developmental biology." *Developmental Biology.* 173: 357–72.

Gilbert, W. (1992). "A vision of the grail." In Kevles, D. and L. Hood, eds., 1992, *The Code of Codes.* Cambridge, MA: Harvard University Press, pp. 83–97.

Gill, M. (1989). *Aristotle on Substance.* Princeton: Princeton University Press.

——— (1997). "Material necessity and *Meteorology 4.12.*" In Kullmann and Föllinger 1997, pp. 145–61.

Gilmour, J. (1940). "Taxonomy and philosophy." In J. Huxley, ed., *The New Systematics,* Oxford: Oxford University Press, pp. 461–74.

Gilson, E. (1930). *Etudes sur le rôle de la formation du système cartésien.* Paris: Vrin.

——— (1965). *Etudes sur le rôle de la pensée médiévale dans la formation du système cartésien,* 4th ed. Paris: Vrin.

Girtanner, C. (1796). *Uber das Kantische Prinzip für die Naturgeschichte.* Göttingen: J. C. Dieterich.

Godfrey-Smith, P. (1994). "A modern history theory of functions." *Nous.* 28: 344–362 (reprinted in Buller 1999, pp. 199–220).

Goldschmidt, R. (1940). *The Material Basis of Evolution.* New Haven: Yale University Press (reprinted with an introduction by S. J. Gould, 1982).

Gordon, D. (1992). "Wittgenstein and ant-watching." *Biology and Philosophy.* 7: 13–25.

Gotthelf, A. (1987). "First principles in Aristotle's *Parts of Animals.*" In Gotthelf and Lennox 1987, pp. 167–85.

——— (1997). "The elephant's nose: Further reflections on the axiomatic structure of biological explanation in Aristotle." In Kullmann and Föllinger 1997, pp. 85–9.

Gotthelf, A. and J. Lennox, eds. (1987). *Philosophical Issues in Aristotle's Biology.* Cambridge: Cambridge University Press.

Gould, S. J. (1977). *Ontogeny and Phylogeny.* Cambridge, MA: Harvard University Press.

—— (1980). "Is a new and general theory of evolution emerging?" *Paleobiology.* 6: 119–30.

—— (1982). "Darwinism and the expansion of evolutionary theory." *Science.* 216: 380–7.

—— (1983). "The hardening of the modern synthesis." In Grene 1983, pp. 71–93.

—— (1987). "The limits of adaptation: Is language a spandrel of the human brain?" Paper presented to the Cognitive Science Seminar, Center for Cognitive Science, Massachusetts Institute of Technology, Cambridge, MA.

—— (1989). *Wonderful Life: The Burgess Shale and the Nature of History.* New York: Norton.

—— (1997). "Darwinian fundamentalism." *New York Review of Books.* June 12, 1997, pp. 34–7.

—— (2002). *The Structure of Evolutionary Theory.* Cambridge, MA: Harvard University Press.

Gould, S. J. and R. Lewontin. (1979). "The spandrels of San Marco and the panglossian paradigm: A critique of the adaptationist programme." *Proceedings of the Royal Society-London B.* 205: 581–98.

Gould, S. J. and N. Eldredge. (1977). "Punctuated equilibria: The tempo and mode of evolution reconsidered." *Paleobiology.* 3: 115–51.

Gould S. J. and E. Lloyd (1999). "Individuality and adaptation across levels of selection: How shall we name and generalize the unit of Darwinism?" *Proceedings of the National Academy of Science USA.* 96: 11904–11909.

Grant, E. (1971). *Physical Science in the Middle Ages.* New York: Wiley.

Grassé, P. (1944). "'La Biologie.' Texte inédit de Lamarck." *Revue Scientifique.* 5: 267–76.

Gray, A. (1963/1876). *Darwiniana.* Cambridge, MA: Harvard University Press (original edition 1876, New York: Appleton).

Gregory, J. K. (1977). *Scientific Materialism in Nineteenth Century Germany.* Dordrecht: Reidel.

Grene, M. (1961). "Statistics and selection." *British Journal for the Philosophy of Science.* 12: 25–42.

—— (1963). *A Portrait of Aristotle.* Chicago: University of Chicago Press (reprinted 1998, Bristol: Thoemmes Press).

—— (1974). *The Understanding of Nature.* Dordrecht: Reidel.

—— ed. (1983). *Dimensions of Darwinism: Themes and Counterthemes in Twentieth Century Evolutionary Theory.* Cambridge: Cambridge University Press.

—— (1990). "Evolution, typology, and population thinking." *American Philosophical Quarterly.* 27: 237–44.

—— (1991). *Descartes among the Scholastics.* Milwaukee: Marquette University Press.

—— (1993a). "The heart and blood: Descartes, Plemp, and Harvey." In Voss, S., ed. (1993). *Essays on the Philosophy and Science of René Descartes.* New York: Oxford University Press, pp. 324–5.

_____ (1993b). "Aristotelico-Cartesian themes in natural philosophy: Some seventeenth-century cases." *Perspectives on Science*. 1: 66–87.

_____ (1995). *A Philosophical Testament*. Chicago: Open Court.

Griesemer, J. (2000). "Reproduction and the reduction of genetics." In Beurton, Falk, and Rheinberger 2000, pp. 240–85.

Griffin, D. (1984). *Animal Thinking*. Cambridge, MA: Harvard University Press.

_____ (1992). *Animal Minds*. Chicago: University of Chicago Press.

Griffiths, P. (1999). "Squaring the circle: Natural kinds with historical essences." In R.Wilson 1999, pp. 187–208.

Griffiths, P. and R. Gray. (1995). Developmental systems and evolutionary explanation. *The Journal of Philosophy*. 91: 277–304.

Guyer, P. (2001). "From nature to morality: Kant's new argument in the 'Critique of Teleological Judgment'." In H.-F. Fulda and J. Stolzenberg, eds., *Architektonik und System in der Philosophie Kants*, Felix Meiner 2001, pp. 375–404.

Haeckel, E. (1866). *Generelle Morphologie der Organismen*. Berlin: Reimer. 2 vols.

Hahlweg, K. and C. Hooker. (1989). *Issues in Evolutionary Epistemology*. Albany: State University of New York Press.

Hales, Stephen (1727/1969). *Vegetable Staticks*. Reprinted 1969. New York: American Elsevier.

Hamburger, V. 1980. "Embryology and the modern synthesis in evolutionary theory." In Mayr and Provine 1980, pp. 97–112.

Hamilton, W. (1964). "The genetical evolution of social behavior, I–II." *Journal of Theoretical Biology*. 7: 1–52.

Hamy, E. (1893). *Les derniers jours du jardin du roi et la fondation du Muséum d'histoire naturelle*. Paris: Imprimerie nationale.

Hanov, M. C. (1766) *Philosophia Naturalis, Tome 3*. Halle: Libraria Rengeriana. In (1997) *Christian Wolff Gesammelte Werke III*. ABT.BD. 40.3. Hildesheim: Georg Olms.

Hardy, G. (1908). "Mendelian proportions in a mixed population." *Science*. 28: 49–50.

Harvey, W. (1847). *Works*. Translated by R. Willis. For the Sydenham Society, London (Johnson Reprint, 1965). Hereafter W.

Hattab, H. (1998). *The Origins of a Modern View of Causation: Descartes and His Predecessors on Efficient Causes*. Ann Arbor, MI: University of Michigan Dissertation Services.

Hennig, W. (1950). *Grundzüge einer Theorie der phylogenetischen Systematik*. Berlin: Deutscher Zentralverlag.

_____ (1966). *Phylogenetic Systematics*, 1st ed. Urbana, IL: University of Illinois Press

_____ (1979). *Phylogenetic Systematics*, 2nd ed. Urbana, IL: University of Illinois Press.

Herder, J. (1784). *Ideen zur Philosophie der Geschichte der Menschheit* (Ideas for a Philosophical History of Mankind). Riga and Leipzig: Hartknock.

Herrnstein, R. and C. Murray. (1994). *The Bell Curve: Intelligence and Class Structure in American Life*. New York: Free Press.

Herschel, W. (1830). *Preliminary Discourse on the Study of Natural Philosophy*. London: Longmans, Rees, Orme, Brown, and Green (reprinted 1987, Chicago: University of Chicago Press).

—— (1841). "Whewell on the inductive sciences." *The Quarterly Review*. 68: 177–238.

Hippocrates, *Hippocratic Writings* (1978). Edited with an Introduction by G. E. R. Lloyd. Translations by J. Chadwick and W. N. Mann. London: Hammondsworth, Penguin Books.

Hirschmann, D. (1973). "Function and explanation." *Proceedings of the Aristotelian Society*. London: Methuen, pp. 16–38.

Hobbes, T. (1655). *Elementa philosophiae sectio prima de corpore*. London.

Hodge, M. (1977). "The structure and strategy of Darwin's 'long argument'." *British Journal of the History of Science*. 10: 237–46.

—— (1982). "Darwin and the laws of the animate part of the terrestrial system (1835–1837): On the Lyellian origins of his zoonomical explanatory program." *Studies in the History of Biology*. 6: 1–106.

—— (1985). "Darwin as a lifelong generation theorist." In D. Kohn 1985, pp. 207–43.

—— (1990). "Darwin studies at work: A re-examination of three decisive years, 1835–1837." In T. Levere and W. Shaw 1990, *Nature, Experiment and the Sciences*, Dordrecht: Kluwer, pp. 249–74.

—— (1992). "Biology and philosophy (including ideology): A study of Fisher and Wright." In S. Sarkar, ed., 1992, *The Founders of Evolutionary Genetics*, Dordrecht: Klewer, pp. 231–93.

—— (2000). "Knowing about evolution: Darwin and his theory of natural selection". In R. Creath and J. Maienschein 2000, pp. 27–47.

Hodge, M. and D. Kohn (1985). "The immediate origins of natural selection." In Kohn 1985, pp. 185–206.

Hubbard, R. and E. Wald (1993). *Exploding the Gene Myth*. Boston: Beacon.

Hull, D. (1970a). "Contemporary systematic philosophies." *Annual Review of Ecology and Systematics*. 1: 19–54.

—— (1970b). *Philosophy of Biological Science*. Englewood Cliffs, NJ: Prentice-Hall.

—— (1976). "Are species really individuals?" *Systematic Zoology*. 25: 174–91.

—— (1978). "A matter of individuality." *Philosophy of Science*. 45: 335–60.

—— (1980). "Individuality and selection." *Annual Review of Ecology and Systematics*. 11: 311–32.

—— (1988a). *Science as a Process. An Evolutionary Account of the Social and Conceptual Development of* Science. Chicago: University of Chicago Press.

—— (1988b). "Progress in ideas of progress." In M. Nitecki, ed., (1988). *Evolutionary Progress*, Chicago: University of Chicago Press, pp. 27–48.

—— (1994). "Species, races, and genders: Differences are not deviations." In Weir, R., S. Lawrence, and E. Fales (1994). *Genes and Human Self-knowledge*, Iowa City: University of Iowa Press, pp. 207–31.

—— (2001). *Science and Selection: Essays on Biological Evolution and the Philosophy of Science*. Cambridge: Cambridge University Press.

Hull, D., R. Langman and S. Glenn. (2001). "A general account of selection: Biology, immunology and behavior." In Hull 2001, pp. 49–93.

Hume, D. (1739). *A Treatise of Human Nature*. London: John Noon (Oxford: Clarendon Press edition, 1955).

_____ (1779). *Dialogues Concerning Natural Religion*. London: no publisher identified.

_____ (1935). *Dialogues Concerning Natural Religion*. Oxford: Clarendon Press.

Hurlbutt, R. (1985). *Hume, Newton, and the Design Argument*. Lincoln: University of Nebraska Press.

Hutchison, G. (1965). *The Ecological Theatre and the Evolutionary Play*. New Haven: Yale University Press.

Hutton, J. (1795). *Theory of the Earth with Proofs and Illustrations*. Edinburgh: William Creech.

Huxley, J. (1942). *Evolution: The Modern Synthesis*. London: Allen and Unwin (American edition, New York: Harper and Brothers, 1943). All references are to the American edition.

Huxley, J., A. Hardy, and E. Ford, eds. (1954). *Evolution As a Process*. London: Allen and Unwin.

International Human Genome Sequencing Consortium (2001). "Initial sequencing and analysis of the human genome." *Nature*. 409: 860–921.

Irwin, T. (1988). *Aristotle's First Principles*. Oxford: Oxford University Press.

Jacob, F. (1977). "Evolution and tinkering." *Science*. 196: 1161–66.

_____ (1982). *The Possible and the Actual*. Seattle: University of Washington Press.

James, W. (1875). Review of *Grundzüge der physiologischen Psychologie* by Wilhelm Wundt. *North American Review* 121: 195–201.

Johannsen, W. (1909). *Elemente der exakten Erblichkeitslehre*. Jena: Fischer.

_____ (1913). *Über Erblichkeit in Populationen in reinen Linien: Ein Beitrag zur Beleuchtung schwebender Selektionsfragen*. Jena: Fischer.

Joysey, K. and A. Friday (1982). *Problems of Phylogenetic Reconstruction*. London: Academic Press.

Kant. I (1908–13). *Kants gesammelte Schriften*. Herausgegeben von der königlich preussischen, bzw. Deutschen Akademie der Wissenschaften. Berlin: George Reimer. (Hereafter Ak.).

_____ (1757). Entwurf und Ankunding eines Collegii der physischen Geographie. Ak. II, pp. 1–11.

_____ (1775/1777). "Of the different races of mankind" (Uber den verschiedenen Rassen der Menschen). Ak II, pp. 427–44 Revised edition, 1777, translated by J. M. Mikkleson, in Bernasconi and Lott, 1999, pp. 8–22).

_____ (1781). *Kritik der reinen Vernunft*. 1st [A] edition. Riga: Hartknoch.

_____ (1783). *Prolegomena to Any Future Metaphysics (Prolegomena zu einer jeden künftigen Metaphysik, die als Wissenschaft wird auftreten können)*. Ak. IV, pp. 255–383.

_____ (1785a). "Determination of the concept of a human race" (Bestimmung des Begriffs einer Menschenrasse). Ak. VIII, pp. 89–106.

_____ (1785b). Review of J. G. Herder's *Ideas for a Philosophy of the History of Mankind* (Recenzion von J. G. Herders *Ideen zur Philosophie der Geschichte*). Ak. VIII, pp. 43–66.

—— (1786a). A conjectural beginning of the human race (Mutmasslicher Anfang der Menschengeschichte) Ak. VIII, pp. 107–23.

—— (1786b). *Metaphysical Principles of Natural Science. Metaphysische Anfangsgrunde der Naturwissenschaft.* Ak. IV, pp. 465–565.

—— (1787). *Kritik der reinen Vernunft.* 2nd [B] edition. Riga: Hartknoch.

—— (1788). *On the Use of Teleological Principles in Philosophy (Über den Gebrach teleologischer Principien in der Philosophie).* Ak. VIII, 157–84.

—— (1793). *Critique of Judgment (Kritik der Urteilskraft,* 2nd ed. 1st ed. 1790). Ak. V, pp. 165–485. Passages quoted are from Kant 1987.

—— (1794). "Metaphysik–Vorlesung K2." Ak. XXVIII, pp. 753–75. Translated by K. Americks and S. Naragon, *Lectures on Metaphysics,* Cambridge: Cambridge University Press, pp. 395–413.

—— (1929). *Critique of Pure Reason.* Translated by Norman Kemp Smith. New York: Macmillan.

—— (1936/1993). *Opus Postumum.* Convolt I–IV. Ak. XXI. Berlin: de Gruyter. Translated by Eckart Forster, Cambridge and New York: Cambridge University Press, 1993.

—— (1987). *Critique of Judgment.* Translated by W. S. Pluhar. Indianapolis: Hackett Publishing Company, 1987.

Keller, E. (2000). *The Century of the Gene.* Cambridge, MA: Harvard University Press.

Kingsolver, J. and M. Koehl. (1985). "Aerodynamics, thermoregulation and evolution of insect wings: Differential scaling and evolutionary change." *Evolution.* 39: 488–504.

Kitcher, P. (1984). "Species." *Philosophy of Science.* 51: 308–33.

—— (1993). "Function and design." *Midwest Studies in Philosophy* 18: 379–97 (reprinted in Allen et al., 1998, pp. 479–503, and in Buller 1999, pp. 159–83).

Kohn, D. (1980). "Theories to work by: Rejected theories, reproduction, and Darwin's path to natural selection." *Studies in the History of Biology.* 4: 67–120.

—— ed. (1985). *The Darwinian Heritage.* Cambridge: Cambridge University Press.

Kohn, D (1989). Darwin's ambiguity: Secularization of biological meaning. *British Journal for the History of Science* 22: 215–39.

Koshland, D. (1989). "Sequences and consequences of the human genome." *Science.* 246: 39.

Krell, D. (1985). "The oldest program toward a system in German idealism." *Owl of Minerva.* 17: 5–19.

Kullmann, W. (1998). *Aristoteles und die moderne Wissenschaft.* Stuttgart: Franz Steiner Verlag.

Kullmann, W. and S. Föllinger, eds. (1997). *Aristotelische Biologie.* Stuttgart: Franz Steiner Verlag.

Lamarck, J. (1809/1984). *Zoological Philosophy.* Edited by R. W. Burkhardt, Jr. Chicago: University of Chicago Press.

Larson, J. (1971). *Reason and Experience: The Representation of Natural Order in the Work of Carl von Linné.* Berkeley: University of California Press.

Latour, B. and S. Woolgar (1979). *Laboratory Life: The Social Construction of Scientific Facts*. Beverly Hills and London: Sage Publications.

Laudan, R. (1987). *From Mineralogy to Geology: Foundations of a Science 1650–1830*. Chicago: University of Chicago Press.

Le Guyader, H. (1988). *Théories et histoire en biologie*. Paris: Vrin.

——— (1998/2004). *Geoffroy Saint-Hilaire, 1772–1844, Un Naturaliste vision-naire*. Paris: Belin. English translation 2001 by M. Grene, *Geoffroy, Saint-Hilaire: Visionary Naturalist*. Chicago: University of Chicago Press.

Le Matieu, D. (1976). *The Mind of William Paley*. Lincoln: University of Nebraska Press.

Lee, H. (1948). "Place names and the date of Aristotle's biological works." *Classical Quarterly*. XLII: 61–7.

Lennox, J. (1982). "Teleology, chance, and Aristotle's theory of spontaneous generation." *Journal of the History of Philosophy*. 20: 21–38 (reprinted in Lennox 2001, pp. 229–49).

——— (1985). "Theophrastus on the limits of teleology." In Fortenbaugh 1985, *Theophrastus of Eresus: His Life and Work*. New Brunswick, NJ: Rutgers University Press. 2 vols. pp. 143–63 (reprinted in Lennox 2001, pp. 259–79).

——— (1987). "Divide and explain: The *Posterior Analytics* in practice." In Gotthelf and Lennox 1987, pp. 90–119 (reprinted in Lennox 2001, pp. 7–38).

——— (1994). "The disappearance of Aristotle's biology: A Hellenistic mystery." *Apeiron*. 27: 7–24 (reprinted in Lennox 2001, pp. 110–25).

——— (2001). *Aristotle's Philosophy of Biology*. Cambridge: Cambridge University Press.

Lenoir, T. (1980). "Kant, Blumenbach, and Vital Materialism in German Biology." *Isis*. 71: 77–108.

——— (1981). "Development of Transcendental *Naturphilosophie*." In Coleman and Limoges, eds., 1981, *Studies in the History of Biology*. 5: 111–205. Baltimore: Johns Hopkins University Press.

——— (1989). *The Strategy of Life: Teleology and Mechanics in Nineteenth-Century German Biology*, 2nd ed. Chicago: University of Chicago Press.

Levere, T. and W. Shaw, eds. (1980). *Nature, Experiment and the Sciences*. Dordrecht: Kluwer.

Lewis, E. (1978). "A gene complex controlling segmentation in *Drosophila*." *Nature*. 276: 565–70.

Lewontin, R. (1970). "The units of selection." *Annual Review of Ecology and Systematics*. 1: 1–18.

——— (1974). *The Genetic Basis of Evolutionary Change*. New York: Columbia University Press.

——— (1978). "Adaptation." *Scientific American*. 239: 212–230.

——— (1991a). "Facts and the factitious in natural science." *Critical Inquiry*. 18: 140–53.

——— (1991b). *Biology as Ideology*. New York: Harper Perennial.

Lewontin, R., and L. Dunn (1960). "The evolutionary dynamics of a polymorphism in the house mouse." *Genetics*. 45: 705–22.

Lewontin, R., S. Rose, and L. Kamin (1984). *Not in Our Genes: Biology, Ideology, and Human Nature*. New York: Pantheon.

Linnaeus, C. (1751). *Philosophia Botanica*. Stockholm: Kiesewetter.

—— (1754). *Reflections on the Study of Nature*. Translated by J. E. Smith, Dublin, 1786.

—— (1758). *Systema Naturae*, 10th ed. Stockholm: L. Salvii.

Lloyd, E. (1988). *The Structure and Confirmation of Evolutionary Theory*. New York: Greenwood Press.

Lloyd, G. (1996). *Aristotelian Explorations*. Cambridge: Cambridge University Press.

Locke, J. (1690). *An Essay Concerning Human Understanding*. London: Awnsham and Churchill 1690 (Oxford: Oxford University Press, 1973).

Loeb, J. (1912). *The Mechanistic Conception of Life*. Cambridge, MA: Harvard University Press.

Löw, R. (1980). *Philosophie des Lebendigen*. Frankfurt: Suhrkamp.

Lumsden, C. and E. Wilson (1981). *Genes, Mind and Culture: The Coevolutionary Process*. Cambridge, MA: Harvard University Press.

Lyell, C. (1830–33). *Principles of Geology, Being an Attempt to Explain the Former Changes of the Earth's Surface, by References to Causes Now in Operation*. London: John Murray. 3 vols. (facsimile edition, Chicago: University of Chicago Press 1990).

—— (1863). *The Geological Evidences of the Antiquity of Man, with Remarks on the Theories of the Origin of Species by Variation*, 2nd revised edition, London: John Murray.

Lyon, J. and P. Sloan, eds. (1981). *From Natural History to the History of Nature: Readings from Buffon and His Critics*. Notre Dame: University of Notre Dame Press.

Malesherbes, C. G. de Lamoignon de (1798). *Observations sur l'histoire naturelle générale et particulier de Buffon et Daubenton*. Paris: Charles Pougens. 2 vols (reprinted in Slatkine Reprints, Geneva, 1971). Selections appear in English under the title "*Observations* on Buffon's natural history" in Lyon and Sloan 1981, pp. 329–45).

Mansion, S., ed. (1961). *Aristote et les problèms de la méthode*. Louvain.

de Maupertuis, P-L. M. (1745/1751). *Vénus physique* (published anonymously). Republished as *Systéme de la Nature*. In Moreau de Maupertuis 1756, *Oeuvres*. Translated by Lyon, Jean-Marie Bruyset, pp. 137–68 1751.

Mayden, R. (1997). "A hierarchy of species concepts: The denouement in the saga of the species problem." In M. Claridge, H. Dawah, and M. Wilson, eds., 1997, *Species: The Units of Diversity*, London: Chapman-Hall.

Mayr, E. (1942). *Systematics and the Origin of Species*. New York: Columbia University Press.

—— (1954). "Change of genetic environment and evolution." In Huxley et al. 1954, pp. 157–80.

—— (1959). "Where are we?" *Cold Spring Harbor Symposia on Quantitative Biology*. XXIV: 1–14.

—— (1961). "Cause and effect in biology." *Science*. 134: 1501–1506.

_____ (1963). *Animal Species and Evolution*. Cambridge, MA: Harvard University Press.

_____ (1970). *Populations, Species and Evolution*. Cambridge, MA: Harvard University Press.

_____ (1976a). "Teleological and teleonomic: A new analysis." In Mayr 1976b, pp. 383–404.

_____ (1976b). *Evolution and the Diversity of Life*. Cambridge, MA: Harvard University Press.

_____ (1980a). "The role of 'systematics' in the evolutionary synthesis." In Mayr and Provine 1980, pp. 127–32.

_____ (1980b). "Prologue: Some thoughts on the history of the evolutionary synthesis." In Mayr and Provine 1980, pp. 1–48.

_____ (1980c). "G. G. Simpson." In Mayr and Provine 1980, pp. 452–63.

_____ (1980d). "How I Became a Darwinian." In Mayr and Provine 1980, pp. 413–23.

_____ (1981). "Biological classification: toward a synthesis of opposing methodologies." *Science*. 214: 510–16 (reprinted in Mayr 1988, pp. 268–88).

_____ (1982). *The Growth of Biological Thought*. Cambridge, MA: Harvard University Press.

_____ (1985). "How does biology differ from the physical sciences?" In Depew and Weber 1985, pp. 43–63.

_____ (1988). *Toward a New Philosophy of Biology: Observations of an Evolutionist*. Cambridge, MA: Harvard University Press.

_____ (1991). *One Long Argument: Charles Darwin and the Genesis of Modern Evolutionary Thought*. Cambridge, MA: Harvard University Press.

_____ (2001). *What Evolution Is*. New York: Basic Books.

Mayr, E., and W. Provine (1980). *The Evolutionary Synthesis*. Cambridge MA: Harvard University Press.

McFarland, J. (1970). *Kant's Concept of Teleology*. Edinburgh: University of Edinburgh Press.

McKirahan, R. (1992). *Principles and Proofs: Aristotle's Theory of Demonstrative Science*. Princeton: Princeton University Press.

McLaughlin, P. (1990). *Kant's Critique of Teleology in Biological Explanation*. Lewiston, NY: Mellon.

_____ (2002). *What Functions Explain: Functional Explanation and Self-Reproducing Systems* Cambridge: Cambridge University Press.

Mendel, G. (1865). "Versuche über Pflanzenhybriden." *Verhandlungen des naturforschenden Vereins in Brünn*. 4: 3–47 (English translation in C. Stern and E. Sherwood, eds., 1966, pp. 1–46).

Millikan, R. (1989). "In defense of proper functions." *Philosophy of Science*. 56: 288–302. (reprinted in Allen et al. 1998, pp. 295–312).

Mills, S. and J. Beatty. (1979). "The propensity interpretation of fitness." *Philosophy of Science*. 46: 263–286.

Mishler, B. (1999). "Getting rid of species." In R. Wilson 1999, pp. 307–15.

Mishler, B. and R. Brandon. (1987). "Individuality, pluralism and the phylogenetic species concept." *Biology and Philosophy*. 2: 397–414.

Mishler, B. and M. Donoghue. (1982). "Species concepts: A case for pluralism." *Systematic Zoology*. 31: 491–503.

Mitchell, S. (2000). "Dimensions of scientific law." *Philosophy of Science*. 67: 242–65.

Monod, J. (1971). *Chance and Necesssity*. New York: Knopf.

Montaigne, M. de (1580/1582/1588). "Apologie de Raimond Sebond," in *Oeuvres Complètes*, 1962. Editions de la Pleiade. Paris: Gallimard.

—— (1965/1962). "Apologie de Raimond Sebond." *Essais* II, pp. 138–351. Paris: Gallimard.

—— (2003). "Apology for Raymond Sebond," trans. R. Ariew and M. Grene. Indianapolis: Hackett.

Morgan, T. (1911). "The application of the conception of pure lines to sex-limited inheritance and to sexual dimorphism." *American Naturalist*. 45: 65–78.

—— (1916). *A Critique of the Theory of Evolution*. Princeton: Princeton University Press.

Moss, L. (2002). *What Genes Can't Do: Prolegomena to a Philosophy Beyond the Modern Synthesis*. Cambridge, MA: MIT Press/Bradford Books.

Muller, H. (1935). *Out of the Night: A Biologist's View of the Future*. New York: Vanguard.

—— (1864). *Für Darwin*. In Moller, A. (ed). *Fritz Müller: Werke, Briefe, und Leben*. Jena: Gustav Fischer, pp. 200–63.

Müller, F. (1878). "Ituna and Thyridia: A remarkable case of mimicry in butterflies." *Transactions of the Entomological Society of London*, pp. 20–9.

—— (1879). "Ituna and Thyridia." In Moller, op. cit.

—— (1948). "Evidence of the precision of genetic adaptation." *The Harvey Lectures*. Springfield, IL: Thomas, pp. 165–229.

Nagel, E. (1961). *The Structure of Science*. New York: Harcourt Brace and World.

Nagel, T. (1974). "What is it like to be a bat?" *Philosophical Review*. 83: 435–50.

Neander, K. (1991). "Functions as selected effects: The conceptual analyst's defence." *Philosophy of Science*. 58: 168–84 (reprinted in Allen et al., 1998, pp. 313–33).

Newton, I. (1718). *Opticks: or, A Treatise of the Reflections, Refractions, Inflections and Colours of Light*, 2nd ed. London: W. and J. Innys.

Newton, I. (1740). *La Méthode des Fluxions et les suites infinites*. Translated by Buffon. Paris: Chex de Bure L' âiné.

Norton, B. (1983). "Fisher's entrance into evolutionary science: The role of eugenics." In Grene, ed., 1983, pp. 19–29.

Nussbaum, M. (1982). "Saving Aristotle's appearances." In M. Nussbaum and M. Schofield, eds., 1982, *Logos and Language*, pp. 267–93. Cambridge: Cambridge University Press.

Olby, R. (1985). *Origins of Mendelism*. 2nd ed. Chicago: University of Chicago Press.

Owen, G. E. L. (1961). "*Tithenai ta phainomena*." In S. Mansion, *Aristotle et les problèmes de la méthode Louvain*, pp. 83–101 (reprinted in Owen, G. E. L., *Logic, Science, and Dialectic*, M. Nussbaum, ed., Ithaca: Cornell University Press, pp. 239–51).

Owen, R. (1992). *The Hunterian Lectures in Comparative Anatomy*. P. Sloan, ed. Chicago: University of Chicago Press.

Owens, J. (1957). *The Doctrine of Being in the Aristotelian Metaphysics*. Toronto: Pontifical Institute of Mediaeval Studies.

Oyama, S., P. Griffiths, and R. Gray, eds. (2001). *Cycles of Contingency: Developmental Systems and Evolution*. Cambridge, MA: MIT Press/Bradford Books.

Pagel, W. (1967). *Harvey's Biological Ideas*. Basel/New York: Karger.

Paley, W. (1802). *Natural Theology*. London: R. Faulder.

Pantin, C. F. A. (1954). "The recognition of species." *Science Progress*. 42: 587–98.

———— (1968). *The Relations Between the Sciences*. Cambridge: Cambridge University Press.

Passmore, J. (1958). "William Harvey and the Philosophy of Science." *Australasian Journal of Philosophy*. 36: 85–94.

Paterson, H. (1985). "The recognition concept of species." In E. S. Vrba, ed. *Species and Speciation. Transvaal Museum Monographs*. 4: 21–9.

Pauly, P. (1987). *Controlling Life: Jacques Loeb and the Engineering Ideal in Biology*. New York: Oxford University Press.

Pearson, K. (1892/1899/1937). *The Grammar of Science*. London: Scott. 2nd ed. London: Black 1899; London: Dent 1937.

Pellegrin, P. (1985). *Aristotle's Classification of Animals*. Translated by A. Preus. Berkeley: University of California Press.

Pennisi, E. (2001). "Linnaeus's last stand?" *Science*. 291: 2304–7.

Piattelli-Palmarini, M. (1989). "Evolution, selection, and cognition: From 'learning' to parameter setting in biology and the study of language." *Cognition*. 31: 1–44

Pinker, S. (1994). *The Language Instinct: How the Mind Creates Language*. New York: William Morrow.

Pinker, S. and P. Bloom (1990). "Natural language and natural selection." *Behavioral and Brain Sciences*. 13: 707–84 (reprinted in Barkow et al. 1992, pp. 451–93). Page numbers in text refer to the 1992 reprint.

Pittendrigh, C. (1958). "Adaptation, natural selection and behavior." In A. Roe and G. Simpson, eds., 1958, *Behavior and Evolution*, New Haven: Yale University Press, pp. 390–416.

Playfair, J. (1802). *Illustrations of the Huttonian Theory of the Earth*. Edinburgh: William Creech.

Pluche, N. (1740). *The History of the Heavens*. Translated by J. B. de Freval. London: no publisher identified.

Polanyi, M. (1958). *Personal Knowledge*. Chicago: University of Chicago Press.

Popper, K. (1972). *Objective Knowledge: An Evolutionary Approach*. Oxford: Oxford University Press.

———— (1978). "Natural selection and the emergence of mind." *Dialectica*. 32: 339–55.

Provine, W. (1971). *The Origins of Theoretical Population Genetics*. Chicago: University of Chicago Press.

———— (1986). *Sewall Wright and Evolutionary Biology*. Chicago: University of Chicago Press.

Raff, R. (1996). *The Shape of Life: Genes, Development and the Evolution of Animal Form*. Chicago: University of Chicago Press.

Ray, J. (1693). *Three Physico-theological Discourses*. London: no publisher identified.

Rensch, B. (1960). *Evolution above the Species Level*. New York: Columbia University Press.

Rey, R. (1992). "Buffon et le vitalisme." In Gayon 1992a, pp. 399–414.

Rheinberger, H-J. (1990). "Buffon: Zeit, Veränderung und Geschichte." *History and Philosophy of the Life Sciences*. 12: 203–33.

Richards, R. (1987). *Darwin and the Emergence of Evolutionary Theories of Mind and Behavior*. Chicago: University of Chicago Press.

—— (2000). "Kant and Blumenbach on *Bildungstrieb*: A historical misunderstanding." *Studies in History and Philosophy of Biological and Biomedical Sciences*. 31: 11–32.

Robin, L. (1944). *Aristote*. Paris: Presses Universitaires de France.

Rodis-Lewis, G. (1971). *L'oeuvre de Descartes*. Paris: Vrin. 2 vols.

Roe, S. (1981). *Matter, Life and Generation*. Cambridge: Cambridge University Press.

Roger, J., ed. (1968). *Buffon, Les Epoques de la Nature (1751)*. Paris: Editions du Muséum (1st ed. 1962).

Roger, J. (1989/1997). *Buffon: Un philosophe au Jardin du Roi*. Paris: Librarie Arthème Fayard. English translation, 1997, *Buffon: A Life in Natural History*, trans. S. L. Bonnefoi. Ithaca, NY: Cornell University Press.

Rohault, J. (1978). "Entretiens sur la philosophie." In *Cahiers d'équipe de recherche*, 75: *Recherches sur le XVIIe Siècle*, 3. Paris: CNRS, pp. 109–54.

Rosenberg, A. (1985). *The Structure of Biological Science*. Cambridge: Cambridge University Press.

—— (1994). *Instrumental Biology or the Disunity of Science*. Chicago: University of Chicago Press.

—— (2000). *Darwinism in Philosophy, Social Science and Policy*. Cambridge: Cambridge University Press.

Rudwick, M. (1964). "The inference of function from structure in fossils." *British Journal for the Philosophy of Science*. 15: 27–40 (reprinted in Allen et al., eds. 1999, pp. 101–15). Quotations are from the reprint.

—— (1972). *The Meaning of Fossils*. New York: Science Publications USA. Revised 1976.

—— (1985). *The Great Devonian Controversy*. Chicago: University of Chicago Press.

—— (1997). *Georges Cuvier: Fossil Bones and Geological Catastrophes*. Chicago: University of Chicago Press.

Rupke, N. (1994). *Richard Owen: Victorian Naturalist*. Yale University Press.

Ruse, M. (1975). "Darwin's debt to philosophy: An examination of the influence of the philosophical ideas of John F. W. Herschel and William Whewell on the development of Charles Darwin's theory of evolution." *Studies in History and Philosophy of Science*. 6: 159–83.

—— (1977). "William Whewell and the argument from design." *The Monist*. 60: 244–68.

_____ (1979). *The Darwinian Revolution.* Chicago: University of Chicago Press.

_____ (1988). *Philosophy of Biology Today.* Albany, NY: State University of New York Press.

_____ (1996). *Monad to Man: The Concept of Progress in Evolutionary Biology.* Cambridge, MA: Harvard University Press.

_____ (2000). "Darwin and the philosophers: Epistemological factors in the development and reception of the theory of the *Origin of Species.*" In Creath and Maienschein 2000, pp. 3–26.

Ruse, M. and E. Wilson. (1985). "The evolution of ethics." *New Scientist.* 17: 50–2.

Russell, E. S. (1916). *Form and Function.* London: John Murray (reprinted 1982, University of Chicago Press).

Ryan, M. (1985). *The Túngara Frog: A Study in Sexual Selection and Communication.* Chicago: University of Chicago Press.

Sahlins, M. (1976). *The Use and Abuse of Biology.* Ann Arbor, MI: University of Michigan Press.

Salthe, S. (1985). *Evolving Hierarchical Systems.* New York: Columbia University Press.

_____ (1998). *Genetics and Reductionism.* Cambridge: Cambridge University Press.

Schaffner, K. (1993). *Discovery and Explanation in Biology and Medicine.* Chicago: University of Chicago Press

_____ (1998). "Genes, behavior, and developmental emergentism." *Philosophy of Science.* 65: 209–52.

Schelling, F. (1797/1988). *Ideas for a Philosophy of Nature.* 1st ed. 1797; 2nd ed. 1803. Translated by E. Harris and P. Heath. Cambridge: Cambridge University Press, 1988.

Schmalhausen, I. (1949). *Factors of Evolution: The Theory of Stabilizing Selection.* Philadelphia: Blakeston.

Schwartz, J. (1999). *Sudden Origins: Fossils, Genes, and the Emergence of Species.* New York: Wiley.

Scotus, John Duns. *Opus Oxoniense.* L. Wadding, ed. (1968 reprint, Hildesheim: George Olms).

Scriven, M. (1959). "Explanation and prediction in evolutionary theory." *Science.* 130: 477–82.

Secord, J. (2000). *Victorian Sensation: The Extraordinary Publication, Reception, and Secret Authorship of Vestiges of Natural Creation.* Chicago: University of Chicago Press.

Sedgwick, A. (1831). "Presidential address to the geological society." *Proceedings of the Geological Society.* I: 281–316.

_____ (1845). "Review of *Vestiges of Natural History of Creation.*" *The Edinburgh Review.* 82: 1–85.

Segerstrale, U. (2000). *Defenders of the Truth.* Oxford: Oxford University Press

Sekia, M. et al. (2001). "Monitoring the expression pattern of 1,300 *Arabidopsis* genes under drought and cold stresses by using a full-length cDNA microarray." *Plant Cell.* 13: 61–72.

Shapin, S. (1994). *The Social History of Truth*. Chicago: University of Chicago Press.

Simpson, G. (1944). *Tempo and Mode in Evolution*. New York: Columbia University Press.

——— (1949). *The Meaning of Evolution*. New Haven: Yale University Press.

——— (1953). *The Major Features of Evolution*. New York: Columbia University Press.

——— (1961). *Principles of Animal Taxonomy*. New York: Columbia University Press.

——— (1967). "The crisis in biology." *The American Scholar*. 36: 363–77.

——— (1969). *Biology and Man*. New York: Harcourt Brace and World.

Sinsheimer, R. C. (1967). *The Book of Life*. Reading, MA: Addison-Wesley.

Sloan, P. R. (1972). "John Locke, John Ray, and the problem of the natural system." *Journal of the History of Biology*. 5: 1–53.

——— (1976). "The Buffon–Linnaeus controversy." *Isis*. 67: 356–75.

——— (1979). "Buffon, German biology, and the historical interpretation of biological species." *British Journal for the History of Science*. 12: 109–53.

——— (1985). "Darwin's invertebrate program, 1826–1836: Preconditions of transformism." In Kohn 1985, pp. 71–120.

——— (1987). "From logical universals to historical individuals: Buffon's idea of biological species." *Histoire du concept d'espèce dans les sciences de la vie*. Paris: Fondation Singer-Polignac. pp. 101–40.

——— (1992). "Organic molecules revisited." In Gayon 1992a, pp. 415–38.

——— (1995). "The gaze of natural history". In Fox et al. 1995, pp. 112–51.

——— (2000). "Mach's phenomenalism and the British reception of Mendelism." In *Comptes Rendus de l'Académie des Sciences, Série III, Sciences de la Vie*. 323: 1069–79.

——— (2001). "The sense of sublimity: Darwin on nature and divinity." *Osiris* 16: 251–69.

——— (in press). "Natural history." In Haakonssen, K., ed., *Cambridge Handbook to Eighteenth-Century Philosophy*. Cambridge: Cambridge University Press.

Smokovitis, V. (2000). "The 1959 Darwin Centennial celebration." In P. Abir-Am and C. Elliott, eds., *Commemorative Practices in Science: Historical Perspectives on the Politics of Collective Memory. Special Issue of Osiris, Second Series*, Vol. 14, pp. 274–323.

Sneath, P. and R. Sokal (1961). "The construction of taxonomic groups." *Symposium of the Society for General Microbiology* 12: 289–332.

Sober, E. (1984). *The Nature of Selection*. Cambridge, MA: MIT Press.

——— (1993). *Philosophy of Biology*. 1st ed. Boulder, CO: Westview Press.

——— (2000). *Philosophy of Biology*. 2nd ed. Boulder, CO: Westview Press.

Sober, E. and D. Wilson (1994). "A critical review of philosophical work on the units of selection problem." *Philosophy of Science*. 61: 534–5.

——— (1998). *Unto Others: The Evolution of Altruism*. Cambridge, MA: Harvard University Press.

Sokal, R. (1962). "Typology and empiricism in taxonomy." *Journal of Theoretical Biology*. 3: 230–67.

Sokal, R. and T. Crovello (1970). "The biological species concept: A critical evaluation." *American Naturalist*. 104: 127–53.

Sokal, R. and P. Sneath (1963). *Principles of Numerical Taxonomy*. San Francisco: W. H. Freeman.

Stadler, L. (1954). "The gene." *Science*. 120: 811–9.

Stanley, S. (1975). "A theory of evolution above the species level." *Proceedings of the National Academy of Science USA*. 72: 646–50.

Stebbins, G. (1969). *The Basis of Progressive Evolution*. Chapel Hill: University of North Carolina Press.

———(1980). "Botany and the synthetic theory of evolution." In Mayr and Provine 1980, pp. 139–152.

Steinle, F. (2002). "Challenging established concepts: Ampère and exploratory experimentation." *Theoria*. 17: 291–310.

Steno, N. (1666). *Discours sur l'anatomie du cerveau* (reprinted in Steno, *Opera Philosophica*, Copenhagen, 1910).

Sterelny, K. (2001). *The Evolution of Agency and Other Essays*. Cambridge: Cambridge University Press.

Sterelny, K. and P. Griffiths. (1999). *Sex and Death: An Introduction to the Philosophy of Biology*. Chicago: University of Chicago Press.

Sterelny, K. and P. Kitcher (1988). "The return of the gene." *Journal of Philosophy*. 85: 339–60.

Stern, C. and E. Sherwood, eds. (1966). *A Mendel Source Book*. San Francisco: Freeman.

Stevens, P. (2001). "Review of M. Ereshevsky, *The Poverty of the Linnean Hierarchy*." *Journal of the History of Biology*. 34: 600–2.

Suarez, F. (1998). *Disputationes metaphysicae*. Hildesheim: Georg Olms. 2 vols.

Tattersall, I. and J. Schwartz (2000). *Extinct Humans*. Boulder, CO: Westview.

Templeton, A. (1989). "The meaning of species and speciation: a genetic perspective." In D. Otte and J. Endler, eds., *Speciation and its Consequences*. Sunderland, England: Sinauer Associates, pp. 3–27.

———(2002). "Out of Africa, again and again." *Nature*. 416: 45–51.

Thompson, P. (1989). *The Structure of Biological Theories*. Albany: State University of New York Press.

Tinbergen, N. (1963). "On the aims and methods of ethology." *Zeitschrift für Tierpsychologie*. 20: 410–33.

Todhunter, L. (1876). *William Whewell: An Account of His Writings*. London: Macmillan. 2 vols.

Toletus, L. (1589). *Commentarium in Octo Libros Physicae Aristotelis*.

Tooby, J. and L. Cosmides (1992). "The psychological foundations of culture." In Barkow et al. 1992, pp. 19–136.

Topinard, P. (1885). *Elements d'anthropologie générale*. Paris: Delahaye et Lecrosnier.

Treviranus, G. R. (1802–22). *Biologie, oder Philosophie der lebenden Natur für Naturforscher und Ärtze*. Göttingen: J. F. Röewer. 6 vols.

Tschermak, E. (1900). "Über künstliche Kreuzung bei *Pisum sativum*." *Berichte der deutschen botanischen Gesellschaft*. 18: 232–9.

Van Fraassen, B. (1980). *The Scientific Image*. Oxford: Oxford University Press.

Van Valen, L. (1973). "A new evolutionary law." *Evolutionary Theory*. 1: 1–30.

——— (1976). "Ecological species, multispecies, and oaks." *Taxonomy*. 25: 233–239.

Venter, J. C., M. D. Adams, E. W. Myers, et al. (2001). "The sequence of the human genome." *Science*. 291: 1304–51.

Von Dassow, G., E. Meir, E. Monro, and M. Odell (2000). "The segment polarity network is a robust developmental module." *Nature*. 406: 188–92.

Von Hoghelande, C. (1676) (1st ed. 1646). *Cogitationes*. Leyden: no publisher identified.

Von Staden, H. (1989). *Herophilus: The Art of Medicine in Early Alexandria*. Cambridge: Cambridge University Press.

——— (1997a). "Character and competence: Personal and professional conduct in Greek medicine." In Von Staden 1997b.

——— (1997b). *Médicine et morale dans l'antique*. Geneva: Fondation Hardt.

——— (1997c). "Teleology and mechanism: Aristotle's biology and early Hellenistic medicine." In Kullmann and Föllinger, eds. 1997, pp. 183–208.

Vrba, E. ed. (1985). *Species and Speciation*. Monograph no. 4. Pretoria: Transvaal Museum.

Waddington, C. (1942). "Canalization of development and the inheritance of acquired characters." *Nature*. 150: 563–5.

——— (1953a). "Genetic assimilation of an acquired character." *Evolution*. 7: 118–26.

——— (1953b). "Epigenetics and evolution." In J. F. Danielli and R. Brown, eds., 1953, *Society for Experimental Biology Symposium: Evolution*, Cambridge: Cambridge University Press, pp. 186–99.

——— (1956). *Principles of Embryology*. New York: Macmillan.

——— (1957). *The Strategy of the Genes*. London: Allen and Unwin.

——— (1960). "Remarks on Panel Three: Man as an Organism." In Tax S. and C. Callendar, eds., 1960, *Evolution After Darwin, Vol. III: Issues in Evolution*. Chicago: University of Chicago Press, pp. 148–9.

——— (1961). "Genetic assimilation." *Advances in Genetics*. 10: 257–90.

Wade, M. (1978). "A critical review of the models of group selection." *Quarterly Review of Biology*. 53: 101–14.

Wagner, G. (1996). "Homologues, natural kinds, and the evolution of modularity." *American Zoology*. 36: 36–43.

Wake, D. (1994). "Comparative terminology." *Science*. 265: 268–9.

Waller, J. (2001). "Gentlemanly men of science: Sir Francis Galton and the professionalization of the British life sciences." *Journal of the History of Biology*. 54: 83–114.

Walsh, D. (1996). "Fitness and function." *British Journal of the Philosophy of Science*. 47: 533–74.

Walsh, D. and A. Ariew (1996). "A taxonomy of functions." *Canadian Journal of Philosophy*. 26: 493–514 (reprinted in Buller 1999, pp. 257–80).

Watkins, E. (1997). "The laws of motion from Newton to Kant." *Perspectives in Science*. 5: 311–48.

——— (1998). "Kant's justification of the laws of mechanics." *Studies in the History of the Philosophy of Science*. 29: 539–60.

Wear, A. (1983). "Harvey and the way of the anatomists." *History of Science*. 21: 223–49.

Weber, B. and D. Depew, eds. (2003). *Evolution and Learning: The Baldwin Effect Reconsidered*. Cambridge, MA: Bradford Books/MIT Press.

Webster, W. and B. Goodwin (1996). *Form and Transformation: Generative and Relational Principles in Biology*. Cambridge: Cambridge University Press.

Weinberg, W. (1908). "Ueber den Nachweis der Vorerbung beim Menschen." *Jahreshefte des Vereins für Vaterländische Naturkunde in Würtemburg*. 61: 368–82.

Weismann, A. (1875). Über den Saison-Dimorphismus der Schmetterlinge." In A. Weismann 1875, *Studien zur Descendenz-Theorie*, Leipzig: Engelmann.

_____ (1904). *The Evolution Theory*, vol. 2: *The Biogenetic Law*. London: Edward Arnold.

Wheeler, Q. and R. Meier, eds. (2000). *Species Concepts and Phylogenetic Theory: A Debate*. New York: Columbia University Press.

Wheeler, Q. and N. Platnick (2000). "The phylogenetic species concept, *sensu* Wheeler and Platnick." In Wheeler and Meier 2000, pp. 55–69.

Whewell, W. (1831). [Review of] *Preliminary Discourse* by J. W. Herschel. *Quarterly Review* XLV, pp. 374–407.

—— (1832). "Lyell's geology." *Quarterly Review*. XLVII, pp. 103–32.

_____ (1833). *Astronomy and General Physics Considered with Reference to Natural Theology* (Bridgewater Treatise). 1st ed. London: Pickering. All quotations are from the 4th, 1839 edition.

_____ (1837). *History of the Inductive Sciences from the Earliest to the Present*. London: Parker. 3 vols.

—— (1840/1967). *Philosophy of the Inductive Sciences Founded on Their History*. London: John W. Parker. Reprint of 2nd edition, 1967, New York: Johnson Reprint Co. All quotations are from the 2nd edition reprint.

_____ (1845). *Indications of the Creator*. London: John W. Parker.

_____ (1850). "Criticism of Aristotle's account of induction." *Transactions of the Cambridge Philosophy Society*. 9: 63–72.

Wiley, E. (1981). *Phylogenetics: The Theory and Practice of Phylogenetic Systematics*. New York: Wiley-Interscience.

Wiley, E. and R. Mayden (2000a). "The evolutionary species concept." In Wheeler and Meier 2000, pp. 70–89.

_____ (2000b). "A critique from the evolutionary species concept perspective". In Wiley and Mayden 2000a, pp. 146–58.

Williams, G. (1966). *Adaptation and Natural Selection*. Princeton: Princeton University Press.

_____ (1992). *Natural Selection: Domains, Levels and Challenges*. Oxford: Oxford University Press.

Wilson, D. (1989). "Levels of selection: An alternative to individualism in biology and the social sciences." *Social Networks*. 11: 257–72.

Wilson, D. and E. Sober (1989). "Reviving the superorganism." *Journal of Theoretical Biology*. 136: 332–56.

Wilson, E. (1898). "Cell lineages and ancestral reminiscence." *Biological Lectures from the Marine Biology Laboratories, Woods Hole, Massachusetts.* Boston: Ginn, pp. 21–42.

Wilson, E. O. (1975). *Sociobiology: The New Synthesis.* Cambridge, MA: Harvard University Press.

—— (1978). *On Human Nature.* Cambridge, MA: Harvard University Press.

—— (1994). *Naturalist.* Washington D. C.: Island Press.

—— (1998). *Consilience: The Unity of Knowledge.* New York: Knopf.

Wilson, L., ed. (1970). *Sir Charles Lyell's Scientific Journals on the Species Question.* New Haven and London: Yale University Press.

Wilson, P. (1980). *Man, the Promising Primate.* New Haven: Yale University Press.

Wilson, R., ed. (1999). *Species: New Interdisciplinary Essays.* Cambridge, MA: MIT Press.

Wimsatt, W. (1980). "Reductionist research strategies and their biases in the units of selection controversy." In T. Nickles 1980, *Scientific Discovery, Vol. II: Case Studies,* Dordecht: Reidel, pp. 213–59.

—— (1981). "The units of selection and the structure of the multi-level genome." In Asquith and Giere 1981, pp. 122–83.

Winther, R. G. (2001). "Darwin on variation and heredity." *Journal of the History of Biology.* 33: 425–55.

Wolf, U. (1995). "Identical mutations and phenotypic variation." *Human Genetics.* 100: 305–21.

Wolff, C. (1730). *Philosophica Prima seu Ontologia.* Publisher not identified.

Woltereck, R. (1909). Weitere experimentelle Untersuchungen über Artveränderung, speziell über das Wesen quantitativer Artunderschiede bei Daphniden. *Versuch. Deutsch. Zool. Ges.* 1009: 110–72.

Woodger, J. H. (1937). *The Axiomatic Method in Biology.* Cambridge: Cambridge University Press.

Woodward, J. (1695). *A Natural History of the Earth.* London: R. Wilkin.

Wright, L. (1958). "Functions." *Philosophical Review.* 82: 139–68 (reprinted in Allen et al., 1998, pp. 51–78, and in Buller 1999, pp. 29–55).

—— (1967). *Teleological Explanations.* Berkeley and Los Angeles: University of California Press.

Wright, S. (1930). "Review of *The Genetical Theory of Natural Selection* by R. A. Fisher." *Journal of Heredity.* 21: 349–56 (reprinted in Wright, 1986, pp. 80–7).

—— (1931). "Evolution in Mendelian populations." *Genetics.* 16: 97–159 (reprinted in Wright 1986).

—— (1932). "The roles of mutation, inbreeding, crossbreeding, and selection in evolution." *Proceedings of the Sixth International Congress of Genetics.* 1: 356–66 (reprinted in Wright 1986, pp. 161–71).

—— (1948). "On the role of directed and random changes in gene frequency in the genetics of populations." *Evolution.* 2: 279–94.

—— (1964). "Biology and the philosophy of science." *Monist.* 48: 265–90.

—— (1982). "Character change, speciation, and the higher taxa." *Evolution.* 36: 427–43 (reprinted in Wright 1986, pp. 622–38).

—— (1986). *Evolution: Selected Papers.* Edited by W. Provine. Chicago: University of Chicago Press.

Wynne Edwards, V. (1962). *Animal Dispersion in Relation to Social Behavior.* Edinburgh: Oliver and Boyd.

Zammito, J. (1992). *The Genesis of Kant's Critique of Judgment.* Chicago: University of Chicago Press.

—— (2002). *Kant, Herder, and the Birth of Anthropology.* Chicago: University of Chicago Press.

Zumbach, C. (1984). *The Transcendent Science.* Boston: Martinus Nijhoff.

Zuckerkandl, E. (1987). On the molecular evolutionary clock. *Journal of Molecular Evolution* 26: 34–46.

Index

Achinstein, Peter, 317
acquired characteristics, heritability of, 150, 223, 230, 283
Adams, M., 257
Adanson, Michel, 73, 75
adaptation, 150–151, 179–180
  Modern Evolutionary Synthesis on, 200, 258–261
  polymorphic, 258
adaptationism, 268, 334, 336, 340, 342. *See also* Modern Evolutionary Synthesis, "hardening of the"
adaptive radiation, 168, 287, 328
Adickes, E., 113–114
affordances, 357. *See also* psychology, ecological; Gibson, J. J.
Alberti, S., 228
Alexander, Richard, 330–331
Alexandria, 32, 33
altruism. *See* cooperative behavior in nature
Amundson, Roald, 318, 350–351
anagenesis, 257, 276, 277
anatomy, comparative. See Cuvier, G., comparative anatomy; Geoffroy St. Hilaire, E., comparative anatomy; Darwin, C., comparative anatomy
anthropic principle, 188

anthropology, as discipline, 323, 327–328, 333, 352
  physical, 334
  cultural, 333, 335
Appel, Toby, 128, 137, 148
Aquinas, Thomas, 8, 39–60, 184–185
argument from design, 98, 160, 164, 184, 188, 189, 191, 194–198, 201. *See also* natural religion; natural theology
Ariew, André, 317–318
Ariew, Roger, 38, 40, 44
aristogenesis, 286
Aristotle, xvi–xvii
  art (craft) (*technē*), as contrasted with nature by, 6–7
  biological interests of, as starting point for inquiry, 11, 14, 35, 306, 351
  biological treatises, reception of, 42
  biology, view of as part of physics, 5, 32. *See also* Aristotle, physics
  camel's dentition, 24–25
causes
  kinds of, doctrine concerning, xvii, 6–7, 31, 36
  final, 211–212, 313, 314
chance, view of, 7–8
change, doctrine of, xvii
  substantial vs. accidental, distinction between, 47
character traits (*ethē*), 12

natural purposes, doctrine of, xv,
xviii, 94, 97–98, 102–108, 112,
115, 116, 122. *See also* Kant,
organisms on
"Newton of a blade of grass," on
the idea of, xv, 125, 134
*On the Use of Teleological
Principles in Philosophy*, 125
ontogeny, 94–95, 285
Opus postumum, 107, 113–114,
126
organisms. *See also* Kant, natural
purposes.
causes and effects of, 89, 94, 99
artefacts, relationship to, 97–100,
102
parts of, as mutually causative of,
94, 103
physico-theological proof. *See* Kant,
argument from design,
response to
polygenism, opposition of, 117
predispositions (*Anlagen*), doctrine
of, according to, 96, 119, 120
preformationism. *See* Kant,
epigenesis and preformation
*Prolegomena to Any Future
Metaphysics*, 102
purposiveness (*Zweckmässigkeit*),
conception of, according to,
xviii, 100–101, 102
races of human beings
racism, accused of, 119
reductionism, 92
"Second Analogy," 93–94, 98, 107.
See also Kant, causality
self-formative force, conception of,
99–100. *See also* Blumenbach,
self-formative force; influence
on Kant of
self-organization, on development
as, 100
species, concept of, according to,
118, 125
supersensible, 114, 115
teleology, 92, 101, 120
external vs. internal, 106, 108,
112

universal (or global), 59, 105,
106
theism and, 109
teleological maxim ("Nature does
nothing in vain"), 105
vitalism and, 89
Kielmayer, Carol Friedrich,
123
Kimura, Motoo, 271–272
King, Jack, 272
Kingsolver, J., 318
kin selection. *See* natural selection, kin
selection as form of
Kitcher, Philip, 317
Knox, Robert, 177, 178
Koehl, M., 318, 319–321
Koelreuter, J. G., 240
Köhler, W., 357
Kohn, David, 152, 193, 196, 208–209
Kripke, Saul, 176

lac operon, 284. *See also* Jacob, F.
La Caze, Louis de, 88, 89
La Flèche, 41
Lamarck, Jean Baptiste, xvi, 129
acquired characteristics, inheritance
of, according to, 150, 230
adaptationism of, 151
biology, origins of term and, 123,
149
classification. *See* classification,
Lamarck on
extinction, view of, 150, 177
influence in Britain of, xviii, 146,
153, 178–179, 180–182
materialism of, 178–179
progress, evolutionary, doctrine of,
333
scientific interests of, 149
spontaneous generation, on, 150
transformism of, xviii, 146, 148,
177, 194
language, evolution of, xxi, 340–343
Laplace, Pierre Simon, Marquis de,
134, 135
Latour, Bruno, 354
Latreille, Pierre-André, 144
Lauden, Rachel, 157, 175